PROTON-EMITTING NUCLEI

GIOTTO di Bondone (b. 1267, Vespignano, d. 1337, Firenze)

Cappella degli Scrovegni (Arena Chapel), Padua, Italy

In about 1305 and 1306 Giotto painted a notable series of 38 frescoes in the Arena Chapel in Padua. The frescoes illustrate the lives of Jesus Christ and of the Virgin Mary. Over the archway of the choir is a scene of the Court of Heaven, and a Last Judgment scene faces it on the entrance wall. In the Last Judgment, below the cross, on the left, is the dedicatory scene, in which Enrico Scrovegni kneels before the Virgin and two saints, offering a model of the Arena Chapel upheld by an Augustinian friar.

This image was provided courtesy of Assessorato alla Cultura, Comune di Padova.

PROTON-EMITTING NUCLEI

Second International Symposium
PROCON 2003

Legnaro, Italy   12-15 February 2003

EDITORS
Enrico Maglione
Università di Padova
Padova, Italy

Francesca Soramel
Università di Udine
Udine, Italy

SPONSORING ORGANIZATIONS
Istituto Nazionale di Fisica Nucleare (INFN)
Università di Padova
Università di Udine

Melville, New York, 2003
AIP CONFERENCE PROCEEDINGS ■ VOLUME 681

Editors:

Enrico Maglione
Dipartimento di Fisica
Università di Padova
Via F. Marzolo, 8
I-35131 Padova
ITALY

E-mail: maglione@pd.infn.it

Francesca Soramel
Dipartimento di Fisica
Università di Udine
Via Delle Scienze, 208
I-33100 Udine
ITALY

E-mail: soramel@fisica.uniud.it

L.C. Catalog Card No. 2003110298
ISBN 0-7354-0150-0
ISSN 0094-243X
Printed in the United States of America

CONTENTS

EXPERIMENTS ON ONE PROTON EMISSION

THEORY ON ONE PROTON EMISSION

TWO PROTON EMISSION

ALPHA EMISSION

BETA DELAYED PROTON EMISSION AND GAMMA SPECTROSCOPY

NUCLEAR ASTROPHYSICS

EXPERIMENTAL TECHNIQUES

PREFACE

The "Procon 2003 - International Symposium On Proton Emitting Nuclei" was held from February 12 to 15, 2003 at the Legnaro National Laboratory near Padova, Italy. This was the second of this kind of meetings following the one held in Oak Ridge in September 1999.

The field of exotic decays at the proton drip-line has had an enormous development in the past decade making it necessary to its community to meet periodically to exchange ideas and results.

The meeting in Legnaro has been very interesting and in a six half a day sessions we had the chance to update our knowledge with new results and developments, both experimental and theoretical. The participation has been quite good and we would like to underline the presence of a group of young and enthusiastic physicists.

We are grateful to Dr. G. Fortuna, Director of the Laboratories, for its kind hospitality. We warmly thank Mrs. A. Spalla, conference secretary, for her precise and professional work which greatly contributed to the conference success. Our thanks to Mr. M. Deanna who prepared the conference poster.

Finally we would like to acknowledge the sponsorship of Istituto Nazionale di Fisica Nucleare (INFN), University of Padova, University of Udine, Comune di Padova, Banca di Roma, and Alitalia.

On behalf of the organizing committee

E. Maglione and F. Soramel   May 2003

COMMITTEES

International Advisory Board

J. Äystö (Jyväskylä)
B. Blank (Bordeaux)
R. Bonetti (Milano)
M. J. Borge (Madrid)
R. Boyd (Ohio)
T. Faestermann (Garching)
S. Hofmann (GSI)
S. G. Kadmensky (Voronezh)
D. Rudolph (Lund)
T. Vertse (Debrecen)
P. J. Woods (Edinburgh)
S. W. Xu (Lanzhou)

Organizing Committee

Jon Batchelder (ORNL)
Cary Davids (Argonne)
Giacomo De Angelis (LNL)
Lídia S. Ferreira (Lisbon)
Enrico Maglione (Padova)
Francesca Soramel (Udine)

FUNDING AGENCIES AND SPONSORS

Istituto Nazionale di Fisica Nucleare
Università di Padova
Università di Udine
Comune di Padova
Banca di Roma
Alitalia

EXPERIMENTS ON ONE PROTON EMISSION

Proton Emitter Studies at Argonne

P.J. Woods

School of Physics, The University of Edinburgh, Edinburgh EH9 3JZ, UK

Abstract. The paper reports on experimental developments for studies of decays of proton radioactive nuclei on the Argonne Fragment Mass Analyzer. The results illustrate that there is a clear proton decay rate sensitivity to the unpaired neutron orbital in highly deformed odd-odd nuclei consistent with the Influential Spectator Model of Ferreira and Maglione. No such sensitivity is observed for the proton decay rates of spherical odd-odd nuclei studied here. The first ever study of proton decay from a nucleus, ^{135}Tb, produced via 1p6n fusion evaporation is also reported.

INTRODUCTION

At the first Procon Conference [1] the recent discoveries at Argonne of proton decay from the highly deformed nuclei ^{131}Eu and ^{141}Ho [2] were reported. Just prior to the meeting we confirmed the highly deformed nature of ^{131}Eu with the first discovery of proton decay fine structure [3], and the Recoil Decay Tagging (RDT) technique [4] was used to demonstrate the more modest deformation of ^{141}Ho [5, 6]. These results, combined with a successful modelling of proton decay spectroscopic factors in near spherical nuclei [7, 8], represented a significant broadening and deepening of our knowledge of proton emitters and proton emission. The continuing richness of the field is reflected by the increasing number of theoreticians and experimentalists being drawn to its study. The present paper will report on some of the more recent experimental developments, and proton radioactivity measurements performed at Argonne using the Fragment Mass Analyzer (FMA). The results of RDT studies performed at Argonne on proton emitters are reported in a separate paper published in this volume [9].

EXPERIMENTAL DEVELOPMENTS ON THE FMA

The development of a new intense ECR ion source for the ATLAS accelerator facility encouraged us to explore the sensitivity limits of our experimental set-up. In particular, we aimed to increase our sensitivity to small cross-sections (eg 1p5n and 1p6n fusion evaporation channels) in order to access the most remote regions of the proton drip-line. Such an approach has the added benefits of increasing precision where measurements are required to discriminate between different theoretical approaches, and enables more detailed features to be revealed, such as isomers, or fine structure.

A new design of double-sided silicon strip detector (DSSD) was developed to encompass most of the region behind the focal plane of the FMA while maintaining

CP681, *Proton-Emitting Nuclei: Second International Symposium; PROCON 2003*,
edited by E. Maglione and F. Soramel
© 2003 American Institute of Physics 0-7354-0150-0/03/$20.00

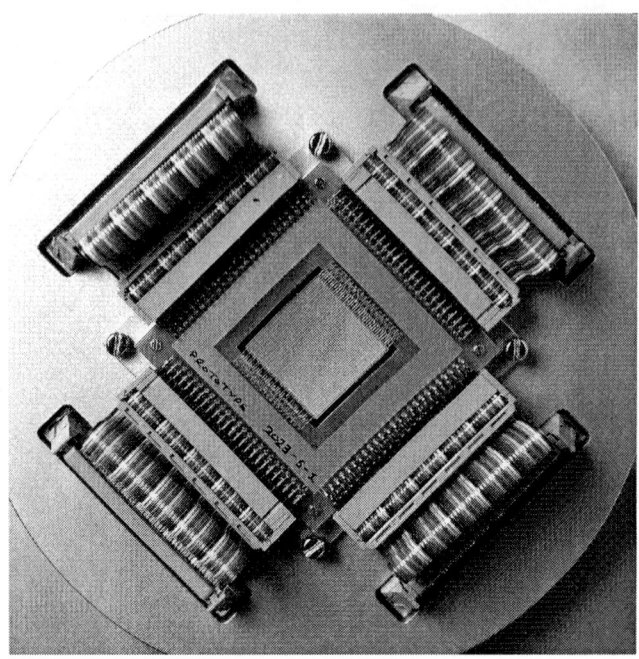

FIGURE 1. The new 80*80 DSSD used behind the focal plane of the Argonne FMA for proton decay studies.

a high granularity for recoil-decay correlation performance. The detector consists of 80*80 strips with a pitch of 400μm (see figure 1) and is consequently larger and more segmented than the original prototype DSSD developed for the Daresbury Recoil Separator [10].

The much higher recoil detection efficiency is particularly valuable for RDT studies where experiments are rate limited by forward angle Ge detector rates. The higher granularity means higher implantation rates can be tolerated for proton decay searches.

The great variation in energy and beam species available from the ATLAS accelerator are a major benefit for our experimental programme. In particular it enables us to choose inverse kinematics to access the nucleus of interest. This can significantly increase the recoil transmission efficiency through the FMA. Inverse kinematics are particularly useful for RDT experiments since the target position has to be displaced upstream of the FMA in order to accommodate Gammasphere - this compensates for what otherwise would be a significant loss in sensitivity. The downside of inverse kinematics is increased levels of scattered beam at the focal plane of the FMA. Following the approach developed at JAERI, a split anode was constructed for the first electric dipole of the FMA. This was successfully commissioned in 2002 resulting in significant reductions in scattered beam rates ensuring that even in extreme kinematical scenarios the recoil rate is dominated by evaporation residues.

The standard electronic response of the Edinburgh/RAL pre-amplifier shaping amplifier system [11] used to instrument the DSSD is limited to decays occurring $\geq 6\mu s$ after heavy ion implantation onto the same strip. This compares with a recoil flight time $\leq 1\mu s$. While in most cases of proton decay this is not a key issue, there are instances where improved detection efficiency for a few short-lived ($t_{1/2}\sim\mu s$) proton emitters can significantly enhance a measurement [12]. A delay line shaping system [13] was successfully introduced to instrument all 160 strips of the DSSD in parallel with the existing instrumentation system. Proton decays from ^{113}Cs could be seen down to $\sim 1.0\mu s$ following implantation. This system was then used to make a high precision half-life measurement of the short-lived isomeric proton transition from ^{150m}Lu [14, 15]. The energy resolution of the delay line system was found to be comparable, but slightly inferior, to that obtained with the Edinburgh/RAL electronics system. The delay line system gives at least equivalent performance to the digital signal processing approach [16] in this application. The delay line system has a significantly lower per channel cost, requires no specialist expertise for setting-up, and there is no limit imposed on the search time range. Our experiments now utilise the delay line and Edinburgh/RAL systems in parallel for search experiments where the energy and half-life of the proton decay is unknown.

PROTON DECAY MEASUREMENTS ON THE FMA

^{117}La

At the first Procon meeting evidence was reported by Soramel et al. for the ground state proton decay of ^{117}La [17]. Subsequently an additional high spin isomeric transition was reported from the same experiment [18]. We made a higher precision measurement of the proton decay of ^{117}La and confirmed the existence of the ground state transition indicating a high degree of prolate deformation, $\beta\sim 0.3$, with the decay occurring from either a $3/2^+$ or $3/2^-$ configuration [19]. However, despite the better statistics and experimental running conditions, no evidence was found for the reported proton decay of a $9/2^+$ isomer.

^{130}Eu

The nucleus ^{131}Eu is currently thought to be the most deformed known proton emitter. Recently, a theoretical framework has been developed for proton decays from highly deformed odd-odd nuclei by Ferreira and Maglione [20]. This demonstrates that the proton decay rate can be affected by the nature and orientation of the neutron orbit even though it remains inert during the transition. With the permission of the originators one may refer to this as being the Influential Spectator Model (ISM). ^{130}Eu is predicted to

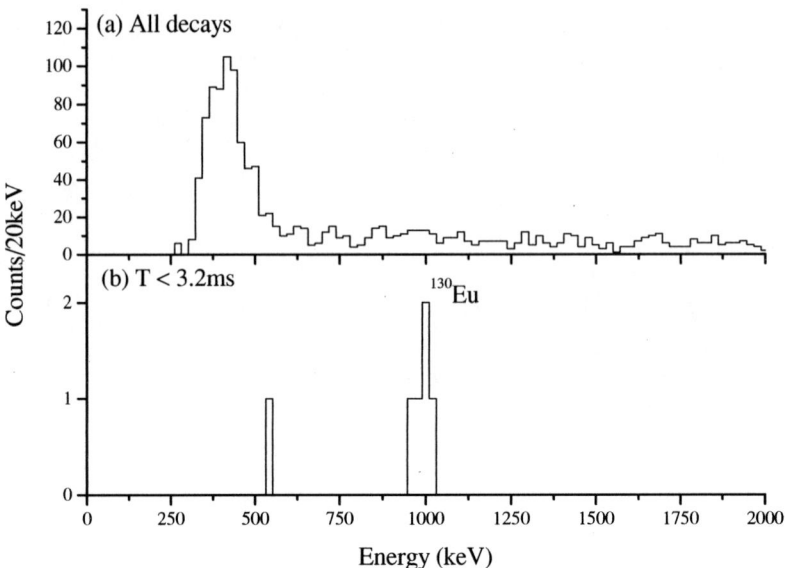

FIGURE 2. (a) All decay events from the reaction ^{78}Kr + ^{58}Ni, restricted by slits to A = 130 recoils. (b) Decay events occurring within 3.2 ms of an implant in the same DSSD pixel. The proton decay peak of ^{130}Eu is clearly visible.

have the same proton configuration and high deformation as its neighbour ^{131}Eu [21] and is an ideal candidate to explore this theoretical model. Figure 2 shows a proton decay energy spectrum indicating the existence of ^{130}Eu produced via a 1p5n fusion evaporation channel [22].

Figure 3 shows a comparison of the measured half-life with calculations performed by Davids based on the ISM. The $K_p=3/2^+$ proton orbital can couple with either the $K_n=1/2^+$ or $7/2^-$ neutron orbitals. It can be seen that the $7/2^-$ neutron orbital is not in agreement with theory but both K = 1^+ and 2^+ states produced by coupling to the $K_n=1/2^+$ orbital are in good agreement with theoretical predictions. Ironically, the proton decay rate gives information on the neutron Fermi surface beyond the proton drip-line even though the neutron itself remains undisturbed in the transition. This is a further illustration of the unique physics insights that can be obtained through proton decay studies.

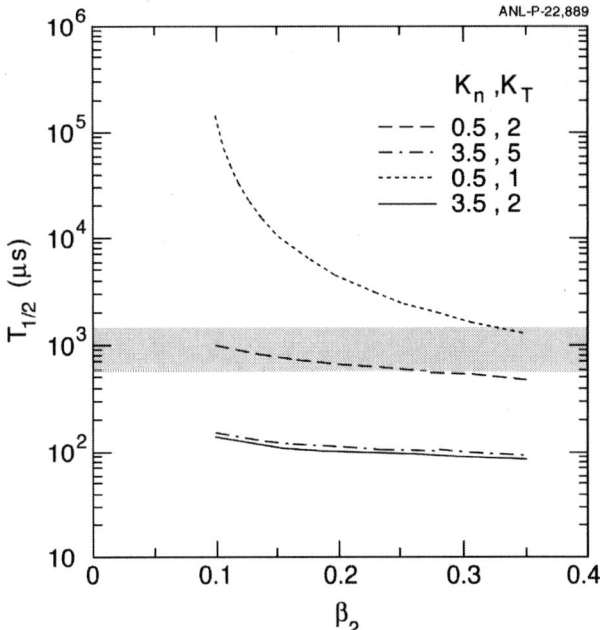

ANL-P-22,889

FIGURE 3. The measured proton decay half-life of ^{130}Eu compared with calculations for proton emission from deformed odd-odd nuclei

^{135}Tb

^{135}Tb was produced by bombarding a ^{92}Mo target with ^{50}Cr beams. Figure 4 shows clear evidence for the proton decay of ^{135}Tb representing the first example of a proton decaying isotope produced via the 1p6n evaporation channel. This data is still being analysed [23] - Figure 5 shows a comparison of the measured half-life with different proton orbital configurations assuming a high prolate deformation.It can be seen that good agreement can be obtained for a 7/2$^-$ proton configuration. The success of this experiment suggests it will be possible to access all of the proton emitting elements between Z=51-83 with the present experimental set-up.

^{164m}Ir and ^{170m}Au

Proton decays were observed from ^{164m}Ir and ^{170m}Au by bombarding ^{92}Mo and ^{96}Ru targets with ^{78}Kr beams. In each case the proton emitter was produced via a 1p3n evaporation channel. The proton half-lives were in both cases consistent with emission from $h_{11/2}$ orbitals, and the spectroscopic factors were in excellent agreement with low seniority shell model calculations [22, 7]. In the case of ^{164m}Ir similar results were also obtained by Kettunen et al. [24]. The results are therefore consistent with a spherical

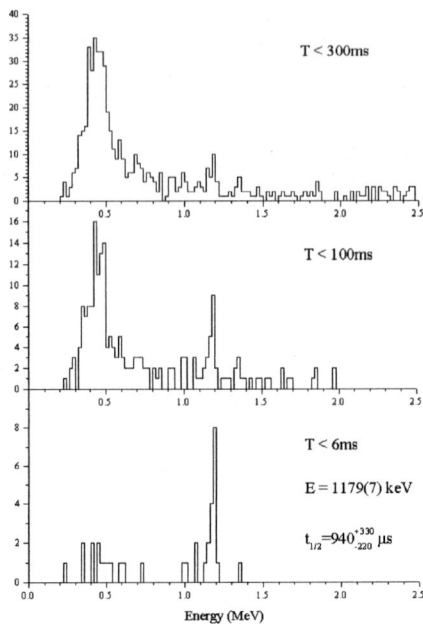

FIGURE 4. These spectra show the clear emergence of the proton decay of ^{135}Tb at short time intervals. This is the first observation of proton decay from a nucleus produced via a 1p6n evaporation channel.

shape and show no clear sensitivity to the neutron configuration.

SUMMARY

A number of changes to the FMA proton decay set-up have been successfully implemented. The increased sensitivity of the system has been used to explore small cross-section and short-lived proton decays. The emphasis in the coming period will be to explore the structure of proton emitters using the RDT technique on the Gammasphere array. As I write this article, exciting new results have been obtained for ^{145,146}Tm including the first observation of γ-decays observed in coincidence with proton radioactivity [25]. Such experiments entail a holistic approach to the study of the structural and decay properties of proton emitters. Tempting as it is, I conclude it would be best to leave these latest results for the next Symposium.

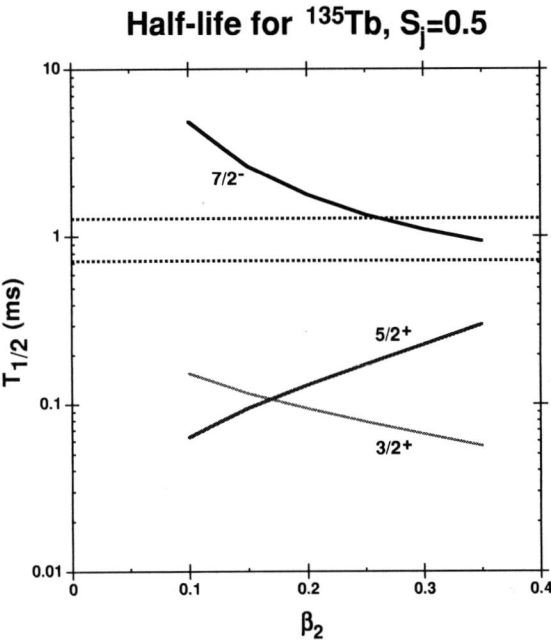

FIGURE 5. A comparison of the measured proton decay half-life of ^{135}Tb with calculations assuming a highly prolate deformed shape.

ACKNOWLEDGMENTS

The author represents here the proton decay collaboration at Argonne. The measurements referred to have been performed over a period of time and have involved different people, all of whom I wish to thank. In particular I would thank Cary Davids, Darek Seweryniak, Andreas Heinz, Bill Walters, Tom Davinson, Hassan Mahmud, Andrew Robinson and Peter Munroe.

REFERENCES

1. P.J. Woods, Procon '99 Proceedings, 34 (1999).
2. C.N. Davids *et al.*, Phys. Rev. Lett. **80**, 1849 (1998).
3. A.A. Sonzogni *et al.*, Phys. Rev. Lett. **83**, 1116 (1999).
4. E.S. Paul *et al.*, Phys. Rev. C **51**, 78 (1995).
5. D. Seweryniak *et al.*, Procon '99 Proceedings, 112 (1999).
6. D. Seweryniak *et al.*, Phys. Rev. Lett. **86**, 1458 (2001).
7. C.N. Davids *et al.*, Phys. Rev. C **55**, 2255 (1997).
8. P.J. Woods *et al.*, Annu. Rev. Nucl. Part. Sci. **47**, 541 (1997).
9. D. Seweryniak, to be published in these Proceedings (2003).
10. P.J. Sellin *et al.*, Nucl. Inst. Meth. **A311**, 217 (1992).
11. S.L. Thomas *et al.*, Nucl. Inst. Meth. **A288**, 212 (1998).

12. M. Karny *et al.*, Phys. Rev. Lett. **90**, 012502 (2003).
13. P. Wilt, private communication
14. R. Grzywacz *et al.*, to be published in these proceedings (2003).
15. T. Ginter *et al.*, Phys. Rev. C **61**, 014308 (2000).
16. A. Robinson *et al.*, private communication.
17. F. Soramel *et al.*, Procon '99 Proceedings, 68 (1999).
18. F. Soramel *et al.*, Phys. Rev. C **63**, 031304(R) (2001).
19. H. Mahmud *et al.*, Phys. Rev. C **64**, 031303(R) (2001).
20. L.S. Ferreira and E. Maglione, Phys. Rev. Lett. **86**, 1721 (2001).
21. P. Moller *et al.*, At. Data Nucl, Data Tables **66**,131 (1997).
22. H. Mahmud *et al.*, Eur. Phys. J. **A15**, 85 (2002).
23. P. Munroe *et al.*, private communication.
24. H. Kettunen *et al.*, Acta. Phys. Pol. **B32**, 989 (2001).
25. D. Seweryniak *et al.*, private communication.

Fine structure in one-proton emission studied at Oak Ridge

K. P. Rykaczewski[1], R. Grzywacz[1,2], J. C. Batchelder[3], C. R. Bingham[1,4],
D. Fong[5], T.N. Ginter[1,5,6], C. J. Gross[1], J. H. Hamilton[5], D. J. Hartley[4],
J. K. Hwang[5], Z. Janas[2,7], M. Karny[2,7], W. Królas[5,7,8] W.D. Kulp[9],
Y. Larochelle[4], T. A. Lewis[1], K. H. Maier[1,7], J. W. McConnell[1],
A. Piechaczek[10], A. V. Ramayya[5], K. Rykaczewski[11], D. Shapira[1],
M. N. Tantawy[4], J. A. Winger[12], C. -H. Yu[1], E. F. Zganjar[10], K. Hagino[13],
A. T. Kruppa[6,14], W. Nazarewicz[1,4,15], P. Semmes[16] and T. Vertse[6,14]

[1]*Physics Division, Oak Ridge National Laboratory, Oak Ridge, TN 37831, USA*
[2]*Institute of Experimental Physics, Warsaw University, PL 00-681 Warsaw, Poland*
[3]*UNIRIB Oak Ridge Associated Universities, Oak Ridge, TN 37831, USA*
[4]*Dept. of Physics and Astronomy, University of Tennessee, Knoxville, TN 37996, USA*
[5]*Dept. of Physics and Astronomy, Vanderbilt University, Nashville, TN 37235, USA*
[6]*NSCL, Michigan State University, East Lansing, MI 48824, USA*
[7]*Joint Institute for Heavy Ion Research, Oak Ridge, TN 37831, USA*
[8]*H. Niewodniczański Institute of Nuclear Physics, PL 31-342 Kraków, Poland*
[9]*Dept. of Physics, Georgia Institute of Technology, Atlanta, GA 30332, USA*
[10]*Dept. of Physics and Astronomy, Louisiana State University, Baton Rouge, LA 70803, USA*
[11]*Oak Ridge High School, Oak Ridge, TN 37830, USA*
[12]*Dept. of Physics and Astronomy, Mississippi State University, Mississippi State, MS 39762, USA*
[13] *Yukawa Institute for Theoretical Physics, Kyoto University, Kyoto 606-8502, Japan*
[14]*Institute of Nuclear Research, Hungarian Academy of Sciences, H 4001 Debrecen, Hungary*
[15]*Institute of Theoretical Physics, Warsaw University, PL 00-681 Warszawa, Poland*
[16]*Dept. of Physics, Tennessee Technological University, Cookeville, TN 38505, USA*

Abstract. Two observations of fine structure in proton emission are reported : 3-μs ^{145}Tm and 4-ms ^{141}Ho. These experiments were performed using the Recoil Mass Separator (RMS) at the Holifield Radioactive Ion Beam Facility at Oak Ridge. The signals from the RMS detectors were digitally processed using Digital Gamma Finder modules. The fine structure branching ratios, 9.6±1.5% and 0.70±0.15%, and measured energies of the 2^+ excited levels in daughter nuclei, 330 and 202 keV, respectively, helped us to determine the deformations and wave functions of proton-emitting states.

There are over thirty proton-radioactive nuclei identified up to date, with about forty proton-emitting ground- and isomeric-states [1, 2, 3, 4]. However, fine structure in proton emission was observed only in few cases. A pioneering experiment on the decay of odd-Z even-N isotope ^{131}Eu was performed at the Fragment Mass Analyzer (FMA) at Argonne. Proton transitions to the 0^+ ground-state and to the 2^+ excited state in ^{130}Sm were observed [5].

The first observation of the fine structure in the proton emission from an odd-odd nucleus, ^{146gs,m}Tm, was made at the Recoil Mass Separator (RMS) at the Holifield Radioactive Ion Beam Facility (HRIBF) in Oak Ridge [6]. Reinvestigation of odd-odd

CP681, *Proton-Emitting Nuclei: Second International Symposium; PROCON 2003*,
edited by E. Maglione and F. Soramel

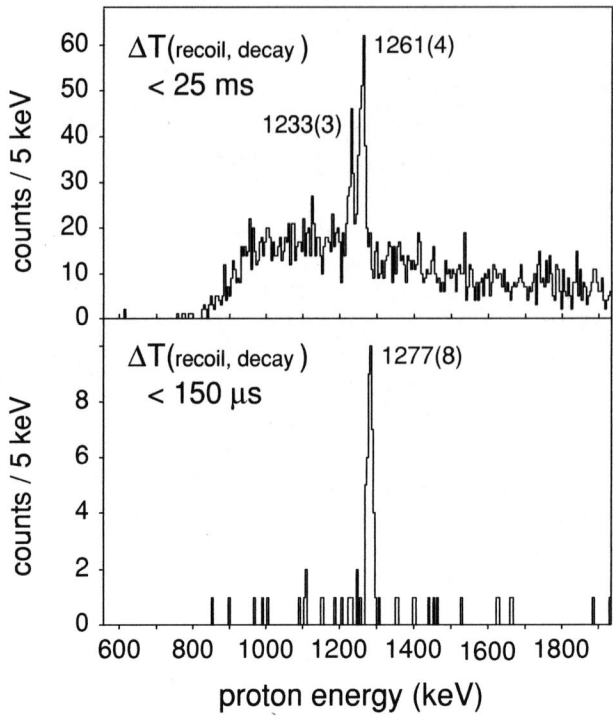

FIGURE 1. The energy spectra of proton events collected within 25 ms (upper panel) and 150 μs (lower panel) after an implantation of A=150 recoils into the DSSD. The recoiling nuclei were produced using a 292 MeV [58]Ni beam on a 0.54 mg/cm^2 [96]Ru target [6, 7]. The charge reset foil was placed 10 cm behind the target. It restores the charge state of a recoiling ion, changed after an isomeric deexcitation involving a conversion electron [8]. Observed increase in the yield of [150m]Lu proton events suggests the presence of such short-lived isomeric levels on the deexcitation path leading to [150m]Lu [6, 7]. Use of a converging solution for the RMS optics [8] resulted in the collection of [150]Lu ions in two charge states. The slits were set in a less restrictive way in comparison to the previous RMS study of [150]Lu [9]. A small fraction of neighboring activity [151gs]Lu (E_p=1233 keV) can be seen near the [150gs]Lu line (E_p=1261 keV) - upper panel. Better energy calibration and improved counting statistics yielded more accurate values for the energy (E_p=1277(8) keV) and half-life ($T_{1/2}$=39(+8,-6) μs) for proton emission from [150m]Lu. No evidence has been obtained for the fine structure in proton emission from [150m,gs]Lu. However, the peak-like weak structures, e.g., around 1.11 - 1.12 MeV, are worth reinvestigation.

[150gs,m]Lu decay did not offer a clear evidence for fine structure, see the spectra in Fig.1.

Here, we report the results of two experiments at the RMS. The decays of [145]Tm [10, 11] and of [141gs]Ho [12, 13, 14] were remeasured. In both RMS experiments, the proton transitions to the 0^+ ground-state and to the 2^+ excited state were found.

A description of the RMS and experimental techniques used for proton radioactivity studies are given in [8, 15, 16, 17]. Briefly, the radioactive nuclei are produced at the target using a beam accelerated by the 25 MV Tandem. Recoiling products of fusion-

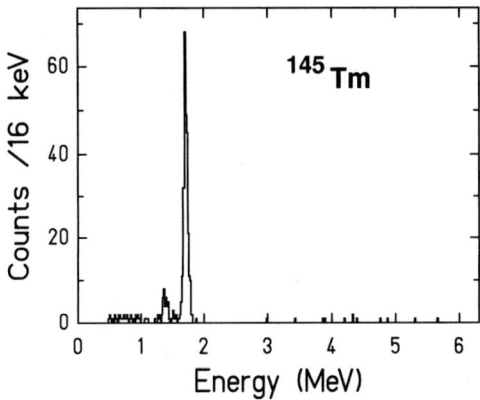

FIGURE 2. The energy spectrum of proton events collected within 0.5 to 10 μs after implantation of A=145 recoils into the DSSD. Similar halflives of 3.1(3) μs and 2.7(10) μs were obtained from the analysis of the decay pattern of 1.73 MeV and 1.40 MeV lines, respectively.

evaporation reactions are separated by magnetic and electric fields, according to their kinetic energy (acceptance $\pm$ 10%) and mass-over-charge Q/A value. Primary beam rejection is excellent and allows us to run with over 20 pnA beam intensity. For the beams of 5 MeV/u ^{54}Fe or ^{58}Ni impinging on a 1 mg/cm^2 ^{92}Mo target, almost no beam particles are reaching the detectors at the RMS final focus. The typical time of flight of fusion-evaporation products through the RMS is about 2.2 μs, and the mass separation is around 1:400. Before implantation, the recoils trigger a position-sensitive Microchannel Plate Detector (MCP) and can be slowed down by a degrader foil. The MCP [18] was designed to determine the position and intensity of radioactive ion beams at the HRIBF. However, this detector has nearly 100% efficiency for recoiling fusion products separated by the RMS. The recoiling ions in two neighboring charge states, e.g., ^{145}Tm^{+26} and ^{145}Tm^{+27}, can be implantated in the 40mm by 40 mm Double-sided Silicon Strip Detector (DSSD). The RMS transmission efficiency is about 5% for such converging ion optics.

The signals from MCP and DSSD detectors are digitally processed using Digital Gamma Finder (DGF4C) modules [19, 15, 16, 17]. The development of this novel pulse processing technique was critical for a search for weak proton transitions, in particular for the short-lived activity ^{145}Tm [11]. Two data acqusition modes can be selected for the DGF module. The 25-μs section of the preamplifier pulse around its rising edge can be recorded and analyzed further off-line. The digital image of the signal is obtained with 25 nanosecond sampling rate. To reduce the data stream, only the pulses with the implantation and decay signals overlapping within a 10 μs time interval are selected by the DGF on-board processors. This, so called "proton-catcher" mode was used during the ^{145}Tm study [11]. In a standard DGF operation mode, the energy and arrival time of the signal is derived on-board, using a trapezoidal filter algorithm. This operation mode is used to study longer-lived proton emitters like the 4-ms activity of ^{141gs}Ho.

The proton energy spectrum correlated with mass 145 recoils measured by the DSSD is shown in Fig. 2. It was derived from the analysis of DSSD preamplifier traces containing overlapping recoil and decay signals [11]. Two proton lines were observed in microsecond time correlations with mass A=145 recoils. The 1.73 MeV line was assigned earlier to the decay of ^{145}Tm [10]. The new line at 1.40 MeV has a similar halflife, of about 3 μs, suggesting the assignment to the activity of ^{145}Tm. Since the daughter activity is an even-even nucleus ^{144}Er, the interpretation of the 1.73 MeV and 1.40 MeV proton lines as the transitions populating the 0^+ ground-state ($I_p(0^+)$ of 90%) and the 2^+ excited state ($I_p(2^+)$=9.6(15)%) is obvious. The derived energy of 0.33(1) MeV is close to the $E(2^+)$ for ^{138}Sm, ^{140}Gd and ^{142}Dy, which are less exotic N=76 isotones. The valence correlation scheme N_pN_n [20] suggests similar $E(2^+)$ for ^{144}Er and its "N=82 mirror" nucleus ^{156}Er. Indeed, 0.33(1) MeV is close to known energy of 0.343 MeV for the 2^+ state in ^{156}Er. Similarly good agreement with N_pN_n systematics was found after the observation of fine structure in proton emission from ^{131}Eu [5]. The derived 2^+ state energy of 121±7 keV was close to the predicted value of 120±20 keV [21]. However, sometimes the shapes and level energy spectra of even-even nuclei near the proton drip line may be more complex than is accounted by the valence correlation scheme. The $E(2^+)$ value predicted as 160±20 keV [21] for ^{140}Dy was found to be 202.2 keV by the studies of ^{140m}Dy decay [22, 23]. Such energy difference, about 200 keV, was also found for two lines in the proton energy spectrum measured for ^{141gs}Ho activity at the RMS, see Fig. 3. It is a first observation of a very weak proton transition (I_p=0.70±0.15%) to the 2^+ excited state in ^{140}Dy, in addition to the dominating ground-state-to-ground-state decay. This small I_p value is in agreement with an upper limit of $I_p(2^+)$=1% obtained from the FMA study of ^{141gs}Ho decay [24].

The measured values of 2^+ energies can be used to estimate the deformation of the even-even nuclei [25, 26]. Using recent energy level systematic analyzed by Raman *et al.* [26], one obtains the deformation parameters of β_2=0.18 for ^{144}Er and β_2 about 0.23-0.24 for ^{140}Dy. The latter value is close to the estimation of β_2=0.25(4) derived from the observed level scheme of ^{141}Ho [24].

The wave function of ^{145}Tm can be composed from single-particle proton orbitals coupled to the ground and excited states of the even-even core [11]. To explain the observed partial-halflives of proton transitions within the spherical picture [27], the I^π=11/2$^-$ wave function should contain about 67% of the $\pi h_{11/2} \otimes 0^+$ and about 3.7% of the $\pi h_{7/2} \otimes 2^+$ configurations. The remaining 30% admixture is created from negative parity proton orbitals (l=3,5,7) coupled to the 2^+ core vibration. Since there is a non-negligible deformation expected from the 2^+ energy value as well as from recent data on neighbouring nuclei such as the proton-emitter ^{147}Tm [28, 29], the spherical estimations should be taken as "spherical reference values". More recently, models based on particle-core vibration coupling and accounting for non-spherical shapes have been developed [30, 31]. Here, the structure of the ^{145}Tm wave function and its decay process analyzed within Hagino's model [30, 11] are presented in Fig. 4. Similarly to the spherical reference description, the $\pi h_{11/2} \otimes 0^+$ forms about 56% of the wave function. The 1.4 MeV transition results from 3% admixture of $\pi h_{7/2} \otimes 2^+$. The $\pi h_{11/2}$ orbital dominates the function, with an additional 40% of $\pi h_{11/2}$ coupled to 2^+ core vibration.

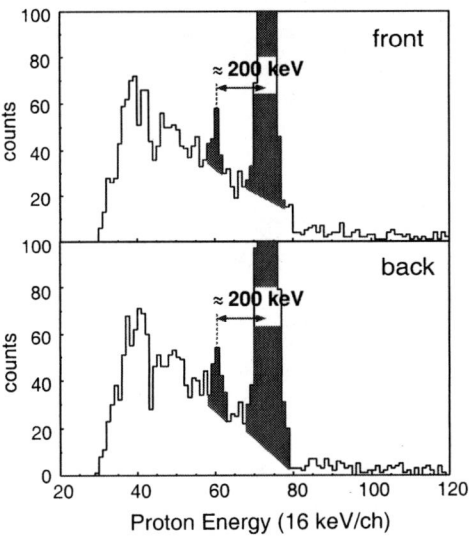

FIGURE 3. Fine structure in proton emission from ^{141gs}Ho. The 0.97 MeV proton lines, which are 0.2 MeV below dominating known transitions at 1.17 MeV, were observed in the energy spectra collected at the front and at back strips of the DSSD. The energy difference betweem proton lines fits perfectly to the excitation energy of the 2^+ state in ^{140}Dy [22, 23].

The $\pi h_{11/2} \otimes 2^+$ component does not contribute substantially to the decay width. The calculated halflife of 3.0(4) μs is in an excellent agreement with the measured value of 3.1(3) μs and the predicted branching ratio of 5.7(3)% is slightly below the measured value of 9.6(15)%. Calculations of Davids and Esbensen [31] reproduce the $I_p(2^+)$ value and halflife for ^{145}Tm decay. The contribution of $\pi f_{7/2} \otimes 2^+$ to the wave function is calculated to be 4%, as in the spherical and Hagino's descriptions. However, the $\pi h_{11/2} \otimes 0^+$ component which dominates the wave function in the above models, is only about 33%. This discreapance might be caused by different parametrizations of the optical potential, resulting in the potential more "transparent" for tunelling protons.

The interpretation of ^{141}Ho decay is more complex. It was anticipated that ^{141}Ho is a highly deformed nucleus, with a β_2 value near 0.3 [12, 13]. The equilibrium deformations calculated within the macroscopic-microscopic approach [32] in the $[\beta_2, \beta_4]$ space for the low-lying one-quasiproton states in ^{141}Ho are very similar, yielding deformations of $\beta_2 \approx 0.27$ and $\beta_4 \approx -0.07$ [13]. Such deformations were also obtained in ref.[33]. Lower β_2 deformation values were suggested by the analysis of the level schemes of ^{141}Ho (β_2=0.25(4) [24]) and of ^{140}Dy (β_2=0.23-0.24 [22, 23]). However, even with β_2 set to 0.24 and β_4 around 0.04-0.05, the halflife of the $7/2^-$[523] ^{141}Ho ground-state calculated within the non-adiabatic model [34, 35] is about 10 times longer in comparison to the observed 4 ms [22]. Also, the $I_p(2^+)$ branching ratio is about 2%, above the observed value 0.70(15)%.

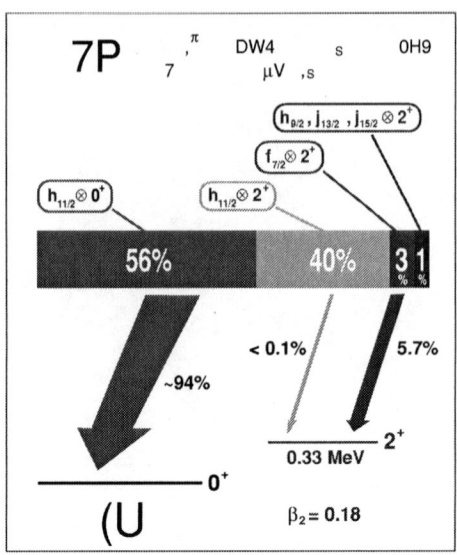

FIGURE 4. The wave function and proton emission widths for [145]Tm according to particle - core vibration coupling model of Hagino [11].

Theoreticcally, the non-adiabatic model for proton radioactivity provides the microscopic description of the emission process and the underlying nuclear structure. It fully takes into account the coupling of the single-particle motion of the proton to the deformed nuclear core. It describes well the decay of the highly deformed emitter [131]Eu [5, 34, 35]. Likely, the assumption of the prolate shape described within the $[\beta_2,\beta_4]$ deformation space might not be adequate for [141]Ho. A new approach to the proton emission from [141]Ho was presented at this meeting by A.Kruppa [36]. Better agreement with experimental data is achieved, when the coupling of the [140]Dy ground-state band and an excited γ-band starting at about 500-600 keV is included in the non-adiabatic description. A different scenario, within the adiabatic model, was presented by Davids [37] pointing to the role of the deformed spin-orbit term and neglecting the strong Coriolis coupling term for the $7/2^-[523]$ state. More complete experimental data on the [140]Dy level scheme, and in particular the observation of fine structure in proton emission from the $1/2^+[411]$ isomeric state in [141]Ho should help the intepretation of the decay process.

This work was supported by the U.S. D.O.E. through Contracts No. DE-FG02-96ER40983, DE-FG05-88ER40407, DE-FG02-96ER40978, DE-FG02-96ER41006, DE-AC05-76OR00033, DE-FG02-96ER40963, by the Polish KBN Grants No. 2P03B 08617 and 2P03B 04516, by the Hungarian OTKA Grants No. T026244 and T029003, and by NATO Grant No. PST.GLG.977613. The Joint Institute for Heavy Ion Research is supported by its members, University of Tennessee, Vanderbilt University and Oak Ridge National Laboratory. ORNL is managed by UT-Battelle, LLC, for the U.S. Department of Energy under Contract No. DE-AC05-00OR22725.

REFERENCES

1. Hofmann, S., *Radiochimica Acta* **70/71**, 93 (1995).
2. Woods, P.J. and Davids, C.N., *Annu. Rev. Nucl. Part. Sci.* **47**, 541 (1997).
3. Rykaczewski, K.P., *Eur. Phys. Jour.* **A 15**, 81 (2002).
4. Maglione, E. and Ferreira, L.S., *Eur. Phys. Jour.* **A 15**, 135 (2002) and refs. therein.
5. Sonzogni, A. A. et al., *Phys. Rev.* **C65**, 031303 (2002).
6. Ginter, T.N. et al., in *Proc. of First Int. Symposium on Proton-Emitting Nuclei (PROCON'99)*, ed. J.C. Batchelder, AIP **518**, p.83 (2002).
7. Ginter, T.N., PhD Dissertation, unpublished, Vanderbilt University, Nashville, TN (1999).
8. Gross, C.J. et al., *Nucl. Instrum. Methods Phys. Res.* **A450**, 12 (2000).
9. Ginter, T.N. et al., *Phys. Rev.* **C61**, 014308 (2000).
10. Batchelder, J.C. et al., *Phys. Rev.* **C57**, R1042 (1998).
11. Karny, M. et al., *Phys. Rev. Lett.* **90**, 012502 (2003).
12. Davids, C.N. et al, *Phys. Rev. Lett.* **80**, 1849-1852 (1998).
13. Rykaczewski, K. et al., *Phys. Rev.* **C60**, 011301 (1999).
14. Rykaczewski, K. P. et al., in *Proc. of Int. Conf. on Nuclear Structure "Mapping the Triangle"*, Wyoming, May 2002, AIP **638**, p. 149 (2002).
15. Grzywacz, R.K. et al., in *Proc. of 3rd Int. Conf. on Exotic Nuclei and Atomic Masses (ENAM 2001)*, Hameenlinna, Finland, July 2001 Springer Verlag, p. 453 (2002)
16. Grzywacz, R.K., in *Proc. of Int. Conf. on Electromagnetic Separation of Nuclei (EMIS14th)*, British Columbia, May 2002, Nucl. Instr. Meth. in Phys. Res. **B204**, 649 (2003).
17. Grzywacz, R.K., in *Proc. of Second Int. Symposium on Proton-Emitting Nuclei (Procon'03)*, Legnaro, Italy, February 2003, E. Maglione and L.Ferreira (eds.)
18. Shapira, D., Lewis, T.A., and Hulett, L.D, *Nucl. Instrum. Methods Phys. Res.* **A454**, 409 (2000).
19. Hubbard-Nelson, B. et al., *Nucl. Instrum. Methods Phys. Res.* **A422**, 411 (1999) and *http://www.xia.com*
20. Casten,R.F., "Nuclear Structure from a Simple Perspective", 2nd ed., Oxford, NY, p. 297-330 (2000) and earlier refs therein.
21. Zamfir, V.N., private communication (1998).
22. Królas, W. et al., *Phys. Rev.* **C65**, 031303 (2002).
23. Cullen, D.M. et al., *Phys. Lett.* **B529**, 42 (2002).
24. Seweryniak, D. et al., *Phys. Rev. Lett.* **86**, 1458-1461 (2001).
25. Grodzins,L., *Phys. Lett.* **2**, 88 (1962).
26. Raman, S., et al., *At. Data Nucl. Data Tables* **78**, 1 (2001).
27. S. Åberg *et al.*, *Phys. Rev.* **C56**, 1762 (1997), and *Phys. Rev.* **C58**, 3011 (1998).
28. Klepper, O., et al., *Zeit. Phys.* **A305**, 125 (1999).
29. Seweryniak, D., et al., *Phys. Rev.* **C55**, R2137 (1997).
30. Hagino, K., *Phys. Rev.* **C64**, 041304R (2001).
31. Davids, C.N. and Esbensen, H., *Phys. Rev.* **C64**, 034317 (2001).
32. Nazarewicz, W., *et al.*, *Nucl. Phys.* **A512**, 61 (1990).
33. Möller, P., et al., *At. Data Nucl. Data Tables* **66**, 131 (1997).
34. Kruppa, A.T., Barmore, B., Nazarewicz, W. and Vertse, T., *Phys. Rev. Lett.* **84**, 4549 (2000).
35. Barmore, B., Kruppa, A.T., Nazarewicz, W. and Vertse, T., *Phys. Rev.* **C62**, 054315 (2000).
36. Kruppa, A. and Nazarewicz, W., in *Proc. of Second Int. Symposium on Proton-Emitting Nuclei (Procon'03)*, Legnaro, Italy, February 2003, E. Maglione and L.Ferreira (eds.)
37. Davids, C.N., in *Proc. of Second Int. Symposium on Proton-Emitting Nuclei (Procon'03)*, Legnaro, Italy, February 2003, E. Maglione and L.Ferreira (eds.)

Studies Of Proton Emitting Nuclei At Legnaro National Laboratories

F. Soramel[*], A. Guglielmetti[¶], R. Bonetti[¶], M. Gernetti[¶], L. Stroe[§], M. Mazzocco[°], C. Signorini[°], M. Ivaşcu[^], C. M. Petrache[+]

[*]Dipartimento di Fisica and INFN, University of Udine, via delle Scienze 208, I-33100 Udine, Italy, E-mail: soramel@fisica.uniud.it
[¶]Istituto di Fisica Generale and INFN, University of Milano, I-20133 Milano, Italy
[§]INFN, Laboratori Nazionali di Legnaro, 35020 Legnaro, Padova, Italy
[°]Dipartimento di Fisica and INFN, University of Padova, I-35131 Padova, Italy
[^]NIPNE, P.O. Box MG-6, R.O.-76900 Bucharest - Magurele, Romania
[+]Dipartimento di Fisica e Matematica and INFN, University of Camerino, I- Camerino, Italy

Abstract. Since few years at the Legnaro National Laboratories we have recently started a program to study exotic decays at the proton drip-line with particular attention to proton radioactivity in the rare earth region. In fact p-emitting nuclei with $54 < Z < 64$ are expected to be quite deformed in their ground state and their study is an important input for the theoretical models that describe this kind of decay. Our studies have brought, as first result, the discovery of the decay by p-emission of ^{117}La where to p-decaying levels have been found. Our experimental program continued with the study of the decay of ^{126}Pm. For this nucleus we can only conclude that our low statistics data show no evidence of p emission.

INTRODUCTION

In the second half of the last decade a large amount of proton emitters have been studied and their decay characteristics have been determined [1,2]. First examples of p-emitters, produced via (p,2n) and (p,3n) evaporation channels and located in the region $64 < Z < 82$, have been interpreted by means of simple WKB calculations being these nuclei spherical. However, in the region $50 < Z < 64$ the predicted p-emitters are expected to be quite deformed in their ground state ($\beta_2 \sim 0.3$) [3], thus making their interpretation more difficult due to the complexity of the emitted proton wave function. Since a complete explanation of the p-emission process should be able to describe this decay as a function of the deformation parameter, it is very important to study and characterize the p-emission from deformed nuclei.

Recent experiments [4,5] have measured the p-decay of well deformed nuclei such as ^{140,141}Ho and ^{131}Eu. Proton emitters below these nuclei and with $Z > 54$ are expected to have largely deformed ground states, among them ^{117}La is predicted to have $\beta_2 = 0.29$ [3] and ^{126}Pm nucleus is predicted to have $\beta_2 = 0.33$ [3]. Their decays would constitute a stringent test for the recently developed models of proton emission from deformed nuclei [6,7]. ^{117}La Decay is expected to have a Q_p ranging between 471 keV [8] and 1021 keV [9]. ^{126}Pm was studied [10] almost 20 years ago at the GSI on-line

CP681, Proton-Emitting Nuclei: Second International Symposium; PROCON 2003,
edited by E. Maglione and F. Soramel
© 2003 American Institute of Physics 0-7354-0150-0/03/$20.00

mass separator. The nucleus was populated via the p5n evaporation channel from the compound nucleus ^{132}Sm produced in the ^{40}Ca + ^{92}Mo reaction. A total number of 980 nuclei were produced, but no evidence for p-emission was found. The calculated Q_p values for ^{126}Pm range from 920 keV [11] to 1010 keV [3]. When our studies strted La and Pm nuclei, together with Pr and Tb, were the only odd Z species between tin and lead for which the proton drip-line had not been identified yet.

EXPERIMENT

At Legnaro National Laboratories we have recently installed a Double-sided Silicon Strip Detector (DSSD) ((40x40) strips, 1 mm wide and 60 µm thick) one meter downstream the RMS (Recoil Mass Spectrometer) focal plane. Combining the RMS M/q selection, the spatial and energy information given by the DSSD and an absolute clock (4 MHz) signal for half-life measurements, we are able to measure the decay properties of many nuclei with half-lives in the range 50 µs - 1 s. The reaction for ^{117}La study used a ^{58}Ni beam, delivered by the Tandem-LINAC (ALPI) accelerator at 310 MeV and with an average intensity of 1.5 pnA, on a ^{64}Zn target (1 mg/cm^2 self-supporting foil). To reduce the high recoil energy that would saturate our amplifiers, we inserted a degrader foil of 2 mg/cm^2 natural Ni in front of the DSSD. Total measuring time for ^{117}La was 36 hours. DSSD energy calibration was obtained from the known ^{147}Tm p-decay (E_p = 1.051 MeV and 1.131 MeV, $T_{1/2}$ = 560 ms and 360 µs) [12] produced through ^{58}Ni (E_{beam} = 261 MeV) + ^{92}Mo (during this calibration run we removed the natural Ni absorber). The data were affected by a high noise background on y signal that inhibited pixel analysis of the DSSD, consequently the analysis refers to correlated recoil-decay events occurring in the same strip. This fact restricts our capability to measure half-lives to values below 160 ms for ^{117}La.

The same setup used for measuring the p-decay of the ^{117}La nucleus was used to investigate a possible decay via proton emission from the odd-odd ^{126}Pm. Improvements were made in what concerns the treatment of the y-DSSD position signals: instead of using a single delay-line (2 ns per strip) and thus having problems with the noise, we now used two delay-lines (20 strips for each), so that the signals are less attenuated, making it possible to use them in the bi-dimensional identification of the position of each event.

A first attempt in studying the possible p-decay of the above mentioned nucleus was made at the ALPI accelerator of the Legnaro National Laboratories. We used a 330 MeV ^{58}Ni beam and an isotopically enriched (96 %) ^{74}Se target. The ^{126}Pm nuclei, populated after the p5n evaporation from the ^{132}Sm compound nucleus, were separated using the RMS and then implanted into the DSSD. Here their decay could be detected, and then correlations could be made between the implanted recoil and the subsequent decay occurring in the same detector pixel.

Unfortunately, due to the ^{74}Se very low melting point, this first experiment gave no conclusive results. Anyway, the measurement revealed that the use of the two delay lines for the Y position signals of the DSSD is seriously improving the quality of the

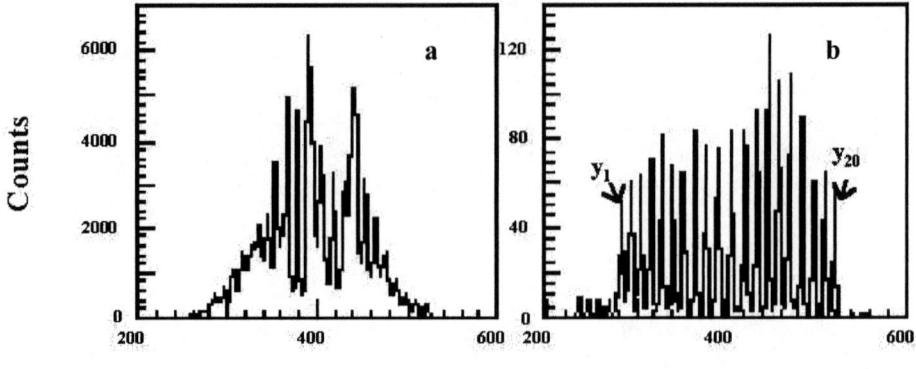

FIGURE 1. Y-DSSD position spectrum taken from one of the two delay-lines used in the present experiments for the ^{58}Ni + ^{92}Mo reaction. Shown are all the events (a) and all the decay events (b) seen by the detector.

Y-DSSD spectra, as shown in Fig. 1.

A second experiment was performed using the same setup and an inverse kinematics reaction, impinging a 425 MeV ^{74}Se beam onto a 1 mg/cm^2 self-supporting ^{58}Ni target. The calculated cross-section for this channel is about 10 nb, very similar to the one (~20 nb) reported [5] for the population of the ^{140}Ho nucleus, using the same (i.e. p5n) evaporation channel. Assuming the above mentioned cross-section, with a beam intensity of 1pnA and taking into account the RMS transmission plus the DSSD efficiency, we estimated to collect about 30 protons, in 7 days of measurement. We actually measured for a total of 94 hours with an average beam current of 0.5 pnA, therefore we expect about 10 events.

DATA ANALYSIS

^{117}La

During the ^{117}La run RMS fields were chosen to focus A = 117 and q = 30$^+$ recoils. Figure 2 shows the data from the ^{58}Ni + ^{64}Zn reaction. Panel (a) presents all decay events in the DSSD. Already from this spectrum with no selection, a peak is clearly emerging at low energy. The peak becomes evident in Figure 2b where a time requirement of $\Delta t \leq 100$ ms between the implanted recoil and the subsequent decay event has been imposed. Finally, in panel (c) a condition on the recoil A/q value has been added.

Despite the high background, the peak at 783 keV becomes more and more pronounced increasing the number of conditions. We therefore attribute this peak to

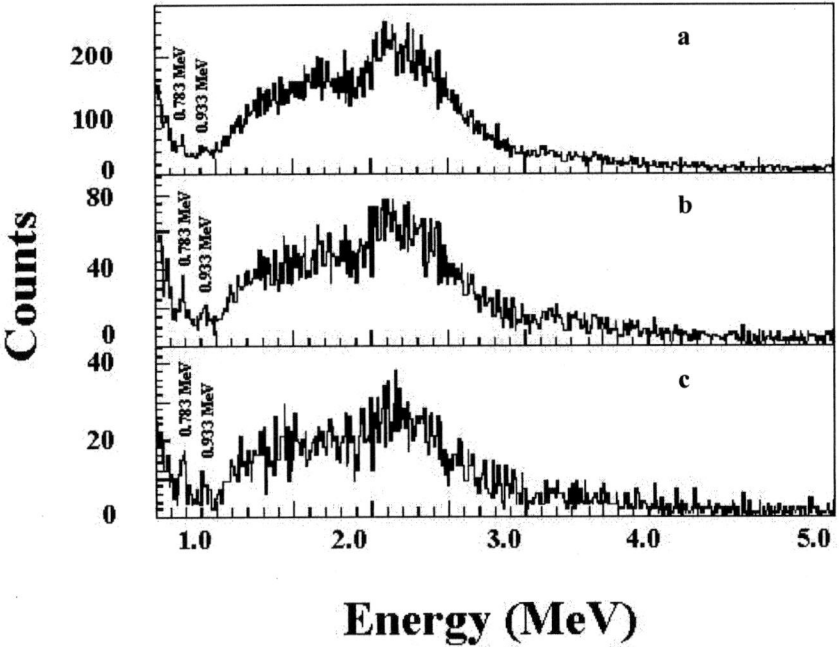

FIGURE 2. Decay events collected in the DSSD during the 310 MeV ^{58}Ni + ^{64}Zn run. Fig. 2a shows all the decays. Fig. 2b displays decay events occurring in a 100 ms time interval after a recoil implantation in a given strip. In Fig. 2c an additional condition M/q = 117/30$^+$ for the recoils is required.

the decay by proton emission of ^{117}La ground state to the ground state of ^{116}Ba, other possibilities being ruled out since they correspond to nuclei closer to the stability line than ^{117}La. After correcting for pulse-height defect and electron screening we get a Q_p-value of (800 ± 6) keV. Proton peak time analysis gives $T_{1/2}$ = (20 ± 5) ms. A correction of 10%, due to random coincidences in the DSSD strips, has to be applied to this value, therefore our final result for the decay half-life is $T_{1/2}$ = (22 ± 5) ms.

In addition, considering an overall RMS transmission of 5-10 %, a 60 % efficiency for the DSSD, and the 75 events detected for this decay, we get a cross section of ~ 200 nb for the ^{117}La ground state, in excellent agreement with other experimental cross sections for (p,4n) evaporation channels leading to p-emitters in this region [4,5].

In Figures 2b and 2c a second proton peak could be seen in the region immediately above 900 keV. Its analysis gives E_p = (933 ± 10) keV and $T_{1/2}$ = (10 ± 5) ms (Q_p = (951 ± 10) keV). This peak has about 1/3 of the 783 keV peak cross section. Our data do not show evidence of α lines. ^{117}La nucleus is predicted to be strongly deformed [3] with $β_2$ = 0.29 and $β_4$ = 0.1 and spherical calculations should be used only to fix a range for the possible half-lives. Using the code PREM [13], based on a simple WKB spherical model, the $Q_p = 800$ keV should correspond either to $T_{1/2} = 0.33$ ms, if the ground state were a $d_{5/2}$, or to $T_{1/2} = 110$ ms, if it were a $g_{7/2}$ level.

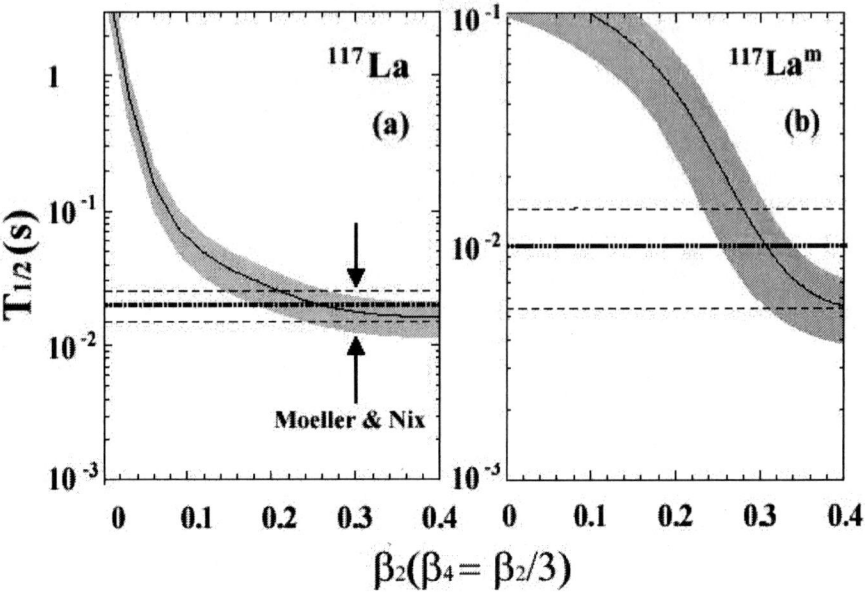

FIGURE 3. Proton partial half-life calculations as a function of the deformation parameter for 783 keV (a) and 933 keV (b) proton decay levels of ^{117}La. Grey bands represent the experimental error for the Q_p values, dashed lines correspond to experimental half-life errors.

A calculation able to take into account the ground state deformation becomes necessary to understand ^{117}La structure. Using the approach of Maglione et al. [6] it is possible to establish the ^{117}La Fermi surface: at large deformation ($\beta_2 = 0.3$ and $\beta_4 = 0.1$) the ground state is likely a $K = 3/2^+$ from the $d_{5/2}$ level (this result agrees with what expected from [3]). Performing a calculation for this level we get the result shown in figure 3a. Calculations include Q_p experimental error (grey band) and calculated [7] spectroscopic factor. Agreement with experimental results is good for $\beta_2 > 0.16$ with $\beta_4 = \beta_2/3$. Similar calculations performed for the second peak assuming that it is a $K = 9/2^+$ from the $g_{9/2}$ level, give the result of figure 3b: the deformation for this level is $\beta_2 > 0.24$ ($\beta_4 = \beta_2/3$). A more complete discussion of these data has been published in [14]. A subsequent experiment [15] performed at Argonne did confirm only the first of these two peaks.

^{126}Pm

The RMS fields during this experiment were set for a central trajectory with A = 126 and q = 36$^+$. As usual, prior to the run, we calibrated in energy the DSSD strips using the very-well known case of the ^{147}Tm.

Due to the low cross section, the ^{126}Pm data analysis requires stringent cuts on the measured parameters to enhance the proton events with respect to the background. Therefore we conditioned all decay events with an anti-coincidence signal from a veto detector positioned behind the DSSD: the veto detector (a (4x4) mm^2 Si detector) collects decay events which did not stop inside the DSSD (whose thickness is only 60 μm), but released only a part of their energy in it contributing to the strong background at E_{decay} ≤ 1 MeV, i.e. in the energy region where proton events are expected. Beside this general condition, we could clean our data acting on other two parameters, namely the TOF between the focal plane detector (MWPC) and the DSSD, and the recoil mass signal coming from the MWPC. The TOF signal allows separation of recoil events from scattered beam events, while the MWPC recoil mass signal allows selection of recoils on the A/q basis (in our case 126/36$^+$).

Last, but not least, we could select the decay events according to the time passed from the recoil implantation in a specific DSSD cell [(1x1) mm^2]. Fig. 4 shows the time evolution of the low energy part of the decay events spectrum in correspondence to a mass gate A/q = 126/36$^+$ and requiring both anti-coincidence with the veto detector and a TOF signal corresponding to recoils.

The result cannot be conclusive: Fig. 4 shows some possible candidates at E_p ~ 900 keV and 1100 keV. They have different half-lives, but in both cases they are too long (> 1 s) in comparison with the β-decay half-life which is expected to be around 200 ms. On the other hand the expected events are only 10 and the odd-odd nature of ^{126}Pm together with a strong deformation might open more than one decay channel, spreading the total cross section on several channels.

CONCLUSIONS

In conclusion ^{117}La has two p-decaying levels that populate the ground state of ^{116}Ba: the ground state with J^π = 3/2$^+$ decays via a (783 ± 6) keV proton with $T_{1/2}$ = (22 ± 5) ms, while the excited level, with E_x = 150 keV and J^π = 9/2$^+$, decays with E_p = (933 ± 10) keV and $T_{1/2}$ = (10 ± 5) ms. A direct γ decay of the ^{117}La excited level to the ground state would proceed through an M3 transition whose partial half-life would be 3.5 s (Weisskopf estimation, corrected for internal conversion), i.e. 350 times slower than the measured p-decay. Since our data do not show evidence of α decays and a possible β decay would have $T_{1/2}{}^\beta$ ~ 400 ms, we attribute 100% branching to each of the two p-decays.

For ^{126}Pm, our first conclusion is that we need a higher statistics run to be able to draw a firm conclusion on its decay nature. Of course a fast decay is always possible and can cot be descarted.

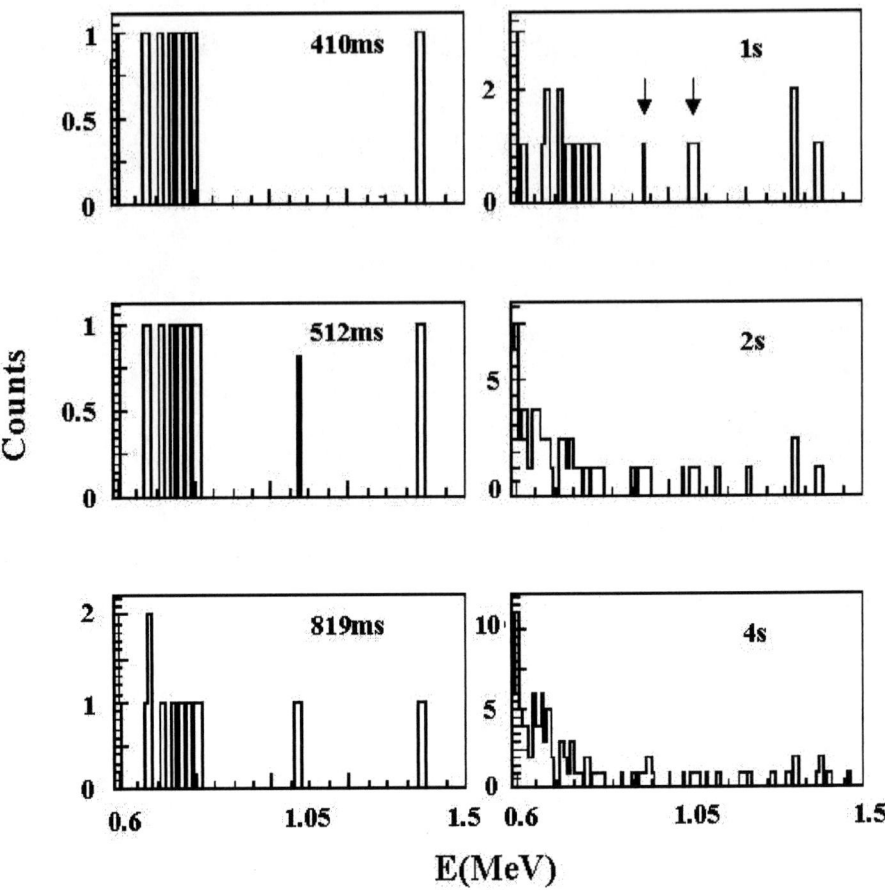

FIGURE 4. Experimental decay spectra for ^{126}Pm. Different spectra correspond on different time between the recoil implantation event and the decay event. The time gates are all starting at zero. All spectra are in anti coincidence with the veto detector, and correspond to recoil events requiring TOF and mass 126 conditions.

REFERENCES

1. Hoffmann, S., in *Nuclear Decay Modes*, edited by Poenar, D.N., IOP, Bristol, 143 (1996)
2. Woods, P.J., and Davids, C.N., *Ann. Rev. Nucl. Part. Sci.* **47**, 541 (1997)
3. Möller, P., et al, *At. Data Nucl. Data Tables* **59**, 185 (1995)
4. Davids, C.N., et al, *Phys. Rev. Lett.* **80**, 1849 (1998)
5. Rykaczewski, K., et al, *Phys. Rev.* **C60**, 011301 (1999)
6. Maglione, E., Ferreira, L.S., and Liotta, R.J., *Phys. Rev.* **C59**, R589 (1999)
7. Ferreira, L.S., and Maglione, E., *Phys. Rev.* **C61**, 021304(R) (2000)

8. Audi, G., et al., *Nucl. Phys.* **A624**, 1 (1997)
9. Jänecke, J., and Masson, P.J., *At. Data Nucl. Data Tables* **39**, 289 (1988)
10. Larsson, P.O., et al., *Z. Phys.* **A314**, 9 (1983)
11. Liran, S., and Zeldes, N., *At. Data Nucl. Data Tables* **17**, 431 (1976)
12. Sellin, P.J., et al, *Phys. Rev.* **C47**, 1933 (1993)
13. Poli, G.L., PhD thesis, University of Milan, Milan, Italy, (1998)
14. Soramel, F., et al., *Phys. Rev.* **C63**, 031304(R) (2001)
15. Mahmud, H., et al., *Phys. Rev.* **C64**, 031303(R) (2001)

Proton decay studies at the gas-filled separator RITU

H. Kettunen, S. Eeckhaudt, T. Enqvist, T. Grahn, P. T. Greenlees, P. Jones,
R. Julin, S. Juutinen, A. Keenan, P. Kuusiniemi, M. Leino, A.-P. Leppänen,
P. Nieminen, J. Pakarinen, P. Rahkila, C. Scholey, and J. Uusitalo

Department of Physics, University of Jyväskylä, Finland

Abstract. The gas-filled recoil separator RITU has been used to study neutron-deficient nuclei above the $N=82$ neutron shell closure and beyond the proton drip line. New information about the proton emitting nuclei in Ir, Au and Tl isotopes was obtained including the identification of two new proton emitters. The systematics of low-lying states in odd-mass At isotopes beyond the proton drip line were studied down to [191]At using alpha-decay. A new low-lying state associated with an oblate $\pi(2p-1h)$ configuration was also observed in the corresponding Bi daughter nuclei. In order to realise these studies at the extreme limit of experimental devices, several technical developments at the RITU separator were required.

INTRODUCTION

During the past few years the decay spectroscopy of neutron deficient nuclei above the $N=82$ neutron shell closure and beyond the proton drip line has been one of the main goals of studies performed at the gas-filled recoil separator RITU [1]. Before this extreme limit of the capability of the experimental devices was reached, several technical developments have taken place at the RITU separator. An improved focal plane detector system was constructed taking into account the special needs of proton decay studies [2]. The dipole chamber and the beam stop of the separator were redesigned for better suppression of the beam particles. A beam-windowless helium filling-gas differential pumping system was developed in order to allow the use of higher beam intensities, which were needed especially for the decay studies of nuclei above the $Z=82$ proton shell closure and beyond the proton drip line.

The first test experiments of the modified focal plane detector system were performed for previously known proton emitters [167]Ir and [165]Ir [3]. In the same measurement proton decay of the new iridium isotope [164]Ir was also observed [2]. In the second experiment the decay properties of previously known proton activities [171]Au and [170]Au [3, 4, 5] were studied with high statistics and proton and alpha decay of the ground state in [170]Au were observed for the first time [6]. At the end of the Au-experiment a short test to measure production cross sections of light Tl isotopes was performed and the proton decay of the new isotope [176]Tl was observed. In addition, in-beam gamma-ray spectroscopy has been performed for the proton emitting nucleus [171]Au [7] utilizing the RDT-technique employing the RITU separator in conjunction with the JUROSPHERE germanium detector array.

CP681, *Proton-Emitting Nuclei: Second International Symposium; PROCON 2003*,
edited by E. Maglione and F. Soramel
© 2003 American Institute of Physics 0-7354-0150-0/03/$20.00

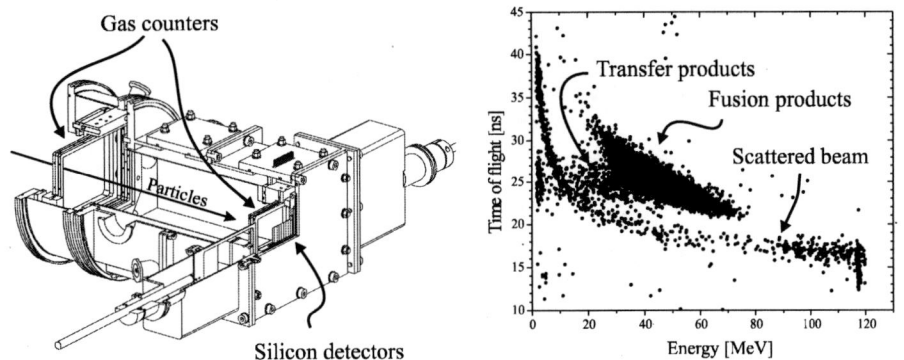

FIGURE 1. A schematic drawing of the focal plane detector system and a two-dimensional energy versus time-of-flight plot of the particles implanted into the silicon strip detector.

One challenging task of the experiments performed at RITU has been the search for proton emitting nuclei in the astatine isotopes. So far, no proton active isotope has been observed, but new information about the systematics of low lying states in odd-mass astatine isotopes [191,193,195]At and in the corresponding daughter nuclei beyond the proton drip-line have been obtained [8, 9].

TECHNICAL DEVELOPMENTS OF THE RITU SEPARATOR

The first improvement of the focal plane detector system was the installation of a multi wire proportional avalanche counter (MWPAC) in front of the position sensitive silicon strip detector shown in Fig. 1. By demanding that the events in the silicon strip detector were in anti-coincidence with events in the gas counter it was possible to separate implantation events from decay events observed in the silicon strip detector. In addition, the energy loss signals in the gas counter usually allowed the discrimination of scattered beam particles from fusion-evaporation products using an $E - \Delta E$ determination. For better discrimination the system was later extended with a second gas counter, allowing the determination of the time-of-flight of implanted particles over a flight path of approximately 300 mm. By combining the time-of-flight with the implantation energy of the particles in the silicon strip detector the fusion products could be easily separated from scattered beam particles and transfer products, as shown on the right hand side of Fig. 1.

Behind the silicon strip detector two quadrant silicon detectors of area $60 \times 60 \, \text{mm}^2$ and of thickness $450 \, \mu\text{m}$ were installed. These detectors were mainly used for the detection of energetic light particles which were able to punch through the first silicon strip detector. By combining the energy loss signals in the silicon strip detector and the second silicon detector, these punch through particles were identified to be protons and alpha particles, as shown with simulated curves in the upper left panel of Fig. 2. The lower left panel of Fig. 2 shows the energy loss spectrum of these particles in the silicon strip detector which was used for spectroscopic measurements. As can be seen

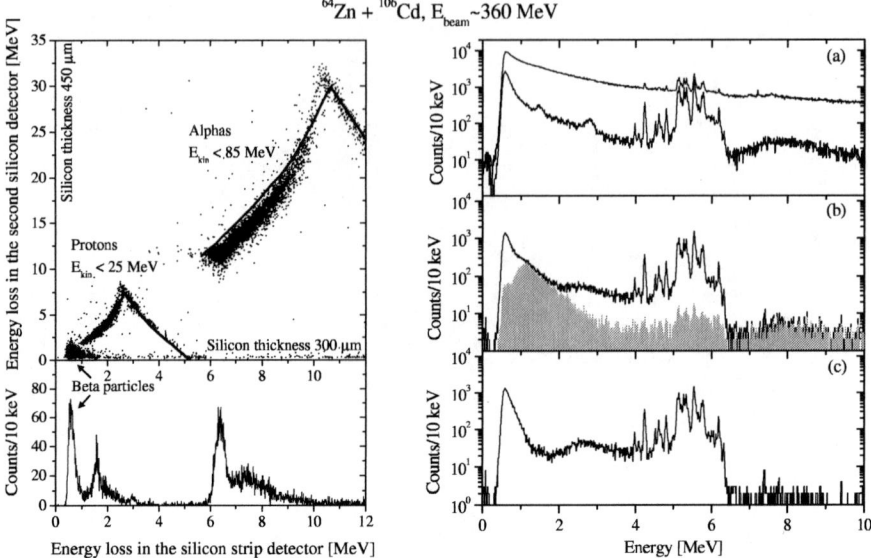

FIGURE 2. Upper left panel: Identification of the punch through particles using energy losses in silicon detectors. Solid lines represent simulated curves. Bottom left panel: The energy loss spectrum of the particles in the silicon strip detector. Right panel shows the decay spectra after various suppression combinations. (a) Total and beam particle suppressed alpha-decay energy spectrum. (b) With punch through suppression. The filled area shows the energy spectrum of the vetoed escape-alpha particles and low energy scattered light particles (alphas) in the region of $E > 6$ MeV. These particles are discriminated from the spectrum shown in panel (c).

the energy losses of the particles overlap the regions where the proton and alpha decay peaks were expected to appear.

For the studies of proton emitting nuclei the second gas counter was placed close (20 mm upstream) to the silicon strip detector such that it could be used to detect and veto so-called escape alpha particles. These alpha particles were emitted at backward angles so that only part of their kinetic energy was deposited into the silicon strip detector. Suppression of the escape alpha particles increases the detection sensitivity of the system especially for proton-emitting nuclei. This is because typical escape alpha particle energies observed in the silicon detector fall into the same region as typical proton decay energies, see the filled spectrum in the right-hand panel of Fig. 2 (b). The other spectra in the right-hand panel of Fig. 2 show the effects of various suppression combinations on a decay spectrum observed in the $^{64}Zn + ^{106}Cd$ reaction with a beam energy of 360 MeV in the middle of the target.

Another major improvement was the modification of the RITU dipole chamber and the beam stop as shown on the left-hand side of Fig. 3. In the recoil separator RITU the beam particles and fusion products are separated in gas using a magnetic dipole field. The beam particles which have smaller magnetic rigidities follow a trajectory with a much tighter radius than fusion products and are thus dumped in the beam stop inside the dipole chamber.

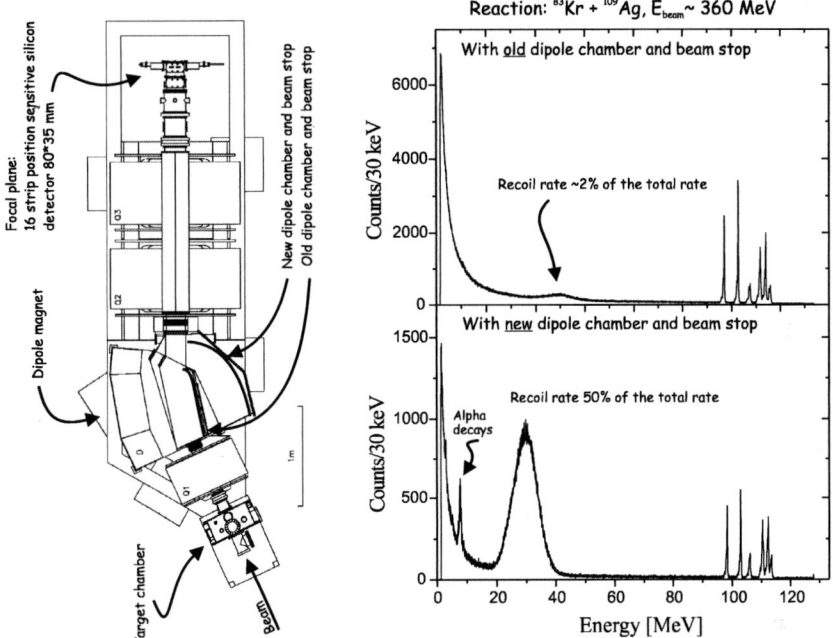

FIGURE 3. A Schematic drawing outlining the modifications to the dipole chamber and beam stop of the RITU separator. At the right hand side of figure is compared the energy spectra of the events observed at the focal plane silicon detector in ^{83}Kr + ^{109}Ag reaction. The upper spectra with original design and the lower with modified dipole chamber and beam stop.

In the original design the beam stop was placed quite close to the optical axis and fully inside the dipole magnet. This allowed the possibility that some of the beam particles scattered from the beam stop were transported to the focal plane. The design of the walls of the old dipole chamber could also be improved. When using symmetric reactions, where the differences between the magnetic rigidities of the beam particles and fusion products are small, beam particles could pass the edge of the old beam stop and scatter from the wall after it to the focal plane.

In the new design the dipole chamber is much larger and care was taken to avoid walls which could scatter particles to the focal plane. The beam stop is placed much further away from the dipole magnet giving more time and space for the separation. This is important especially in symmetric reactions as mentioned above. Also the shape of the new beam stop is designed so that the separated particles are dumped perpendicular to the stop to avoid to scattering the particles back into the separator.

The right-hand panel of Fig. 3 shows the effect of the new design on the total spectrum measured with the silicon strip detector at the focal plane. Both of the spectra are measured in the bombardment of a ^{109}Ag target with a ^{83}Kr beam at an energy of 357 MeV in the middle of the target. The upper panel shows the total spectrum measured with the original dipole chamber. The counting rate of the fusion products was approximately 2% of the total counting rate and their contribution is hardly visible

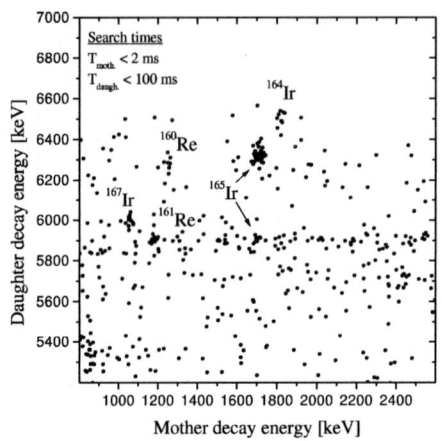

FIGURE 4. Two-dimensional mother and daughter decay energy plot for recoil-mother-daughter decay correlated decay chains in ^{64}Zn+^{106}Cd reaction.

in the spectrum. The lower panel shows the total spectrum from the same reaction but now measured after the installation of the new dipole chamber and beam stop. For this reaction the background is reduced by factor of 25 and the counting rate of fusion products is approximately 50 % of the total rate. Fusion products and even alpha-decay peaks are clearly visible in the spectrum.

PROTON DECAY STUDIES

The focal plane detector system described above was first tested in an experiment performed to study the previously known proton emitting nuclei ^{167}Ir and ^{165}Ir [3]. In the same measurement the proton decay of the new iridium isotope ^{164}Ir was also observed [2]. The experiment was performed using a ^{64}Zn beam with four different bombarding energies varying from 314 MeV to 362 MeV in the middle of a $550\,\mu g/cm^2$ thick ^{106}Cd target. The lowest bombarding energies were used to measure known proton decays of ^{167g}Ir, $E_p = 1064(6)$ keV and ^{165m}Ir, $E_p = 1707(7)$ keV [3], which were used for the proton decay energy calibration.

Figure 4 shows a two-dimensional mother and daughter decay energy plot for correlated decay chains of the type recoil-mother decay-daughter decay ($ER - p_m/\alpha_m - \alpha_d$). Maximum search times for the mother and daughter decays were 2 ms and 100 ms, respectively. The plot includes data from all bombarding energies and five different proton emitters can be identified from the plot, even though the search times used are unsuitable for most of them. Four of the proton emitters were previously known (^{161}Re [10], ^{160}Re [11], ^{167}Ir and ^{165}Ir [3]) and in addition the new proton emitter ^{164}Ir can also be identified. The new radioactivity was seen to be followed by an alpha decay with energy of $E_\alpha = 6493(11)$ keV and half-life $T_{1/2} = (7.5^{+4.2}_{-2.0})$ ms compatible with the decay properties of ^{163}Os presented in ref. [12].

30

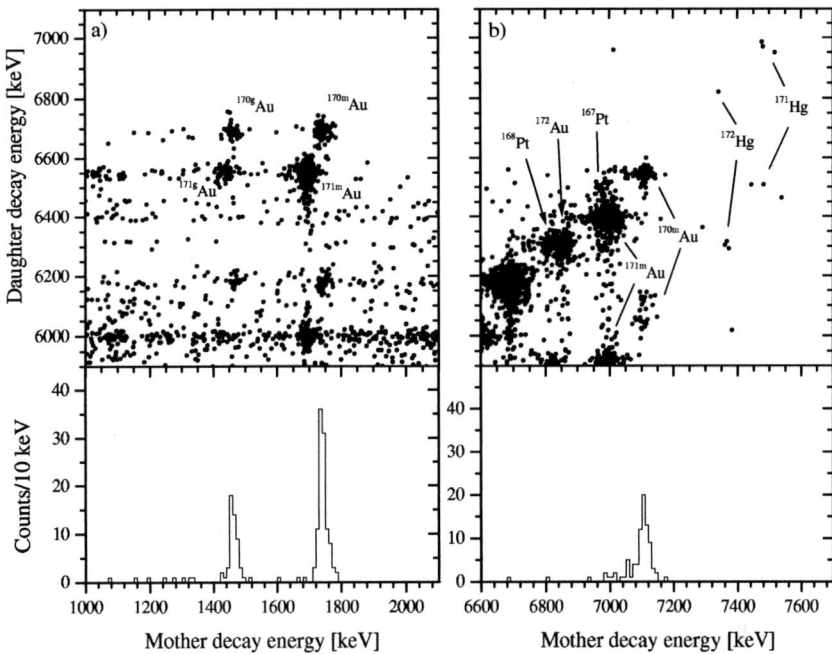

FIGURE 5. Two-dimensional plots for mother and daughter decay energies in ER $-$ p$_m$/α_m $-$ α_d correlated events in the ^{78}Kr$+^{96}$Ru reaction. (a) Correlated events where the proton decay of mother nuclei is followed by an alpha decay of the daughter nuclei (ER $-$ p$_m$ $-$ α_d). (b) Correlated decay chains consist of alpha decays (ER $-$ α_m $-$ α_d). Maximum search times for the mother and daughter decays were 10 ms and 200 ms, respectively. The lower panels show the energy spectra for ^{170}Au decays gated with the corresponding daughter decay in two-dimensional plots above.

The decay energy of $E_p = 1817(9)$ keV and half-life $T_{1/2} = \left(113^{+62}_{-30}\right)\mu$s obtained for ^{164}Ir can only be explained by $l=5$ emission, corresponding to a proton decay originating from an $\pi h_{11/2}$ orbital. The measured spectroscopic factor $S_p^{exp} = 0.19(7)$, based on a WKB approximation through the real part of a Becchetti-Greenlees optical potential [13], is in agreement with the theoretical spectroscopic factor of $S_p^{th} = 0.33$ predicted by low-seniority shell model calculations [3]. Results of an independent measurement of the decay of ^{164}Ir are presented in ref. [5].

Decay studies of the previously known proton active ^{171}Au and ^{170}Au isotopes [3, 4, 5] were performed with improved accuracy and the decay of the ^{170}Au ground state was observed for the first time [6]. A ^{78}Kr beam with seven bombarding energies varying from 361 MeV to 391 MeV in the middle of the ^{96}Ru target with thickness of 500 μg/cm^2 and enrichment of 96.52 % was used. The known proton decay energies of ^{171}Au and ^{177}Tl (see below) were used for the proton decay energy calibration.

Two-dimensional mother and daughter decay energy plots for correlated decay chains of the type ER $-$ p$_m$/α_m $-$ α_d are shown in Figs. 5 (a) and (b). Figure 5 (a) shows the correlated decay chains where the proton decays of the mother nuclei are correlated with alpha decays of the daughter nuclei. Figure 5 (b) shows the correlated decay chains consisting of alpha decays. Maximum search times of 10 ms and 200 ms were used for mother and daughter decays, respectively. The lower panels in Fig. 5 show the energy spectra for ^{170}Au decays gated with the corresponding daughter decay in the two-dimensional plots. In addition, a new alpha active nucleus ^{171}Hg [6] was also observed in the experiment, as shown in Fig. 5 (b).

The spectroscopic factor obtained for the proton decay from the isomeric state in ^{170}Au indicates that the decay originates from the $\pi h_{11/2}$ orbital, which is in agreement with the result presented in ref. [5]. Correspondingly, the ground state proton decay was observed to originate from the $\pi d_{3/2}$ orbital. This would indicate, that the configuration of these states in ^{170}Au are similar to those observed in alpha decay daughter nucleus ^{166}Ir. This assumption is supported by the observation of an unhindered alpha decay of these states in ^{170}Au to the corresponding states in ^{166}Ir.

At the end of the Au-experiment, a short test to produce the new proton emitting nucleus ^{176}Tl was performed by using a ^{78}Kr beam on a ^{102}Pd target [6]. Three bombarding energies, varying from 380 MeV to 389 MeV in the middle of an enriched 800 μg/cm^2 thick ^{102}Pd target were used. For ^{176}Tl one proton decaying branch was observed, corresponding to the ground state decay. Based on the spectroscopic factor examination the proton decay was deduced to originate from the $\pi s_{1/2}$ orbital (low spin state). This would differ from what is observed in the lighter odd-odd proton emitters, where the $\pi d_{3/2}$ orbital is associated with the ground state configurations.

The decay of an isomeric high spin state in ^{176}Tl was not observed in the experiment although high-spin states are expected to be populated much more strongly than low-spin states in heavy-ion fusion-evaporation reactions. Therefore it can be deduced that the half-life of the high-spin state is too short to be observed with the data acquisition system used. The system dead time was approximately 10 μs. This assumption is also supported by the high excitation energy of 807(18) keV of the high-spin state in ^{177}Tl [4]. The production cross-section of the ground state of ^{176}Tl was approximately 3 nb. In addition to ^{176}Tl, a few events corresponding to the proton and alpha decays of the isomeric state in ^{177}Tl were also observed in the experiment. The results of the Au- and Tl-experiments will be discussed in more detail in ref. [6].

DECAY STUDIES OF ODD-MASS ASTATINE ISOTOPES BEYOND THE PROTON DRIP LINE

Alpha-decay properties of the new isotope ^{191}At were investigated for the first time and the decay properties of ^{193}At and ^{195}At were studied with improved accuracy. New information concerning the low lying states in the corresponding daughter nuclei ^{187}Bi, ^{189}Bi and ^{191}Bi was also obtained. The results of the experiments are discussed in more detail in refs. [9, 8].

Three alpha-decaying states were identified for ^{193}At, and two for both ^{191}At and ^{195}At. For each of these isotopes the 1/2$^+$ intruder state was observed to be the ground state (see Fig. 6 (a)), though in ^{193}At also the 7/2$^-$ state with an excitation energy of 7(10) keV could also be the ground state within the accuracy of the measurement. The alpha decays of the excited $7/2^-$ states in ^{195}At and ^{193}At were observed to feed excited $7/2^-$ states at $148.7(5)$ keV and $99.6(5)$ keV, seen for the first time in the corresponding daughter nuclei ^{191}Bi and ^{189}Bi, respectively. The spin, parity and excitation energies of these final states in the bismuth isotopes were determined using the properties of gamma-ray transitions observed in coincidence with the alpha decay of the ^{195}At and ^{193}At isotopes. Identification of the $13/2^+$ state in ^{193}At was also based on alpha-gamma coincidences. In ^{187}Bi the existence of the excited $7/2^-$ state at $63(10)$ keV was deduced based on the shape of the alpha-decay energy spectrum of ^{191}At and the systematics of the heavier odd-mass bismuth isotopes. The spin and parity assignments of the initial states in the astatine isotopes were based on the observations of unhindered alpha decays to the corresponding final states in the bismuth isotopes.

The level schemes suggested for ^{191}At, ^{193}At and ^{195}At were observed to differ from those observed in heavier odd-mass astatine isotopes from ^{197}At to ^{211}At. The intruder $1/2^+$ state, having a $\pi(4p-1h)$ configuration, was observed to become the ground state in ^{195}At. In the heavier odd-mass astatine isotopes, the ground state is the $9/2^-$ state with configuration $(\pi h_{9/2})^3$. Also a $7/2^-$ state rather than a $9/2^-$ state was suggested to represent the first excited state in these light astatine isotopes. The emergence of the $7/2^-$ state over the $9/2^-$ state can be understood by assuming a change in deformation of this three-particle configuration between the ^{197}At and ^{195}At isotopes. Since no sizeable ground state deformation was observed in ^{197}At [14] the last proton in the $\pi h_{9/2}$ orbital creates the $9/2^-$ ground state. According to the Nilsson diagram a $7/2^-$ state, associated with an oblate $7/2^-[514]$ Nilsson state, originating predominantly from the $\pi h_{9/2}$ orbital at sphericity and having a mixed $\pi f_{7/2}/\pi h_{9/2}$ character at oblate deformation, becomes available for the 85th proton in odd-mass astatine isotopes if sufficient oblate deformation is assumed. Based on the results presented in Refs. [8, 9] it was proposed that the deformed three-particle configuration, driving the last proton to the $7/2^-[514]$ Nilsson, state is energetically more favoured than the nearly spherical $(\pi h_{9/2})^3$ configuration in light $A < 197$ odd-mass astatine isotopes. A sudden change in the ground state deformation from a nearly spherical shape to an oblate shape is theoretically predicted to happen between ^{199}At and ^{198}At by Möller et al. in ref. [15]. Correspondingly, the existence of a low-lying $7/2^-$ state in bismuth isotopes can be understood by a single particle $7/2^-[514]$ Nilsson proton state associated with oblate deformed structures.

The proton separation energies of astatine isotopes decrease smoothly with decreasing mass number until the first clearly proton unbound isotope ^{195}At [8] is reached, as can be seen in Fig. 6 (b). The graph behaves like two curves would cross between ^{197}At and ^{195}At. This is indeed true, since the change in the ground state assignment from a $9/2^-$ state to a $1/2^+$ state (see Fig. 6 (a)) causes the bend in the proton separation energy systematics. Surprisingly, a similar, but weaker bend can also been seen in the proton separation energy systematics of bismuth isotopes. In analogy to astatine isotopes it may indicate a change also in the structure of bismuth isotopes. However, a possible inaccuracy in the present atomic mass data on this region of nuclear chart, used for the plot, cannot be fully excluded. Proton separation energies of approximately -560 keV and -1020 keV can be estimated for ^{193}At and ^{191}At, respectively. For ^{191}At this would correspond to a partial half-life of approximately 57 s for an unhindered proton decay

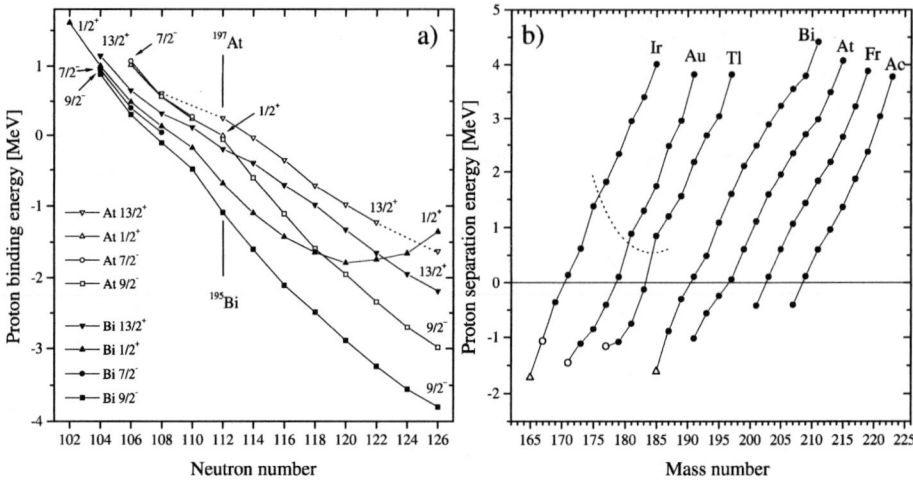

FIGURE 6. (a) Proton binding energies of odd-mass bismuth and astatine isotopes. The energies of the excited states were obtained by simply adding the excitation energies to the corresponding binding energy of the ground state. (b) Proton separation energies for odd-mass Ir, Au, Tl, Bi, At, Fr and Ac isotopes. The filled circles denote energies derived from measured atomic masses presented in refs. [16, 17], except for Ac and Fr isotopes and for Tl, Au and Ir isotopes below the dotted line the masses presented in ref. [18] were used. The open circles represent the measured proton separation energies for the ground state to ground state proton decays [3, 4]. The open triangles represent the measured proton separation energies, but in these cases the proton decay most probably originates from an excited state [19, 20]

from the $\pi s_{1/2}$ orbital. This is calculated using the WKB barrier transmission approximation through the real part of a Becchetti-Greenlees optical potential [13] and using a spectroscopic factor of one. Although the WKB approximation is very rough and works mainly for the spherical nuclei, it shows that the branching ratio of the proton decay compared to the alpha-decay with a half-life of approximately 2 ms would be too small for detection. By extrapolating the systematics of the proton separation energies of the astatine isotopes it can be estimated that the proton separation energy of the next odd-mass astatine isotope [189]At would be approximately -1500 keV, which would surely be negative enough for a proton decay to compete successfully with an alpha decay.

FUTURE DEVELOPMENTS

For future experiments, the focal plane detector system has been upgraded by the U.K. Universities GREAT spectrometer. The main detector of the spectrometer consists of two 60×40 double-sided silicon strip detectors for the implantation of fusion products and decay spectroscopy. At backward angles around the silicon strip detectors a box of 28 silicon PIN diode detectors is installed for the detection of conversion electrons and escaped alpha particles. Approximately 450 mm upstream of the silicon strip detectors is a position-sensitive multiwire proportional counter for energy loss and time-of-flight measurements of the implanted particles. Directly behind the silicon strip detector, inside the focal plane vacuum chamber a segmented planar Ge detector with 24×12 orthogonal strips for low-energy gamma rays, X-rays and beta particles will soon be installed. For the detection of higher energy gamma-rays a Compton-suppressed large volume segmented Ge clover detector will be placed adjacent to the detector chamber. The GREAT spectrometer will increase the sensitivity of the focal plane detector system facilitating the detection of the weakest reaction channels. Due to the high granularity of the spectrometer the reduction of the background events using correlation and coincidence methods will be considerable. An important part of the GREAT project is the development of a new type of data acquisition system known as total data readout (TDR) [21]. The system is based on VXI-electronics and is designed to minimize the dead time of the acquisition process. Instead of a common master trigger for the reading of the ADC-channels, all the channels are running independently. The data is associated and time-stamped by software using a global 100 MHz clock. The GREAT spectrometer and the new data acquisition system have recently been commissioned at RITU.

ACKNOWLEDGMENTS

This work was supported by the Academy of Finland under the Finnish Centre of Excellence Programme (Project No. 44875, Nuclear and Condensed Matter Physics Programme at JYFL), and by the EU under the RTD project EXOTAG (Contract No. HPRI-CT-1999-50017).

REFERENCES

1. M. Leino, *et al.*, *Nucl. Instrum. Methods Phys. Res. B*, **99**, 653–656 (1995).
2. H. Kettunen, *et al.*, *Acta Phys. Pol. B*, **32**, 989–992 (2001).
3. C. N. Davids, *et al.*, *Phys. Rev. C*, **55**, 2255–2266 (1997).
4. G. L. Poli, *et al.*, *Phys. Rev. C*, **59**, R2979–2983 (1999).
5. H. Mahmud, *et al.*, *Eur. Phys. J. A*, **15**, 85–87 (2002).
6. H. Kettunen, *et al.*, *to be published* (2003).
7. T. Bäck, *et al.*, *accepted for publication in Eur. Phys. J. A.* (2003).
8. H. Kettunen, *et al.*, *Eur. Phys. J. A, in press* (2003).
9. H. Kettunen, *et al.*, *submitted for publication in Eur. Phys. J. A.* (2003).
10. R. J. Irvine, *et al..*, *Phys. Rev. C*, **55**, R1621–1624 (1997).
11. R. D. Page, *et al..*, *Phys. Rev. Lett.*, **68**, 1287–1290 (1992).
12. C. R. Bingham, *et al..*, *Phys. Rev. C*, **54**, R20–23 (1996).
13. F. D. Becchetti, *et al..*, *Phys. Rev.*, **182**, 1190–1209 (1969).
14. M. B. Smith, *et al..*, *Eur. Phys. J. A*, **5**, 43–47 (1999).
15. P. Möller, *et al.*, *At. Data Nucl. Data Tables*, **59**, 185–383 (1995).
16. T. Radon, *et al.*, *Nucl. Phys. A*, **677**, 75–99 (2000).
17. Yu. N. Novikov, *et al.*, *Nucl. Phys. A*, **697**, 92–106 (2002).
18. G. Audi, *et al.*, *Nucl. Phys. A*, **624**, 1–124 (1997).
19. C. N. Davids, *et al.*, *Phys. Rev. C*, **76**, 592–595 (1996).
20. G. L. Poli, *et al.*, *Phys. Rev. C*, **63**, 044304 (2001).
21. I. H. Lazarus, *et al.*, *IEEE Trans. Nuc. Sc.*, **48**, 567–569 (2001).

Particle Emission From High-Spin States

D. Rudolph

Department of Physics, Lund University, S-22100 Lund, Sweden

Abstract. The present status of prompt particle emission from high-spin states in the mass $A \sim 60$ region is shortly summarized. The focus lies on recent results from a GAMMASPHERE experiment, which aimed at the combined in-beam high-resolution spectroscopy of light charged particles and γ rays. More detailed studies of known cases as well as the identification of new cases of discrete prompt proton and α-particle emission from highly- and superdeformed states have been possible. Finally, future directions are briefly outlined.

The decay mode of discrete-energy prompt proton- and α-particle emission has been established in nuclei near [56]Ni since 1998 [1]. Two main differences to the ground-state proton- and -α-particle emitters should be noted: At first, the prompt particle emission competes with γ rays instead of β^+ radiation. This places the time scale of the decays into the 10^{-12}–10^{-15} s regime, and allows their study in 'prompt' coincidence with preceeding and subsequent γ rays emitted from the parent and daughter nuclei, respectively. Secondly, the majority of prompt particle decays proceeds from highly- or superdeformed initial states into (near) spherical daughter states. This implies a drastic rearrangement of the nuclear mean field in the course of the decay. Hence, the decay mode may be viewed as a self-regulated two-dimensional quantum tunneling process, which is unique in Nature.

Following the first observation of prompt proton decay in [58]Cu [1], evidence was presented for prompt proton decays in [56]Ni [2] and [59]Cu [3] as well as the first case of prompt α-particle decay in [58]Ni [4]. An experiment with a set-up comprising the GAMMASPHERE Ge-detector array [5], parts of the charged-particle detector system MICROBALL [6], and a wall of four ΔE-E Silicon-Strip telescopes [7, 8] for high-resolution in-beam particle-γ coincidence spectroscopy rendered more precise results in [58]Cu [9]. In addition, more than ten new proton-decay lines and one new α-decay line have so far been observed in the decay-out regime of several rotational bands in [58]Ni and [59]Cu [10]. They imply the first observation of 'fine structure' for the new decay mode [11], involve states with competing α-, proton, and γ-radioactivity, and a near-spherical state in [58]Ni, which reveals a $h_{11/2}$ proton decay branch of approximately 2% [12]. Note that the results concerning [58]Ni have preliminary character.

A recent conference proceeding contains a table with information on all prompt particle decays observed until now (March 2003) [13]. Additional information may also be found in several previous conference proceedings as well as references therein [14, 15, 16, 17].

Given the plain number of prompt particle decays identified until now the new decay mode seems to be a common feature at least in proton rich nuclei near [56]Ni. The

CP681, *Proton-Emitting Nuclei: Second International Symposium; PROCON 2003*,
edited by E. Maglione and F. Soramel
© 2003 American Institute of Physics 0-7354-0150-0/03/$20.00

quantitative continuation of the present studies is clearly the quest for more candidates in the vicinity of ^{56}Ni and nearby, neutron-deficient regions. Qualitatively, a new high-statistics experiment has been approved at the GAMMASPHERE facility, which is going to focus on as precise as possible measurements of the angular distributions of the emitted particles with respect to the spin axis of the emitting nuclei. The spin axis can be determined on an event-by-event basis as it is perpendicular to the plane spanned by the beam axis and the recoil vector. The direction of the latter is indirectly determined from the measured momenta of all evaporated particles. To do this as precisely as possible, not only the forward but also the central section of MICROBALL will be replaced by a total of eight ΔE-E Si-strip telescopes with a total of some 1800 pixels. It should be noted that the spin alignment coefficient is on the order of $\alpha_2 \sim 0.7$-0.8 ($\sigma/J \sim 0.3$) when the particles are emitted from the high-spin states. In this case the spin alignment is a natural consequence of the reaction mechanism, while proton angular distribution measurements of ground-state- or β-delayed proton emitters require low temperature hyperfine interaction techniques [18, 19].

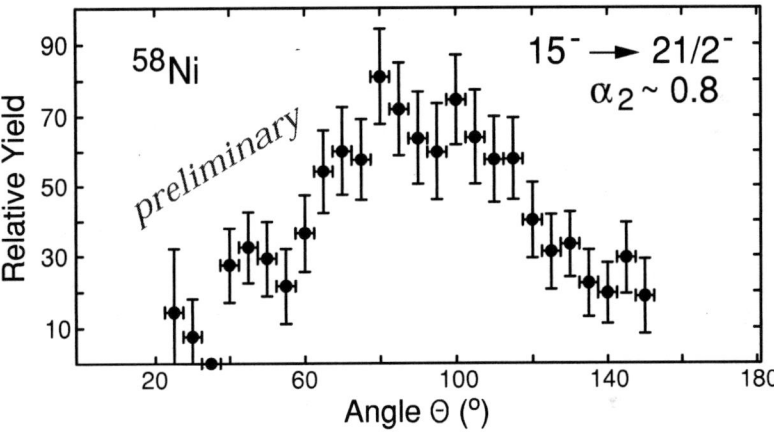

FIGURE 1. Angular distribution relative to the spin axis for one of the $1g_{9/2}$ prompt proton decays in ^{58}Ni (preliminary). See text for details.

Figure 1 shows preliminary results of the proton angular distribution (decay $p8$ in ^{58}Ni, see table in Ref. [12]) relative to the spin axis of emitting ^{58}Ni nuclei from the previous experiment. Only a subset of data is used where the evaporated α-particle and the later-on emitted proton were each detected in one of the four Si-strip telescopes. The overall shape of the distribution is consistent with the expectations for a particle with rather high angular momentum [20, 21, 22]. Interestingly, the peak around 90° indicates an emission from the tips of the axially deformed ^{58}Ni nucleus, which has been suggested as early as 1953 [23] (for α-decay from deformed nuclei). It remains to be seen whether eventual deviations from the basic shape of the angular distribution may provide access to more fundamental issues such as the radial distribution of a $1g_{9/2}$ proton in a deformed nuclear mean field or probing the nuclear time scale of the shape change of the nucleus [24, 25].

ACKNOWLEDGMENTS

D.G. Sarantites and his co-workers from Washington University deserve a lot of credit for their perfect and persistent work concerning the Si-strip high-resolution experiment. I would also like to thank all friends and colleagues who participated in one or several of the mentioned experiments. This research was supported in part by the Swedish Research Council.

REFERENCES

1. D. Rudolph et al., Phys. Rev. Lett. **80**, 3018 (1998).
2. D. Rudolph et al., Phys. Rev. Lett. **82**, 3763 (1999).
3. C. Andreoiu et al., in Proc. International Workshop Pingst 2000 – Selected Topics on $N = Z$ Nuclei, June 2000, Lund, Sweden, Eds. D. Rudolph and M. Hellström, (Bloms i Lund AB, 2000), p. 21.
4. D. Rudolph et al., Phys. Rev. Lett. **86**, 1450 (2001).
5. I.-Y. Lee, Nucl. Phys. A520, 641c (1990).
6. D.G. Sarantites et al., Nucl. Instrum. Meth. **A381**, 418 (1996).
7. MICROBALL, http://wunmr.wustl.edu/~dgs/mball.
8. B. Davin et al., Nucl. Instrum. Meth. **A473**, 302 (2001).
9. D. Rudolph et al., Eur. Phys. J. A **14**, 137 (2002).
10. C. Andreoiu et al., Eur. Phys. J. A **14**, 317 (2002).
11. D. Rudolph et al., Phys. Rev. Lett. **89**, 022501 (2002).
12. D. Rudolph et al., to be published.
13. D. Rudolph, in Proc. Nuclear Structure with Large γ-Arrays: Status and Perspectives, September 2002, Legnaro/Padova, Italy, to be published in Eur. Phys. J. A.
14. D. Rudolph, in Proc. International Conference on Achievements and Perspectives in Nuclear Structure, July 1999, Crete, Greece, Eds. S. Åberg and C. Kalfas, Phys. Script. **T88**, 21 (2000).
15. D. Rudolph, in Proc. The Nucleus: New Physics for the New Millenium, Faure, South Africa, January 1999, Eds. F.D. Smit, R. Lindsay, and S.V. Förtsch, Kluwer Academic / Plenum Publishers, New York, 1999, p. 397.
16. D. Rudolph, in Proc. Symposium on Proton–Emitting Nuclei, Oak Ridge, TN, U.S.A., October 1999, Ed. J.C. Batchelder, AIP Conference Proceedings **518**, 285 (2000).
17. D. Rudolph, in Proc. 3rd International Conference on Exotic Nuclei and Atomic Masses, Hämeenlinna, Finland, July 2001, Eds. J. Äystö, P. Dendooven, A. Jokinen, and M. Leino, Eur. Phys. J. A **15**, 161 (2002).
18. N.J. Stone and J. Rikovska, in Proc. Exotic Nuclei at the Proton Drip Line, Camerino, Italy, September 2001, Eds. C.M. Petrache and G. Lo Bianco, Universitá di Camerino, 2001, p. 229.
19. N.J. Stone, J. Rikovska, Sun Punan, and A. Woehr, in Proc. 12th International Conference on Hyperfine Interactions, Park City, Utah, August 2001, Eds. W. Evenson, H. Jaeger, and M.O. Zacate, Hyp. Int. **136**, 143 (2001).
20. S.G. Kadmensky and A.A. Sonzogni, Phys. Rev. C **62**, 054601 (2000).
21. S.G. Kadmensky, Phys. At. Nucl. **65**, 831 (2002).
22. E. Maglione, priv. comm.
23. D.L. Hill and J.A. Wheeler, Phys. Rev. **89**, 1102 (1953).
24. N. Carjan, P. Talou, and D. Strottmann, in Proc. The Nucleus: New Physics for the New Millenium, Faure, South Africa, January 1999, Eds. F.D. Smit, R. Lindsay, and S.V. Förtsch, Kluwer Academic / Plenum Publishers, New York, 1999, p. 115.
25. P. Talou, in Proc. International Workshop Pingst 2000 – Selected Topics on $N = Z$ Nuclei, June 2000, Lund, Sweden, Eds. D. Rudolph and M. Hellström, (Bloms i Lund AB, 2000), p. 10.

THEORY ON ONE PROTON EMISSION

Fine Structure and Triaxial Deformation in Proton Radioactivity[1]

Cary N. Davids and Henning Esbensen

Physics Division, Argonne National Laboratory, Argonne, IL 60439

Abstract. We report on theoretical calculations in proton radioactivity that have been carried out over the past 2 years at Argonne National Laboratory. This includes calculations of decay rates for near-spherical proton emitters based on particle-vibration coupling [1], fine structure in the proton decay of ^{141}Ho [2], and recent work on the effect of triaxiality on the decay rate of a deformed proton emitter.

FINE STRUCTURE IN PROTON EMISSION

Fine structure in proton radioactivity refers to the presence of a decay branch from the ground state of an odd-Z proton-emitting parent to an excited state of the daughter. For odd-A emitters this means decay to the first excited 2^+ state, as shown in Figure 1. The branching ratio provides information on individual wave-function components and whether or not the single-particle energies have been correctly determined. In addition, the energy difference between the proton groups directly gives the excitation energy of the daughter excited state, information which is usually very difficult to obtain by other means. From this quantity we can in turn derive information on the shape parameters of the daughter nucleus. A recent example of fine structure following decay of a near-

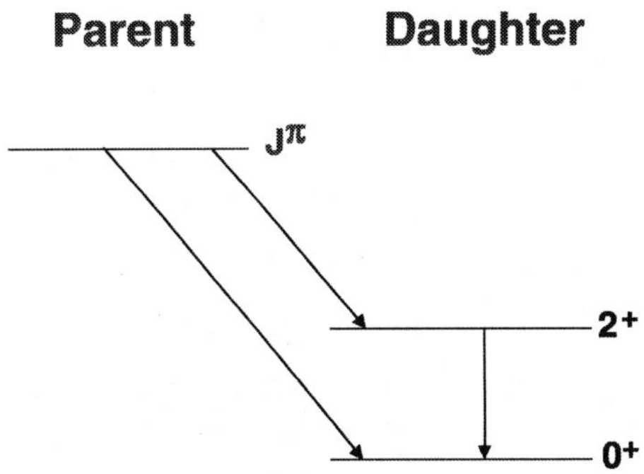

FIGURE 1. Fine structure in proton radioactivity.

spherical proton emitter is that of 3-μs ^{145}Tm [3]. The authors reported a branching ratio

[1] Work supported by the U.S. Department of Energy, Nuclear Physics Division, under Contract W-31-109-ENG-38.

CP681, *Proton-Emitting Nuclei: Second International Symposium; PROCON 2003,* edited by E. Maglione and F. Soramel

of .096(15) to the 0.33(1) MeV 2^+ state in the daughter ^{144}Er, while the ground-state decay rate is consistent with that expected for an $l = 5$ transition in the spherical picture. Under these circumstances, no decay to the 2^+ state is allowed by angular momentum conservation, and thus the observed branch must be due to small admixtures in the ground-state wave function caused either by collective rotational or vibrational effects.

Wave-function admixtures were studied by Davids and Esbensen [1]. Considering first rotational coupling, a quadrupole deformation parameter β_2-value of 0.18 is indicated by the 2^+ excitation energy in ^{144}Er. Using the adiabatic formalism developed in ref. [4], and assuming the same deformation for both parent and daughter, the decay rate and 2^+ branching ratio were calculated for all ^{145}Tm spins between $1/2^\pm$ and $11/2^\pm$, and both oblate and prolate deformations. The results indicate that only $J=5/2^-$ with oblate deformation could come close to explaining the experimental results. Other considerations led to this configuration being an unlikely description of the ^{145}Tm ground state, and attention turned to wave-function admixtures as being responsible for the decay properties of ^{145}Tm.

Another way to generate wave-function admixtures is through particle-vibration coupling. In this mode, the nuclear shape oscillates, but the time-averaged shape is spherical:

$$R(\theta, \phi) = R_0 \left[1 + \sum_{\lambda\mu} \alpha_{\lambda\mu} Y^*_{\lambda\mu}(\theta, \phi) \right],$$

where the $\alpha_{\lambda\mu}$ are time-dependent. The ground-state of an odd-Z odd-A proton emitter contains components produced by coupling a vibrational phonon to the proton single-particle wave function. Restricting to a $\lambda = 2$ phonon, this brings into the wave function of a proton emitter with spin I components like $|j_p \otimes 2^+\rangle$, where $|I - 2| \le j_p \le I + 2$.

We use a Woods-Saxon form for the single-particle potential, with a form factor $f\left(\frac{r-R}{a}\right) = \left[1 + exp\left(\frac{r-R}{a}\right)\right]^{-1}$. Since we expect the size of the $\alpha_{\lambda\mu}$ to be small, we can expand the form factor in powers of $\alpha_{\lambda\mu}$:

$$f\left(\frac{r-R}{a}\right) = f\left(\frac{r-R_0}{a}\right) - R_0 \sum_\mu \alpha_{\lambda\mu} Y^*_{2\mu}(\hat{r}) \frac{d}{dr} f\left(\frac{r-R}{a}\right).$$

The vibrational term in the single-particle potential is then

$$\delta V_{vib}(r, \alpha_{2\mu}) = F_{vib} \sum_{\lambda\mu} \alpha_{\lambda\mu} Y^*_{\lambda\mu}(\hat{r}),$$

$$F_{vib} = -R_N \frac{dV_n(r)}{dr} - R_C \frac{4\pi Z_D e^2}{5} \int dr' r'^2 \frac{d\rho_D(r')}{dr'} \frac{r^2_<}{r^3_>},$$

where $r_< = \min(r, r')$. Fig. 2 shows the wave-function components resulting from a coupled-channel calculation which includes particle-vibration coupling in the proton-core interaction. The small $l = 3$ $f_{7/2}$ wave-function component is responsible for the decay branch to the 2^+ state of the daughter, and Fig. 3 shows how both the observed branching ratio and total decay rate are well reproduced by these calculations. Figure 4 shows the good agreement between experimental and calculated spectroscopic factors

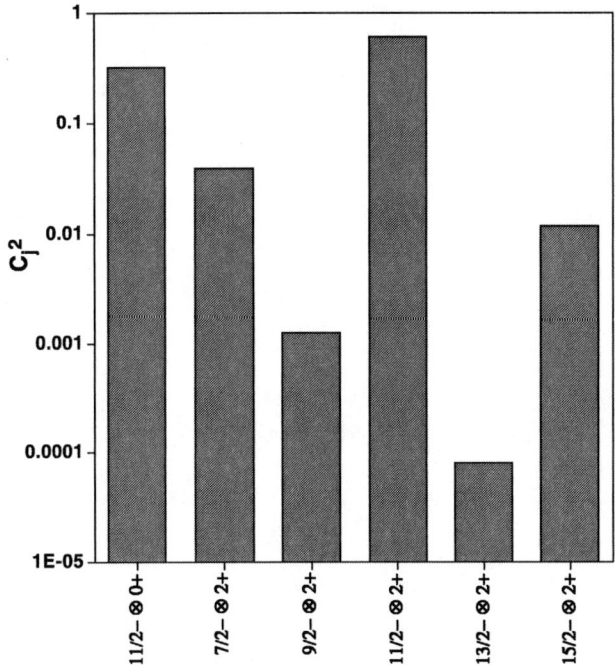

FIGURE 2. Wave function components for the ground state of ^{145}Tm.

obtained with this model for other near-spherical proton emitters, especially for the $d_{3/2}$ emitters, which were poorly fitted in previous treatments.

Fine structure in the decay of a deformed proton emitter was first observed in ^{131}Eu [6]. Decay of the $J = 3/2^+$ ground state populates both the 0^+ ground state of the daughter nucleus ^{130}Sm and the 2^+ excited state at 122 keV, with an observed branching ratio of 0.24(5). Fine structure was searched for in the decay of ^{141}Ho [7], and an upper limit of 1% was found for both proton-emitting states. At that time the excitation energy of the first 2^+ state in the daughter nuclide ^{140}Dy was not known, and a value of 166 keV estimated from $n_p n_n$ systematics [8] was used in calculations. Table I shows a comparison of calculations of the decay rate and branching ratios for ^{141}Ho, using both the adiabatic and non-adiabatic formalisms [4]. It was found that it was important to use a deformed spin-orbit term in the proton-core potential [4]. It can be seen from Table I that the non-adiabatic calculations give too small a decay rate compared with experiment. This is due to the presence of the full Coriolis interaction, which is not present in the adiabatic calculation.

Using the adiabatic formalism, ref. [4] translated the 1% upper limit on the ^{141}Ho branching ratio into a lower limit of 190 keV for the excitation energy of the first 2^+ state in the daughter nuclide ^{140}Dy. In recent experiments at ANL [2] and HRIBF [9], the excitation energy was determined to be 202 keV, leading to a predicted branching ratio

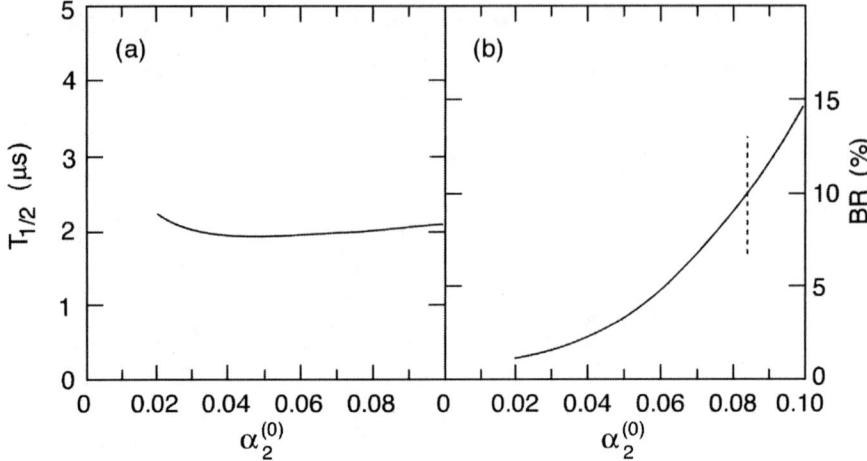

FIGURE 3. (a)Calculated proton decay half-life for ^{145}Tm as a function of $\alpha_2^{(0)}$, for an initial spin of $11/2^-$. (b)Calculated branching ratio $\Gamma(2^+)/[\Gamma(0^+) + \Gamma(2^+)]$. The vertical dashed line indicates $\alpha_2^{(0)}$=0.084. The spectroscopic factors for both branches have been set to unity.

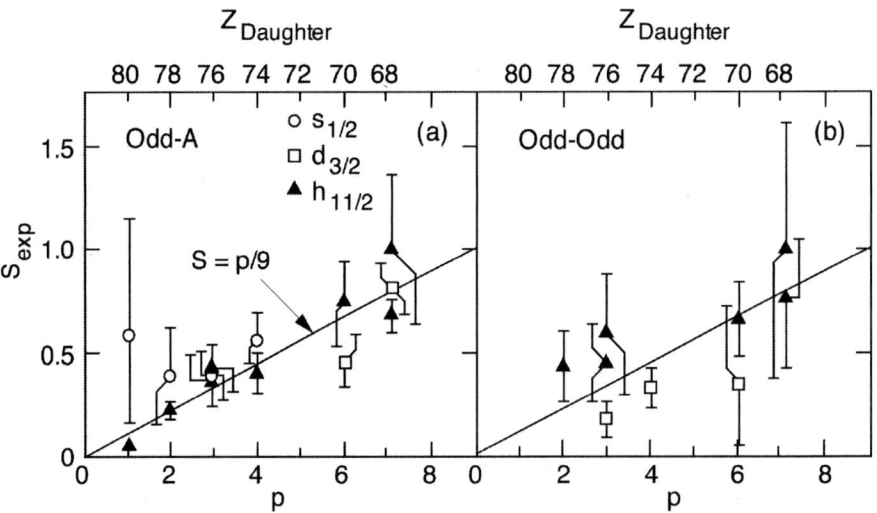

FIGURE 4. (a) Experimental spectroscopic factors S_{exp} calculated with particle-vibration coupling for odd-A proton emitters, plotted as a function of p, the number of pairs of proton holes below Z=82 possessed by the daughter nucleus. (b) Same as (a) except for odd-odd proton emitters. The solid line labelled $S = p/9$ is the predicted spectroscopic factor from a low-seniority shell model calculation [5].

in the adiabatic formalism of 0.7% for the ground state decay [2]. This is in excellent agreement with the value of 0.70(15)% from a recent experiment at HRIBF reported at

TABLE 1. Comparison of Adiabatic and Coupled Channels Calculations for Decay Widths and Branching Ratios in ^{141}Ho.

Method	I^π	Γ_0	Γ_2	Γ_2/Γ_{tot}	$t_{1/2}$
Adiabatic	$7/2^-$	16.5	0.45	0.027	2.7 ms
Coup. Chan.	$7/2^-$	3.38	0.27	0.079	12.5 ms
Experiment*	$7/2^-$	10.9(10)	-	$<.01$	4.2(4) ms
Adiabatic	$1/2^+$	21700	330	0.015	2.1 μs
Coup. Chan.	$1/2^+$	22530	317	0.014	2.0 μs
Experiment†	$1/2^+$	5700(2140)	-	-	8(3) μs
Experiment**	$1/2^+$	7020(1080)	-	$<.01$	6.5(10) μs

* Phys. Rev. Lett. **80**, 1849(1998), Phys. Rev. Lett. **86**, 1458(2001)
† Phys. Rev. C**60**, 011301(R) (1999)
** Phys. Rev. Lett. **86**, 1458(2001)

this conference [10]. The same calculation predicts a value of 0.4% for the branching ratio of the ^{141}Ho^m($1/2^+$) decay, for which only an experimental upper limit of 1% exists [7].

EFFECT OF TRIAXIAL DEFORMATION ON THE PROTON DECAY RATE AND BRANCHING RATIO

Calculations in ref. [7] of the energy levels of ^{141}Ho indicated that one possibility for bringing these calculations into better agreement with experiment was to introduce a small amount of triaxial deformation. Triaxial deformation means that the moments of inertia of the 3 axes of the nucleus are all different. In contrast, all calculations of deformed proton emitters currently assume that the nucleus is axially symmetric.

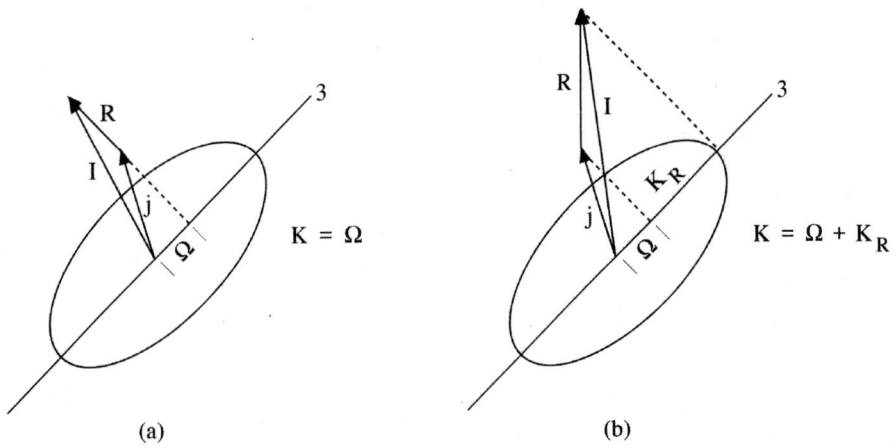

FIGURE 5. (a) Relationship of angular momentum vectors **j**, **R**, and **I** in the axially-symmetric case. (b) Same as (a) except for the non-axially-symmetric case.

In the laboratory system (R-representation) the projection of the total angular momentum $\mathbf{I}$ on the z-axis is M, while its projection along the 3- or long axis in the body-centered system is K. The total angular momentum $\mathbf{I}$ is the vector sum of $\mathbf{j}$, the particle angular momentum, and $\mathbf{R}$, the angular momentum of the core. In the body-centered system the projection of the particle angular momentum $\mathbf{j}$ on the 3-axis is denoted by Ω, the projection of the core angular momentum $\mathbf{R}$ is K_R, and $K = \Omega + K_R$. In the axially-symmetric case, the nucleus is symmetric under rotation around the 3-axis. This means that there can be no component of $\mathbf{R}$ along this axis ($K_R = 0$), and thus $K = \Omega$. This situation is depicted in Fig. 5(a).

The effect of introducing triaxiality into the nuclear shape is illustrated in Fig. 5(b). The core angular momentum $\mathbf{R}$ can have an arbitrary orientation, and since there is no longer a requirement that $K_R = 0$, we have $K = \Omega + K_R$. The nucleus is still symmetric upon a rotation of 180° about either of the 3 axes in the body-centered system, and the particle-core potential will contain a dependence on the orientation angle ϕ.

To proceed we make a multipole expansion of the particle-core interaction, and in the usual way obtain a set of coupled differential equations

$$(h_{lj} + E_R - E)\phi^I_{ljR}(r) = \sum_{l'j'R'} \sum_{\lambda \neq 0, \mu} \langle l(jR)IM|D^\lambda_{\mu 0}(\theta', \phi')|l'(j'R')IM\rangle V_{\lambda\mu}(r)\phi^I_{l'j'R'}(r),$$

where h_{lj} includes the monopole parts of the interaction (see [4] for further details on the notation). This expression differs from the analogous one for the axially-symmetric case by having an extra dependence on the projection quantum number μ and making the expansion in the D-functions instead of Legendre polynomials. Because of symmetry considerations, λ and μ are even numbers.

Since we will be comparing our results in the adiabatic limit, it is more convenient to work in the K-representation [4]. In this case we have

$$\Psi_{IM} = \sum_{lj} \sum_{K>0} \sum_{\Omega} \frac{\phi^{IK}_{lj\Omega}(r)}{r} |lj\Omega, KIM\rangle,$$

where we cast the radial wave functions $\phi^{IK}_{lj\Omega}(r)$ and the spin-angular wave functions $|lj\Omega, KIM\rangle$ in terms of their R-representation equivalents. Symmetry considerations require $K_R = K - \Omega$ to be an even integer $(0, \pm 2, \pm 4\ldots)$[11]. It should be noted that, compared to the axially-symmetric case, the radial wave function is labeled with an extra quantum number, Ω, and there is an additional sum over K.

The decay rate to a given state in the daughter nucleus with spin R is given by

$$\Gamma^I_R = \sum_{lj} \Gamma^I_{ljR} = \frac{\hbar^2 k_R}{\mu} \left| N^I_{ljR} \right|^2,$$

where

$$N^I_{ljR} = -\frac{2\mu}{\hbar^2 k_R} \sum_{\lambda\mu} \sum_{l'j'R'} \langle l(jR)IM|D^\lambda_{\mu 0}(\theta', \phi')|l'(j'R')IM\rangle$$

$$\times \int_0^{r_{int}} dr F_l(k_R r)\left[V_{\lambda\mu}(r) - \delta_{\lambda 0}\frac{Z_D e^2}{r}\right]\phi^I_{l'j'R'}(r).$$

46

TABLE 2. Allowed combinations of j and Ω for $I = K = 7/2^-$ originating from $j = 7/2^- \rightarrow 15/2^-$. The states occurring in the axially-symmetric case are marked with a •.

j	Ω	Axially-Symmetric	j	Ω	Axially-Symmetric
$7/2^-$	$-5/2$		$13/2^-$	$-13/2$	
$7/2^-$	$-1/2$		$13/2^-$	$-9/2$	
$7/2^-$	$3/2$		$13/2^-$	$-5/2$	
$7/2^-$	$7/2$	•	$13/2^-$	$-1/2$	
$9/2^-$	$-9/2$		$13/2^-$	$3/2$	
$9/2^-$	$-5/2$		$13/2^-$	$7/2$	•
$9/2^-$	$-1/2$		$13/2^-$	$11/2$	
$9/2^-$	$3/2$		$15/2^-$	$-13/2$	
$9/2^-$	$7/2$	•	$15/2^-$	$-9/2$	
$11/2^-$	$-9/2$		$15/2^-$	$-5/2$	
$11/2^-$	$-5/2$		$15/2^-$	$-1/2$	
$11/2^-$	$-1/2$		$15/2^-$	$3/2$	
$11/2^-$	$3/2$		$15/2^-$	$7/2$	•
$11/2^-$	$7/2$	•	$15/2^-$	$11/2$	
$11/2^-$	$11/2$		$15/2^-$	$15/2$	

Adiabatic Limit

In the adiabatic limit the coupled equations are diagonal in K, but retain Ω-mixing. This means that, as in the axially-symmetric case, K is a good quantum number, but in addition to the $\Omega = K$ component, the interaction mixes into the wave function components having Ω-values differing from K by ± 2, ± 4, etc., subject to the restriction $|\Omega| \le j$.

In the case of the ^{141}Ho ground state, $K = 7/2^-$. We expand the wave function in spherical components, with $j = 7/2^-, 9/2^-, 11/2^-, 13/2^-$, and $15/2^-$. This gives a total of 30 equations, with Ω taking on values of $-13/2^-, -9/2^-, -5/2^-, -1/2^-$, $+3/2^-, +7/2^-, +11/2^-$, and $+15/2^-$. Table II shows the allowed j-Ω combinations for the $I = K = 7/2^-$ ^{141}Ho ground state.

The resulting wave function amplitudes for the $j = 7/2$ spherical component are shown in Fig. 6, plotted as a function of the Hill-Wheeler parameter $a_{22} = \frac{1}{\sqrt{2}}\beta_2 \sin\gamma$. It can be seen that for axial symmetry (no triaxiality, $a_{22} = 0$), only the $\Omega = K = 7/2$ component is present, and as a_{22} increases, the components with other Ω-values begin to appear.

Fig. 7 shows the total calculated decay width and 2^+ branching ratio, also plotted as a function of a_{22}. For no triaxiality the total width has a value that agrees with experiment, with a spectroscopic factor of 0.73. The decay width to the ground state accounts for about 99% of the total decay width. In the adiabatic approximation, only the $j = 7/2^-$, $\Omega = 7/2$ wave function component can participate in the ground state decay. Its amplitude decreases with increasing a_{22} because of the growth of other Ω-components, which explains the decrease in overall decay rate. For the decay to the 2^+ state, the predicted branching ratio of 0.007 with $a_{22} = 0$ agrees very well with the experimental value of 0.0070(15) [10], and with increasing triaxiality it diverges from the experimental value. Only the $j = 7/2^-$ orbital contributes to this branch as well,

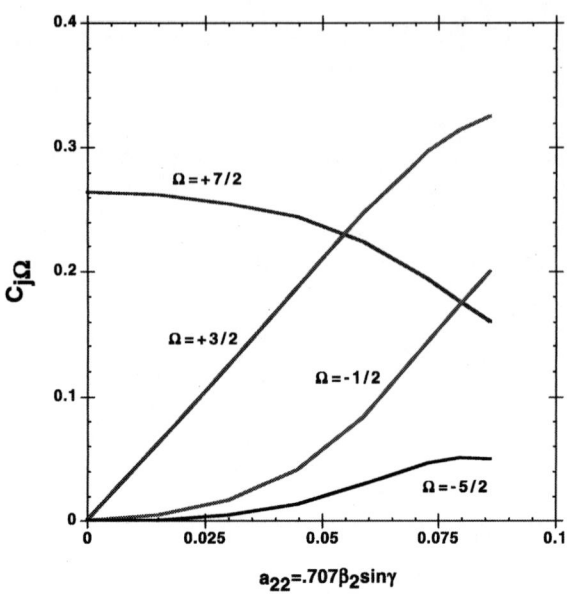

FIGURE 6. Wavefunction components for ^{141}Ho($7/2^-$) with increasing triaxiality.

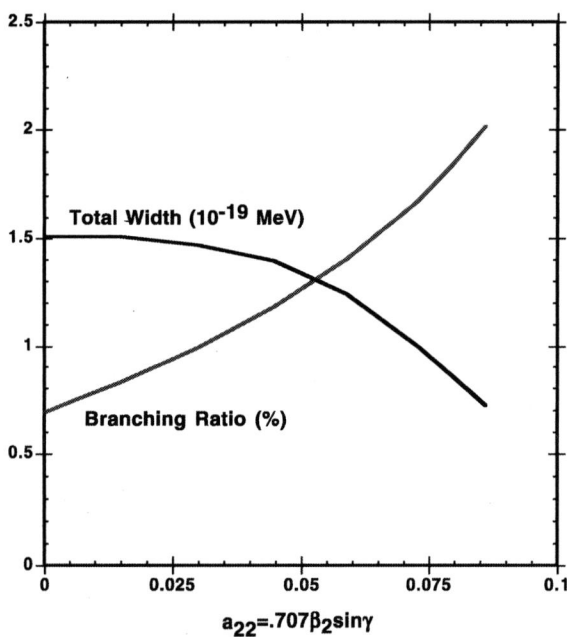

FIGURE 7. Total calculated decay width and 2^+ branching ratio for ^{141}Ho, plotted as a function of a_{22}.

since the higher-j components have $l \geq 5$. However, for $j = 7/2^-$, both $\Omega = 7/2$ and $\Omega = 3/2$ can participate, and since the $\Omega = 3/2$ amplitude increases rapidly with a_{22}, the 2^+ decay width also increases. The main conclusion to be drawn from these calculations is that, in the adiabatic limit, the axially-symmetric description agrees quite well with experiment, and the addition of triaxiality causes the total decay rate to decrease slowly and the 2^+ branching ratio to increase rapidly.

REFERENCES

1. C.N. Davids and H. Esbensen, Phys. Rev. C64, 034317 (2001).
2. D. M. Cullen et al., Phys. Lett. B529, 42 (2002).
3. M. Karny et al., Phys. Rev. Lett. 90, 012502 (2003).
4. H. Esbensen and C. N. Davids, Phys. Rev. C63, 014315 (2000).
5. C.N. Davids et al., Phys. Rev. C55, 2255 (1997).
6. A. A. Sonzogni et al., Phys. Rev. Lett 83, 1116 (1999).
7. D. Seweryniak et al., Phys. Rev. Lett 86, 1458 (2001).
8. V. Zamfir, private communication.
9. W. Królas et al., Phys. Rev. C65, 031303 (2002).
10. K. Rykaczewski et al., this proceedings.
11. A. Bohr, Mat. Fys. Medd. Dan. Vid. Selsk. 26, No. 14 (1952)

Non-adiabatic quasi-particle model for deformed proton emitters

Lídia S. Ferreira* and Enrico Maglione†

* Centro de Física das Interacções Fundamentais, and Departamento de Física,
Instituto Superior Técnico, Av. Rovisco Pais, P-1049-001 Lisboa, Portugal,
email: flidia@ist.utl.pt
† Dip. di Fisica "G. Galilei" and INFN, Via Marzolo 8, I-35131 Padova, Italy,
email: maglione@pd.infn.it

Abstract. Proton radioactivity from deformed nuclei is usually well reproduced using the particle-rotor model in the strong coupling limit. Recent attempts to go beyond the adiabatic approximation, by the addition of the Coriolis interaction, have failed in reproducing the experimental results. We show that the introduction of the pairing residual interaction in the non-adiabatic coupled channel method permits to recover the perfect agreement with data.

PROTON RADIOACTIVITY IN THE STRONG COUPLING APPROACH

Since nuclei on the drip–line have a Fermi level very close or even immersed in the continuum, decay of odd–Z even–N nuclei has been interpreted, as decay from a single particle Nilsson resonance of the unbound core–proton system[1, 2, 3, 4]. The states close to the Fermi surface are the most probable ones for decay to occur. Therefore, one can solve the Schrödinger equation with outgoing wave boundary conditions and determine the half–lives for decay from these s.p. resonances from the imaginary part of the energy, or simply by noticing that the partial decay width is the overlap between the initial and final states. If the second method is used, the Schrödinger equation still has to be solved in order to obtain the proton wave function. Some assumption has to be made for the wave function of the parent nucleus. If the parent nucleus is viewed as one particle plus a rotor with infinite moment of inertia, it will have a frozen excitation spectrum and a degenerate ground state. No Coriolis mixing is then included. This approach corresponds to the strong coupling limit [1]. Due to this collapse of the rotational spectrum into the ground state, the calculations have been called adiabatic.

Calculations in the strong coupling limit were also performed within the coupled-channel Green's function model [1, 2].

In these conditions, the partial decay with in the channel $(l_p j_p)$ is given by,

$$\Gamma^{J_d}_{l_p j_p} = \frac{\hbar^2 k}{\mu} \frac{(2J_d+1) < J_d, 0, j_p, K_i | K_i, K_i >^2}{K_i + 1/2} |N_{l_p j_p K_i}|^2 u^2_{K_i}, \tag{1}$$

where the total momentum of the initial and daughter nucleus K_i and J_d, are related to the one of the escaping proton by the relation $J_d + K_i \geq j_p \geq Max(K_i, |J_d - K_i|)$.

CP681, *Proton-Emitting Nuclei: Second International Symposium; PROCON 2003*,
edited by E. Maglione and F. Soramel
© 2003 American Institute of Physics 0-7354-0150-0/03/$20.00

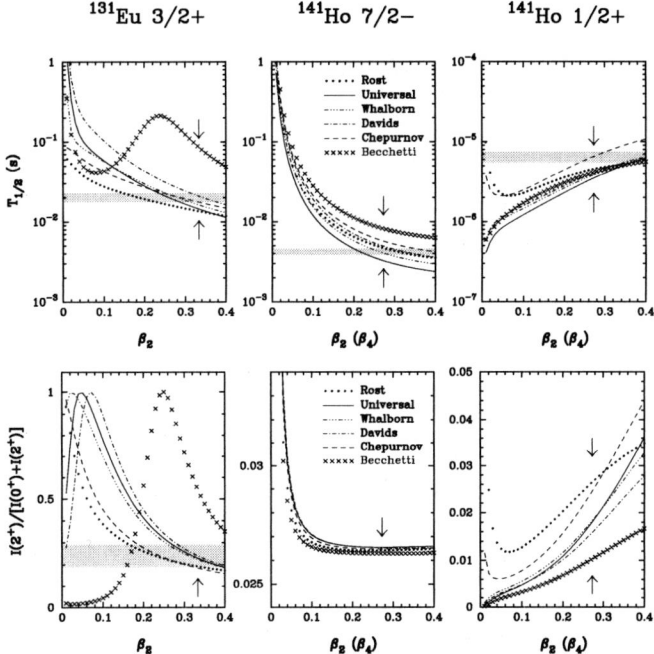

FIGURE 1. Total half–lives for decays from the $3/2^+$ ground state of ^{131}Eu , and from the $[523]7/2^-$ and $[411]1/2^+$ ground and isomeric states of ^{141}Ho respectively, obtained from the available s. p. potential models (upper part of figure). The arrow indicates the deformation predicted by Ref. [5]. The experimental result [2, 6, 7] is within the shaded area. The branching ratios for the fine structure with the same potentials are shown in the lower part.

Beyond the range of the nuclear force the wave function of the proton is just a pure Coulomb wave, therefore, aside from angular momentum recoupling coefficients, the decay width is completely defined by the asymptotic normalization of this function, obtained from the condition, $\lim_{r\to\infty} u_{l_p j_p}(r) = N_{l_p j_p K_i}(G_{l_p}(kr) + iF_{l_p}(kr))$, where F and G are the regular and irregular Coulomb functions, respectively. The residual pairing interaction is taken into by inserting the spectroscopic factor $u^2_{K_i}$, that is, the probability that the single particle level in the daughter nucleus is empty is obtained in the BCS approach. For decay to the ground state only the component of the s.p. wave function with the same angular momentum as the ground state contributes.

Due to energy considerations, one expects proton decay to proceed mainly to the ground state of the daughter nucleus. In this way the proton has the largest possible energy and $J_d = 0$, so that j_p is equal to the total and initial momentum of the parent nucleus, $j_p = J_i = K_i$. This component of the wave function could be very small and quite sensitive to details of the calculation. In the case of decay to excited states, few combinations are permitted for $l_p j_p$ according to angular momentum coupling rules, and consequently different components of the parent wave function are then tested. The decay width obtained from Eq. 1 depends on deformation, and is very sensitive to the

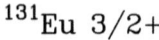

^{131}Eu 3/2+

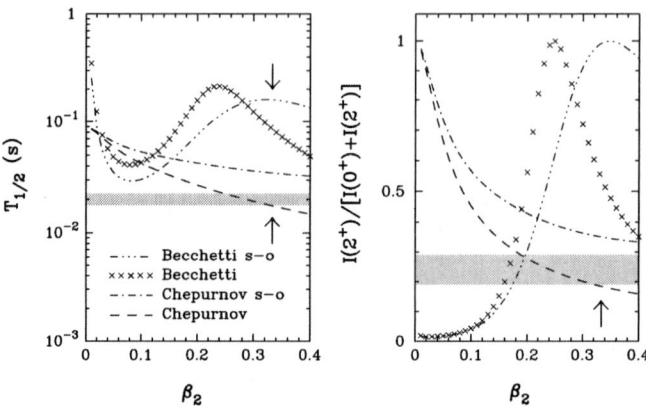

FIGURE 2. Total half–live and branching ratio for decay from the $3/2^+$ ground state of ^{131}Eu calculated with the Becchetti–Greenlees and Chepurnov potentials with deformed and spherical (s–o) spin–orbit interaction. The experimental result[2, 6, 7] is within the shaded area.

wave function of the decaying state. Therefore, a theoretical width that reproduces the experimental value has clear structure information on the deformation of the decaying nucleus, and properties of the decaying state.

The calculation of s.p. resonances relies on the knowledge of the s. p. potential. In Fig. 1, half-lives and branching ratios for decay of ^{131}Eu to ground state and to the 2^+ excited state in ^{130}Sm, and decay from ground and isomeric states of ^{141}Ho to the 2^+ of ^{140}Dy are shown for different s. p. potentials. As it can be seen, it is reasonable [8] to use any of the different parameterizations available in the literature to determine the decay rates, since they lead to quite similar results, inside the limits of experimental uncertainties, with the exception of the Becchetti–Greenlees potential [9], that cannot describe the experimental data. It has a quite small radius parameter and interactions with larger radii were usually adopted uniformly for spherical and deformed systems in a consistent determination of the experimental spectroscopic factors [10]. The ordering of levels for this potential is very odd, therefore the Nilsson wave functions and decay widths become very unreasonable. It was fitted to scattering data around 15 and 40 MeV, mainly from medium weight nuclei, and it is difficult to expect that its extrapolation to low energies involved in proton radioactivity, should work well. Calculations with this interaction are then not very reliable.

The other very important condition to be able to describe the experimental data is the correct treatment of the deformed spin-orbit interaction. If only the monopole part is considered by assuming a spherical spin–orbit interaction, as it was done in Ref. [4, 11] the calculated half-lives are quite different as pointed out in Refs. [8, 12], and differences from the exact deformed treatment become more dramatic with increasing deformation. This is shown in Fig. 2, for decay of ^{131}Eu to ground and fine structure. It is impossible to reproduce the experimental result for a reasonable deformation without a deformed

TABLE 1. Total angular momentum and deformation that reproduce the experimental half–lives for the measured deformed odd–even proton emitters compared with the predictions of [5]. The theoretical results are from Refs. [3, 13, 14]. The label m refers to decays from isomeric states.

	Proton decay		Möller–Nix	
	J	β	J	β
^{109}I	1/2+	0.14	1/2+	0.16
^{113}Cs	3/2+	0.15 ÷ 0.20	3/2+	0.21
^{117}La	3/2+	0.20 ÷ 0.30	3/2+	0.29
^{117m}La	9/2+	0.25 ÷ 0.35		
^{131}Eu	3/2+	0.27 ÷ 0.34	3/2+	0.33
^{141}Ho	7/2–	0.30 ÷ 0.40	7/2–	0.29
^{141m}Ho	1/2+	0.30 ÷ 0.40		
^{151}Lu	5/2–	–0.18 ÷ –0.14	5/2–	–0.16
^{151m}Lu	3/2+	–0.18 ÷ –0.14		

spin–orbit interaction. There are factors 2–3 of difference between them, appearing in a non systematic manner, since the spherical approach sometimes gives a shorter half–life and other times a longer one.

We have applied [3, 13, 14] our model to all measured deformed proton emitters including isomeric decays. The results of the calculations are shown in Table 1, with the "universal" parameterization of the s. p. potential. The experimental half-lives are perfectly reproduced by a specific state, with defined quantum numbers and deformation, thus leading to unambiguous assignments of the angular momentum of the decaying states [3, 13]. Extra experimental information provided by isomeric decay observed in ^{117}La, ^{141}Ho and ^{151}Lu, and fine structure in ^{131}Eu can also be successfully accounted by the model. The experimental half–lives for decay from the excited states were reproduced in a consistent way with the same deformation that describes ground state emission. In general, the half–lives determined with the strong coupling limit, are also perfectly described with a deformation close to the theoretical predictions of Ref. [5].

Emission from deformed systems with an odd number of protons and neutrons can be discussed in a similar fashion [15]. The decaying nucleus is described by a wave function of two particles–plus–rotor in the strong coupling limit. represented in terms of the single particle functions of the odd nucleons. However, in contrast with decay to ground state of odd-even nuclei where the proton is forced to escape with a specific angular momentum, many channels will be open due to the angular momentum coupling of the proton and daughter nucleus, $\vec{J}_d + \vec{j}_p$, giving the total width for decay as a sum of partial widths allowed by parity and momentum conservation,

$$\Gamma^{J_d} = \sum_{j_p=\max(|J_d-K_T|,K_p)}^{J_d+K_T} \Gamma^{J_d}_{l_p j_p} \qquad (2)$$

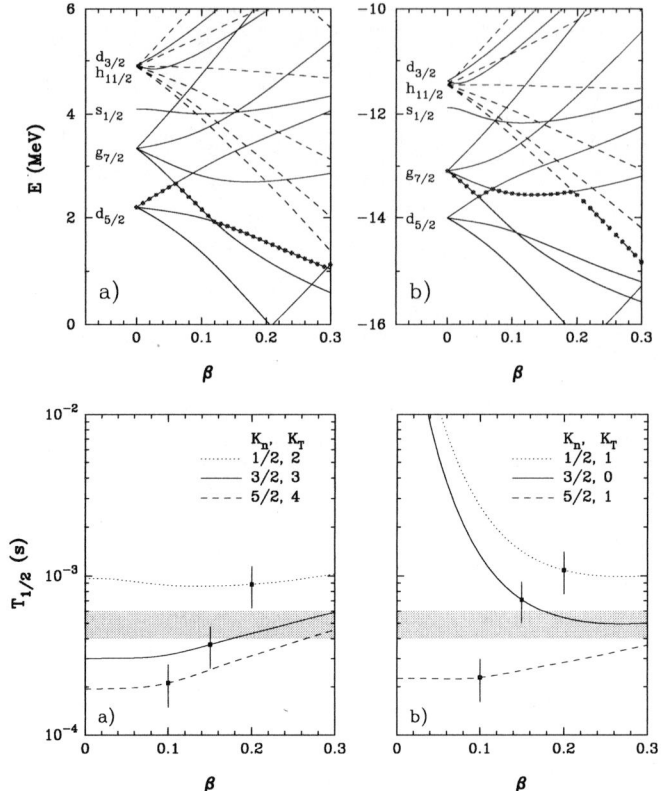

FIGURE 3. Proton and neutron Nilsson levels in ^{112}Cs. The dotted lines denote the Fermi surface (upper part). The lower part displays the half–life for decay from the ground state as a function of deformation, for neutrons in $K_n = 1/2^+, 3/2^+$ and $5/2^+$ states, with $K_T = K_p + K_n$, and $K_T = |K_p - K_n|$. The experimental value[17] is within the shaded area, while the bars drawn at a generic β, indicate typical uncertainties coming from the error in the experimental energy and u^2.

where the width for decay in the channel $l_p j_p$ is given by,

$$\Gamma^{J_d}_{l_p j_p} = \frac{\hbar^2 k}{\mu} \frac{(2J_d+1)}{(2K_T+1)} < J_d, K_n, j_p, K_p | K_T, K_T >^2 \frac{|u_{l_p j_p}(r)|^2}{|G_{l_p}(kr) + iF_{l_p}(kr)|^2} u^2_{K_p}. \quad (3)$$

The quantity in brackets represents a Clebsch-Gordan resulting from the angular momentum coupling of the odd nucleons, and $K_T = |K_p \pm K_n|$ the spin of the bandhead state of the decaying nucleus. Since the neutron intrinsic state does not change during decay $K_d = K_n$. According to Eqs. 2 and 3, the total decay width depends on the quantum numbers of the unpaired neutron state which cannot be considered only a spectator, but contributes significantly, as an "influential spectator"[16] with its angular momentum to the decay.

The proton and neutron single particle Nilsson levels in ^{112}Cs are depicted in Fig. 3. The neutron Fermi level is at levels with $K_n = 1/2^\pm, 3/2^+$ or $5/2^+$ according to

TABLE 2. As in Table 1 for the measured odd–odd proton emitters. Results from the calculation of Ref. [15]. The quantities K_p and K_n are the magnetic quantum numbers of the proton and neutron Nilsson wave functions, J the total angular momentum of the parent nucleus, and β and β_M the deformation coming from the proton decay calculation and the prediction of Ref. [5].

	K_p	K_n	J	β	β_M
^{112}Cs	3/2+	3/2+	0+,3+	0.12 ÷ 0.22	0.21
^{140}Ho	7/2–	9/2–,7/2+	8+,0–	0.26 ÷ 0.34	0.30
^{150}Lu	5/2–	1/2–	2+	–0.15 ÷ –0.17	–0.16
^{150m}Lu	3/2+	1/2–	1–,2–	–0.22 ÷ 0.00	

deformation, whereas for protons we took $K_p = 3/2^+$ since it reproduces[3] the decay of ^{113}Cs. The corresponding half–lives evaluated from Eqs. 2 and 3, are shown in Fig. 3, including the BCS spectroscopic factors calculated according Ref. [13]. They are displayed separately in the two possible coupling cases $K_T = K_p \pm K_n$. Since in each coupling the neutron single particle level leads to quite different factors and intermediate partial widths, the half–lives depend strongly on these quantities. The experimental value is reproduced considering the odd neutron in states with $K_n = 3/2^+, 5/2^+$, with corresponding deformations $\beta > 0.1$ with $K_T = 3^+$, and $\beta > 0.2$ with $K_T = 4^+$, respectively. When the proton and neutron are antiparallel, the same neutron states give $\beta > 0.14$ and $\beta > 0.24$ for $K_T = 0^+, 1^+$, respectively, whereas the $1/2^\pm$ states always have a very long lifetime. The state $K_n = 5/2^+$, should be excluded, since it is the Fermi level only at very low deformation, giving a quite short half–life. Therefore, the only possibility is given by $K_n = 3/2^+$, which is the Fermi level for $0.05 < \beta < 0.19$, and reproduces the experimental half-life in this range of deformations. These deformations are consistent with the ones obtained[3] for ^{113}Cs, giving further support to this calculation.

As in the case of odd-even nuclei, there is a perfect description of experimental decay rates for deformations in agreement with predictions made by other models [5]. The same Nilsson state of the odd proton is used in the calculation of odd–odd and neighbour odd–even nuclei, as can be seen comparing Tables 1 and 2. Also similar deformations were found for the odd-odd and nearby odd-even nuclei. This represents a further consistency check of the model.

THE NON-ADIABATIC QUASI-PARTICLE APPROACH

The effect of a finite moment of inertia of the daughter nucleus on proton decay, was studied within the non-adiabatic coupled channel[11], and coupled-channel Green's function[12] methods. In this way, the Coriolis mixing was taken into account in the total Hamiltonian. In both, decay rates and branching ratios were obtained, but the excellent agreement with experiment found in the adiabatic context was lost. The results differ by factors of three or four from the experiment, and even the branching ratio for fine structure decay is not reproduced[11, 12, 18]. While in the case of the non-adiabatic coupled channel calculations [11, 18], these unreasonable factors were also due[12] to a drastic approximation made by considering only a spherical spin-orbit interaction, in the Green's function approach, they are a matter of concern, since calculations that include the Coriolis mixing should undoubtedly be better. However, decay rates in deformed nuclei, are extremely sensitive to small components of the wave function. Since the Coriolis interaction mixes different Nilsson wave functions, it is not surprising to observe strong changes in the decay widths. The residual pairing interaction can modify this mixing of states. The available calculations[11, 12] do not consider the effect of the residual interaction. It was suggested in Ref.[12] a quenching of the Coriolis matrix elements by the pairing interaction. In order to clarify these points, we decided to perform a calculation [19] where not only the Coriolis mixing is included, but due to the pairing residual interaction in the BCS approach, this mixing is between quasi-particle states instead of particle ones.

In order to take into account the pairing residual interaction, it is more advantageous to employ an intrinsic coordinate frame, and diagonalize the collective Hamiltonian that describes rotations of the inert core with axial symmetry and includes the Coriolis mixing, using the basis of Nilsson quasi-particle wave functions.

Since the intrinsic components of the angular momentum vector commute with the fixed space components, it is possible to choose a representation diagonal in the square of the total angular momentum of the decaying nucleus $\vec{I}$, and on its projections on the symmetry axis of the rotor and on the z of the fixed frame. In this representation, one has a basis of states $|IKM>$, denoted by the corresponding eigenvalues IKM of the commuting operators, that can be expressed in terms of Nilsson single particle and core wave functions. Therefore, the total wave function of the nucleus can be written as a superposition of these states with coefficients a_K^I,

$$\Psi_{IM} = \sum_{K>0} a_K^I < \vec{r}|IKM>, \tag{4}$$

where the single particle Nilsson functions Υ_K are given by,

$$\Upsilon_K = \sum_{lj} \frac{\phi_{lj}^K(r)}{r} \left[\mathbf{Y}_l \times \chi_S \right]_j^K, \tag{5}$$

with φ_{ljR}^I the radial wave function of the relative motion of proton with respect to the core, and $\mathbf{Y}$ and χ_S the orbital angular momentum and spin tensor functions.

The Coriolis mixing can be taken into account diagonalizing the Coriolis Hamiltonian in the $|IKM>$ basis, and determining the coefficients a_K^I. From the properties of operators $I_\pm$ and $j_\pm$ it can be seen that there are only non vanishing off-diagonal matrix elements between states where K differs by one unit,

$$< IK+1M|H_{Cr}|IKM> = -\frac{\hbar^2}{2\mathcal{I}} \sum_{lj} [(I-K)(I+K+1)$$
$$(j-K)(j+K+1)]^{1/2} \int dr \phi_{lj}^{K+1^*}(r) \phi_{lj}^K(r). \tag{6}$$

For symmetry reasons, it is enough to consider only states with $K > 0$. In the particular case of $K = 1/2$, there is also a diagonal contribution and the diagonal matrix elements

56

TABLE 3. Proton decay widths in units os 10^{-20} MeV of $7/2^-$ ground state at 1.190 MeV of ^{141}Ho, to the ground Γ_0 and 2^+(Γ_2) of ^{140}Dy, branching ratio and total half lives calculated in the adiabatic, with Coriolis mixing and with pairing residual interaction, compared with the calculations of Ref. [12].

Method	I^π	Γ_0	Γ_2	Γ_2/Γ_T	$t_{1/2}$
Adiab [12]	$7/2^-$	16.5	0.45	0.027	2.7 ms
Corl [12]	$7/2^-$	3.38	0.27	0.074	12.5 ms
Corl + pairing	$7/2^-$	17.7	0.21	0.012	2.6 ms
Exp. [7]	$7/2^-$	10.9 ± 1.0		<0.01	4.2 ± 0.4 ms

of the full Hamiltonian H become,

$$<IKM|H|IKM>= \epsilon_K+\frac{\hbar^2}{2\mathcal{I}}[I(I+1)-2K^2+\sum_{lj}j(j+1)\gamma_{lj}+\alpha\delta_{K1/2}(-)^{I+\frac{1}{2}}(I+\frac{1}{2})],$$

(7)

where $\alpha = \sum_{lj}(-)^{j-1/2}(j + 1/2)\gamma_{lj}$ is the decoupling parameter, with $\gamma_{lj} = \int dr|\phi_{lj}^K(r)|^2$, the probability that the Nilsson state has angular momenta (l,j) and depends on deformation. The quantity ϵ_K is the Nilsson single particle energy.

The effect of the residual pairing interaction can be included making the usual Bogoliubov transformation from Nilsson single particles to quasi-particles with energy $\tilde{\epsilon}_K = [\Delta^2 + (\epsilon_K - \lambda)^2]^{1/2}$, depending on the pairing gap Δ, and the Fermi energy λ. Under this transformation, the matrix elements of I^2, I_3, j_3 and j^2 in the new basis of quasi-particle states $|IKM>_{qp}$, do not change but the Coriolis contribution of Eq. (6) does, and it has to be multiplied by a factor $f_{uv} = (u_{K+1}u_K + v_{K+1}v_K)$ where v^2 and u^2 are the probability of having the single particle state occupied or empty, respectively. The pairing residual interaction introduces two effects, the quenching factor f_{uv} to Eq. (6) on one side, which was found very small in the present case. On the other hand, in Eq. (7) quasi particle energies $\tilde{\epsilon}_K$ replace the particle ones ϵ_K.

The partial decay width is the overlap between the initial state of the parent nucleus given by Eq. (4), and a final state made of a daughter nucleus left with angular momenta R, and an outgoing proton, in a scattering state $(l_p j_p)$, obtained from the solution of the radial equation with outgoing boundary conditions[20].

For a proton escaping with angular momentum $(l_p j_p)$, it becomes,

$$\Gamma_{l_p j_p}^R(r) = \frac{\hbar^2 k}{\mu}\frac{2(2R+1)}{(2I+1)}\left|\sum_{K>0} u_K^f < j_p KR0|IK > a_K^I \frac{\phi_{l_p j_p}^K(r)}{G_{l_p}(kr)+iF_{l_p}(kr)}\right|^2.$$

(8)

which is analogous to the previous expressions for the decay widths of Eqs. (1) and (3).

Following this procedure, we show the results of the calculated decay widths of the $7/2^-$ ground state of ^{141}Ho to the ground and first 2^+ excited states in ^{140}Dy. It is the example where the Coriolis force is stronger, since it involves a state with high angular momentum, the spherical $h11/2$ state. In order to verify whether our treatment of the Coriolis mixing in the intrinsic frame is valid, we repeated the calculation of Ref.[12] with our formalism, but without the residual pairing force by suppressing the factor

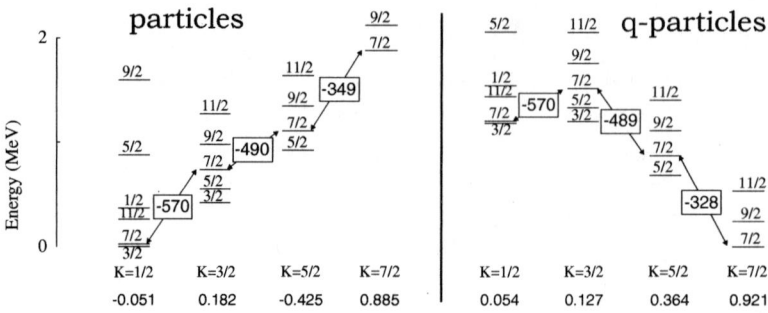

FIGURE 4. a) Level scheme for particle states $|IKM>$ at energies equal to the diagonal matrix elements of the Hamiltonian H. The numbers in the boxes are the off-diagonal matrix elements of Eq. 5 (in keV). States with same angular momentum are connected. b) As in a) for quasi-particles. The numbers in the bottom row are the components of the wave function of the $7/2^-$ decaying state.

u^f_K and the pairing contribution in the calculation of a^I_K. The same excitation energy, $\beta_2 = 0.267$ and $\beta_4 = -0.05$ deformations, and s. p. interaction were used. The results shown in Table I are in complete agreement with the ones of Ref.[12] obtained within the adiabatic approach, and when the Coriolis mixing is taken into account, in spite of the completely different numerical procedure adopted. This proves that the diagonalization of this force in the intrinsic frame is quite adequate.

Including the pairing residual interaction in the diagonalization of the Coriolis mixing, and using for the pairing gap the value $12/A^{1/2}$ MeV [21], the decay width to the ground state increases by a factor of four, whereas the width for decay to the excited state is practically unchanged. The branching ratio is consequently reduced. This can be understood from a careful analysis of the diagonalization procedure. The level scheme corresponding to the basis states $|IKM>$ is displayed in Fig. 4, at energies equal to the diagonal matrix elements of the full Hamiltonian for particles and after the transformation to quasi-particles. The difference between both representations, is an inversion of the level ordering, while the off-diagonal matrix elements are practically unchanged. After diagonalization, the wave function that describes the decaying nucleus corresponds, in the particle case, to the highest state in energy, while in the quasi-particles to the lowest one. This inversion implies a change of sign of the wave function components, with respect to the K=7/2 component.

The sign of a component depends on the difference between the unperturbed energies. For particles there is a destructive interference between these components, that decreases the decay width. For quasi-particles the interference is constructive and the width is enhanced, going back to values close to the ones of the adiabatic approach. In the case of decay to the 2^+ state of ^{140}Dy, the phase of the Clebsch-Gordan of Eq. 8, makes the interference destructive in both cases. The energy of the 2^+ of ^{140}Dy was recently measured[18, 22] at 202 keV. From this quantity one can derive the moment of inertia and, using the Grodzin's formula[23], the quadrupole deformation, getting $\beta_2 = 0.244$. Redoing the calculation with the experimental values for $\mathcal{I}$ and β_2, the half-life for decay to ground state and branching ratio for decay to 2^+ are shown in Fig. 5, as a function of hexadecapole deformation β_4, a quantity not yet known experimentally.

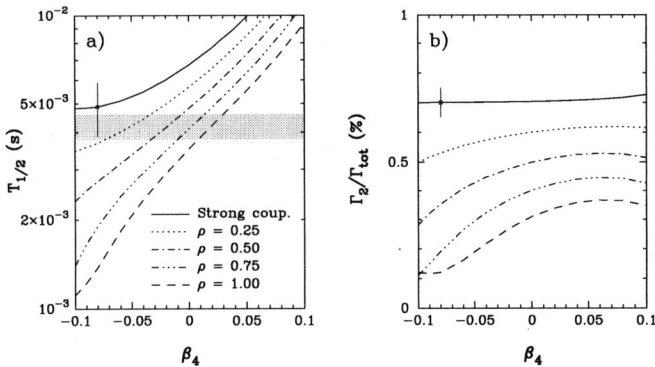

FIGURE 5. a) Total half-life for decay of the $7/2^-$ ^{141}Ho ground state as a function of β_4 for different values of the attenuation factor ρ. The shaded area represents the experimental value[7]. The bars drawn at a generic deformation indicate typical uncertainties due to the error in the experimental energy. b) As in a) for the branching ratio to decay in the 2^+ state of ^{140}Dy.

Quite frequently, to reproduce the experimental spectrum, structure calculations require an attenuation factor $\rho \approx 0.4 - 0.8$[24], that weakens the Coriolis interaction. In order to show the effect that this condition could have in our calculation, we have also introduced such factor. It is possible to observe from Fig. 5, how the effect of the Coriolis force evolves from the strong coupling limit, where ρ is zero, up to its maximum value. Due to various uncertainties in the experimental proton energy, pairing gap, Fermi energy and β_4, one can define a band that can reasonably cover the region between the two extreme lines of the calculation with and without Coriolis. The uncertainty in the energy, for example, brings an error of 20% on the half-life and 7% on the branching ratio. Therefore, there is not so much difference between the results obtained in the strong coupling limit and a correct treatment of Coriolis and residual interaction.

The experimental half-life and branching ratio seem to suggest a very small negative β_4 deformation. The branching ratio is always smaller than the experimental upper limit of 1% for any β_4. This is in contrast with the results of Ref.[18], where such agreement was not found, since the reported branching ratio is between 2 and 3%. The half-lives obtained in this calculation are in very good agreement with experiment. Therefore, the previous disagreement between the calculation with Coriolis[11, 12, 18] and the experimental data were simply due to an inadequate treatment of the residual pairing interaction.

CONCLUSIONS

Proton radioactivity from deformed nuclei is by now well understood. The adiabatic approach based on the exact calculation of proton resonances in a deformed nucleus with deformed spin-orbit interaction accounts for the available experimental data on even–odd and odd–odd deformed proton emitters. Half-lives for decay from ground and isomeric states, as well as data on fine structure, are accurately and consistently reproduced, identifying the decay level and deformation of the decaying nucleus, and also supporting previous predictions made by other models on nuclear structure properties of the decaying nucleus.

An interesting result derived from the study of decay from odd–odd nuclei, is the strong dependence of the final decay widths on the quantum numbers of the unpaired neutron state which cannot be considered only a spectator, but contributes significantly with its angular momentum to decay.

Going beyond the strong coupling limit by the inclusion of the rotational spectra of the daughter nucleus, requires not only the inclusion of the Coriolis mixing, but a consistent treatment of the pairing residual interaction. The inclusion of the rotational degrees of freedom destroys the good agreement with the data, achieved in the strong coupling limit. But quasi–particle excitations in the medium, induce correlations and a new ordering of levels that bring back the good agreement with experiment.

ACKNOWLEDGMENTS

This work was supported by the Fundação de Ciência e Tecnologia (Portugal), Project: POCTI-36575/99.

REFERENCES

1. V. P. Bugrov and S. G. Kadmenskii, Sov. J. Nucl. Phys. **49**, 967 (1989); D. D. Bogdanov, V. P. Bugrov and S. G. Kadmenskii, Sov. J. Nucl. Phys. **52**, 229 (1990).
2. C. N. Davids, *et al.,* Phys. Rev. Lett. **80**, 1849 (1998).
3. E. Maglione, L. S. Ferreira and R. J. Liotta, Phys. Rev. Lett. **81**, 538 (1998); Phys. Rev. **C59**, R589 (1999).
4. K. Rykaczewski *et al.*, Phys. Rev. **C60**, 011301(R) (1999).
5. P. Möller, J. R. Nix, W. D. Myers and W.J. Swiatecki, At. Data Nucl. Data Tables **59**, 185 (1995); P. Möller, R. J. Nix and K. L. Kratz, ibd. **66**, 131 (1997).
6. A. A. Sonzogni, *et al.,* Phys. Rev. Lett. **83**, 1116 (1999).
7. D. Seweryniak *et. al.*, Phys. Rev. Lett. **86**, 1458 (2001).
8. L. S. Ferreira, E. Maglione and D.E.P. Fernandes, Phys. Rev. **C65**, 024323 (2002).
9. F. D. Becchetti and G. W. Greenlees, Phys. Rev. **182**, 1190 (1969).
10. E. Maglione and L. S. Ferreira, European Physical Journal **A15**, 89 (2002).
11. A. T. Kruppa, B. Barmore, W. Nazarewicz and T. Vertse, Phys. Rev. Lett. **84**, 4549 (2000); B. Barmore, A. T. Kruppa, W. Nazarewicz and T. Vertse, Phys. Rev. **C62**, 054315 (2000).
12. H. Esbensen and C. N. Davids, Phys. Rev. **C63**, 014315 (2001).
13. L. S. Ferreira and E. Maglione, Phys. Rev. **C61** (2000) 021304(R); ibd. **C61**, 47307 (2000).
14. F. Soramel *et al.*, Phys. Rev. **C63**, 031304(R) (2001).
15. L. S. Ferreira and E. Maglione, Phys. Rev. Lett. **86**, 1721 (2001).
16. P. J. Woods, in these proceedings.
17. R. D. Page *et al.*, Phys. Rev. Lett. **72**, 1798 (1994).
18. W. Królas *et. al.*, Phys. Rev. **C65**, 031303(R) (2002).
19. G. Fiorin, E. Maglione and L. S. Ferreira, Phys.Rev. **C67**, 054302 (2003).
20. L. S. Ferreira, E. Maglione and R. J. Liotta, Phys. Rev. Lett. **78**, 1640 (1997).
21. A. Bohr and B. R. Mottelson, *Nuclear Structure* (Benjamin, New York, 1969), vol. II.
22. D. M. Cullen et al, Phys. Lett. B **529**, 42 (2002).
23. L. Grodzins, Phys. Lett. **2**, 88 (1962).
24. P. Ring and P. Schuck, *The Nuclear Many-Body Problem*, Springer Verlag, 1980.

Nilsson-orbit and weak-coupling model: non-axial deformation

A.T. Kruppa* and W. Nazarewicz†

*Institute of Nuclear Research, Bem tér 18/c, 4026 Debrecen, Hungary
†Department of Physics and Astronomy, University of Tennessee, Knoxville, Tennessee 37996
Physics Division, Oak Ridge National Laboratory, P.O. Box 2008, Oak Ridge, Tennessee 37831
Institute of Theoretical Physics, Warsaw University, ul Hoża 69, PL-00681 Warsaw

Abstract. Models for deformed proton emitters are reviewed. The non-adiabatic coupled-channel framework with rotational coupling and the adiabatic Nilsson-orbit model are discussed and compared. An improvement of the adiabatic approach is obtained by taking care of the diagonal part of the Coriolis coupling. For the description of the proton radioactivity of ^{141}Ho, we investigate the role of γ-vibrational excitations in the daughter nucleus ^{140}Dy. It is shown that the coupling to triaxial degrees of freedom strongly influences theoretical predictions.

INTRODUCTION

Theoretical models applied to the description of non-spherical proton emitters can be divided into two groups. The core-plus-particle models describe the radioactive parent nucleus in terms of a proton interacting with a core (i.e., the daughter nucleus). Usually, the core is modeled by some phenomenological collective model, e.g., the Bohr-Mottelson (geometric) model. Depending on the structure of the daughter nucleus, rotational [1-3] or vibrational [4, 5] coupling of the proton is assumed. The models belonging to this group employ the coupled-channel reaction theory which have been developed for the description of the elastic or inelastic scattering.

The second group of models uses the framework of the deformed shell model. In the simplest case, the proton resonance corresponds to a single-particle resonant (Gamow) state of a deformed field [6-10]. This model can be generalized to include BCS pairing correlations (see these proceedings).

We may refer to the first group of models as weak coupling models or coupled-channel models. For the second group of models, we reserve the term resonance Nilsson-orbit (or adiabatic) description. The term "adiabatic" requires an explanation. It is very difficult, if not impossible, to relate both groups of models to each other, because they operate on different approximation levels. In special situations, however, the relation between the two types of models can be revealed. For instance, if one considers axial-symmetric nuclei, strong rotational coupling [11], and the degenerate ground-state rotational band in the daughter nucleus, one recovers the resonance Nilsson-orbit model. So one may say that *in this case* the adiabatic model is an approximation of the weak-coupling non-adiabatic model. Generally, however, the relation between the models is not that simple. For example, the resonance Nilsson-orbit model with a non-axial symmetric potential

CP681, *Proton-Emitting Nuclei: Second International Symposium; PROCON 2003*,
edited by E. Maglione and F. Soramel
© 2003 American Institute of Physics 0-7354-0150-0/03/$20.00

(i.e., nonzero γ deformation) cannot be trivially related to a weak coupling model with non-axial deformation.

If the coupled-channel model with the rotational coupling is applied to the nucleus ^{141}Ho, the ground-state decay (half-life time and branching ratio) is poorly described [2, 3]. There are several explanations possible. For example, it may be that the Coriolis mixing is too strong [3]. This can be partly cured if pairing correlation is introduced (see these proceedings). Another possibility, explored in this work, is the coupling to triaxial vibrations. Indeed, in particle-rotor calculations, the best description of the experimentally observed band structures of ^{141}Ho can be explained if γ deformation is introduced [12]. In addition, in the neighboring $N=74$ isotones there are low-lying 2_2^+ and 3^+ levels. We may interpret these states as members of the γ-vibrational band. There are also other indications that in this mass region the nuclei may have triaxial shapes [13, 14].

The ground-state $K = 0$ rotational band of ^{140}Dy has recently been observed [15]. In this work, we assume that ^{140}Dy also has $K=2$ γ-vibrational band. This structure can be coupled to the ground-state band if the proton-daughter interaction in the body-fixed system deviates from the axial symmetry. In our weak-coupling model we do not assume, however, that the daughter nucleus has a permanent γ deformation. Our aim is to investigate the possibility of triaxial vibrations around the deformed axial shape. The experimentally observed rotational band of the parent nucleus is assumed to be a $K = 7/2^-$ band [12]. In the strong-coupling picture, this implies the presence of two addtional rotational bands in ^{141}Ho with quantum numbers $7/2\pm2$, i.e., $K^\pi = 3/2^-$ and $K^\pi = 11/2^-$.

This paper is organized as follows. We will begin with the overview of the weak coupling model in the case of rotational coupling. We will then discuss the Nilsson-orbit model and its relation to the weak coupling approach. In particular, we consider the diagonal part of the Coriolis coupling, which was neglected in earlier calculations. Finally, we present numerical results. Here we show how the position of excited states of the daughter nucleus can influence predictions of the weak-coupling model. We also present preliminary results for the proton emission from ^{141}Ho assuming the presence of the γ-vibrational band in the daughter nucleus. The final conclusion is that the coupling to γ-vibration improves the agreement between the weak-coupling model and experiment.

WEAK COUPLING MODEL: NON-ADIABATIC APPROACH

The model of the parent nucleus describes a single proton interacting with a deformed core. The model Hamiltonian can be written as

$$H_{\rm rot} = H_{\rm d} - \frac{\hbar^2}{2m} \triangle_{\bf r} + V({\bf r},\omega), \tag{1}$$

where $H_{\rm d}$ is the collective Hamiltonian of the daughter nucleus, the second term represents the relative kinetic energy, and V is the proton-core interaction, which depends on the position of the proton ${\bf r}$ and the orientation ω of the core. In the laboratory system, the daughter proton interaction is given by

$$V(\mathbf{r},\omega) = V^{(1)}(\mathbf{r},\omega) + a_2 V^{(2)}(\mathbf{r},\omega)$$
$$= \sum_{\lambda\mu} V_\lambda^{(1)}(r) D_{\mu0}^\lambda Y_{\lambda,\mu}(\hat{r}) + a_2 \sum_{\lambda\mu} V_\lambda^{(2)}(r) \left(D_{\mu2}^\lambda + D_{\mu-2}^\lambda \right) Y_{\lambda,\mu}(\hat{r}), \qquad (2)$$

where the deformation parameters are a_0 and a_2 [$V_\lambda^{(1)}(r)$ depends on a_0]. For the core we take the rotational-vibrational collective model. The daughter states $\phi_{I\mu K}$ are given by the standard ansatz

$$\phi_{I\mu K} = \sqrt{\frac{2I+1}{16\pi^2(\delta_{K,0}+1)}} \left[D_{\mu K}^{I*} + (-1)^I D_{\mu-K}^{I*} \right] \chi_{Kn_2}(a_2)|\text{g.s.}\rangle. \qquad (3)$$

The wave function of the parent nucleus can be written in the weak-coupling form

$$\Psi^{JM} = \sum_{IKlj} \frac{f_{IKlj}^J(r)}{r} \Phi_{JMIKlj}, \qquad (4)$$

where the channel function is given by

$$\Phi_{JMIKlj} = \sum_{\Omega\mu} \langle j\Omega I\mu|JM\rangle \mathcal{Y}_{lj\Omega}\phi_{I\mu K}, \qquad (5)$$

and

$$\mathcal{Y}_{lj\Omega} = \sum_{ms} \langle lm\tfrac{1}{2}s|j\Omega\rangle i^l Y_{lm}(\hat{r})\chi_{1/2}(s) \qquad (6)$$

arises from the coupling of the proton spin with the orbital angular momentum. The unknown radial functions $f_{IKlj}^J(r)$ can be then obtained from the set of coupled-channel equations:

$$\frac{\hbar^2}{2m}\left(-\frac{d^2}{dr^2} + \frac{l(l+1)}{r^2}\right) f_{IKlj}^J + \sum_{\lambda I'l'j'} A_\lambda(Ilj, I'l'j', J)B_\lambda(II'K)V_\lambda^{(1)} f_{I'Kl'j'}^J + \qquad (7)$$
$$\sum_{\lambda I'K'l'j'} A_\lambda(Ilj, I'l'j', J)C_\lambda(IKI'K', a_2)V_\lambda^{(2)} f_{I'K'l'j'}^J = (E - E_{IK})f_{IKlj}^J.$$

Here, the r-independent coupling coefficients can be written in terms of the reduced nuclear matrix elements

$$B_\lambda(II'K) = \langle \Phi_{IK}||D_{;0}^\lambda||\Phi_{I'K}\rangle \qquad (8)$$

and

$$C_\lambda(IKI'K', a_2) = \langle \Phi_{IK}||a_2(D_{;2}^\lambda + D_{;-2}^\lambda)||\Phi_{I'K'}\rangle. \qquad (9)$$

The explicit expressions for geometric coefficients A_λ are given, e.g., in Ref. [16]. The nuclear structure model of the daughter nucleus enters the formalism through the reduced matrix elements B_λ and C_λ.

Weak-coupling model in the body-fixed frame

The ansatz (4) is given in the laboratory frame but the total wave function can be easily transformed to the body-fixed system. Following Ref. [17], where the α-decay was described in the adiabatic limit of the weak-coupling approach, one obtains:

$$\Psi^{JM} = \sum_{K\Omega lj} \frac{g^J_{K,\Omega lj}(r)}{r} \sqrt{\frac{2J+1}{16\pi^2}}$$
$$\left(\mathcal{Y}'_{lj\Omega} D^{J*}_{M,\Omega+K} + (-1)^{J-j}\mathcal{Y}'_{lj-\Omega} D^{J*}_{M,-\Omega-K}\right) \chi_{Kn_2}(a_2)|\text{g.s.}\rangle. \tag{10}$$

The new radial wave functions $g^J_{K,\Omega lj}(r)$ are related to the laboratory-system wave functions through the equation

$$g^J_{K,\Omega lj}(r) = \sqrt{\frac{1}{\delta_{K,0}+1}} \sum_I \frac{\hat{I}}{\hat{j}} \langle j\Omega IK|J\Omega+K\rangle f^J_{IKlj}(r). \tag{11}$$

In the body-fixed frame the proton daughter interaction is

$$V_{\text{def}}(r,\theta'\phi') = V^{(1)}_{\text{def}}(r,\theta') + a_2 V^{(2)}_{\text{def}}(r,\theta'\phi')$$
$$= \sum_\lambda V^{(1)}_\lambda(r) Y'_{\lambda,0}(\theta') + a_2 \sum_\lambda V^{(2)}_\lambda(r)\left[Y'_{\lambda,2}(\theta',\phi') + Y'_{\lambda,-2}(\theta',\phi')\right], \tag{12}$$

i.e., it is given by the triaxial average potential. The action of the Hamiltonian of the daughter nucleus on the laboratory channel function is very simple: $H_d\Phi_{JMIKlj} = E_{IK}\Phi_{JMIKlj}$. On the other hand, the action of the daughter's Hamiltonian on the wave function (10) is more complicated:

$$H_d\Psi^{JM} = \sum_{Klj}\sum_{\Omega\Omega'} A^J_{Klj}(\Omega,\Omega') \frac{g^J_{K,\Omega lj}(r)}{r}\Phi^{\text{body}}_{JMK\Omega'lj}, \tag{13}$$

where

$$A^J_{Klj}(\Omega,\Omega') = \sum_I \frac{2I+1}{2J+1} E_{IK}\langle j\Omega IK|J\Omega+K\rangle\langle j\Omega' IK|J\Omega'+K\rangle \tag{14}$$

and

$$\Phi^{\text{body}}_{JMK\Omega lj} = \sqrt{\frac{2J+1}{16\pi^2}}\left(\mathcal{Y}'_{lj\Omega} D^{J*}_{M,\Omega+K} + (-1)^{J-j}\mathcal{Y}'_{lj-\Omega} D^{J*}_{M,-\Omega-K}\right)\chi_{Kn_2}(a_2)|\text{g.s.}\rangle. \tag{15}$$

For simplicity, we give the coupled-channel equations in the body-fixed system in the case of axial symmetry

$$-\frac{\hbar^2}{2m}\left[\frac{d^2}{dr^2} - \frac{l(l+1)}{r^2}\right]g^J_{0,\Omega lj} + \sum_{l'j'}\langle\mathcal{Y}_{lj\Omega}|V(r,\theta')|\mathcal{Y}_{l'j'\Omega}\rangle g^J_{0,\Omega l'j'}$$
$$+ \sum_{\Omega'} A^J_{0lj}(\Omega\Omega')g^J_{0,\Omega'lj} = E\, g^J_{0,\Omega lj}. \tag{16}$$

We must emphasize that the coupled-channel equations (7) and (16) are completely equivalent. We note that under special circumstances (E_{IK} are given by the pure rotational formula) the third term of the left hand side of Eq. (16) is the rotational energy in the strong coupling limit [3].

NILSSON-ORBIT MODEL: ADIABATIC APPROACH

In the spirit of the deformed shell model, we assume that the state of the emitted proton is the lowest resonance single-particle orbit of a deformed potential. In the body-fixed coordinate system the deformed potential is given by Eq. (12). The Hamiltonian of the resonance Nilsson-orbit model contains only the deformed potential (12), and it can be written as

$$H_{\text{def}} = -\frac{\hbar^2}{2m}\triangle_{\mathbf{r}'} + V_{\text{def}}(r, \theta'\phi').$$ (17)

The wave function of a Nilsson-orbit can be expanded

$$\Psi_{\text{def}} = \sum_{\Omega l j} \frac{\bar{g}_{\Omega l j}}{r} \mathcal{Y}'_{l j \Omega}.$$ (18)

For the radial functions, one obtains a set of coupled-channel equations:

$$\frac{\hbar^2}{2m}\left(-\frac{d^2}{dr^2} + \frac{l(l+1)}{r^2}\right)\bar{g}_{\Omega l j}(r) + \sum_{l'j'}\langle\mathcal{Y}'_{lj\Omega}|V^{(1)}_{\text{def}}|\mathcal{Y}'_{l'j'\Omega}\rangle\bar{g}_{\Omega l'j'}(r)$$
$$+a_2\sum_{\Omega'l'j'}\langle\mathcal{Y}'_{lj\Omega}|V^{(2)}_{\text{def}}|\mathcal{Y}'_{l'j'\Omega'}\rangle\bar{g}_{\Omega'l'j'}(r) = E\bar{g}_{\Omega l j}(r).$$ (19)

For simplicity, let us assume that the daughter nucleus is axially deformed ($a_2 = 0$). The comparison of Eqs. (19) and (16) reveals that if the last term on the left hand side of (16) is neglected, then the Nilsson-orbit model and the weak-coupling model are equivalent. This approximation is severe. The term in question vanishes only if all the energies of the excited states of the daughter nucleus are set to zero. Only in the case of extreme degeneracy (or infinite collective moment of inertia) is the resonance Nilsson-orbit model related to the weak-coupling model.

Corrected Nilsson-orbit model: inclusion of the diagonal part of the Coriolis coupling

In the above section we have demonstrated that the Nilsson-orbit model (an extension of the classical Nilsson-orbit picture to narrow resonances) is an approximation to the weak-coupling model. Nevertheless, it has proved to be a fairly useful approach for the description of proton-emitting nuclei. There is also a very practical reason why it is useful to study connections between the non-adiabatic approach and the Nilsson-orbit description. The number of coupled-channel equations in the weak-coupling model quickly increases with the number of active states of the daughter nucleus.

One of the drawbacks of the Nilsson-orbit model is that the excitation energies of the daughter nucleus are neglected. This is clearly an unphysical assumption. The excitation energies come into play through the action of the operator H_d, which is given by Eqs. (13) and (14). Since our aim is to avoid the increase of the number of coupled-channel equations, we neglect the non-diagonal part of the Ω coupling and approximate (13) by

$$H_d \Psi^{JM} \approx \sum_{Klj} \sum_{\Omega} \tilde{A}^J_{Klj}(\Omega, \Omega) \frac{g^J_{K,\Omega lj}(r)}{r} \Phi^{body}_{JMK\Omega lj}. \tag{20}$$

Taking into account the above expression, the coupled-channel equation (16) turns into

$$\frac{\hbar^2}{2m} \left(-\frac{d^2}{dr^2} + \frac{l(l+1)}{r^2} \right) h^J_{0,\Omega lj}(r) \;+\; \sum_{l'j'} \langle \mathcal{Y}'_{lj\Omega} | V^{(1)}_{def} | \mathcal{Y}'_{l'j'\Omega} \rangle h^J_{0,\Omega l'j'}(r)$$
$$= \left[E - A^J_{0lj}(\Omega, \Omega) \right] h^J_{0,\Omega lj}(r). \tag{21}$$

In this way we have introduced an effective excitation energy dependence in each (lj) channel. Consequently, Eq. (21) introduces the J-dependence into the Nilsson-orbit model. We have achieved an interesting result. Namely, using the coupled equations (21) for a fixed Ω, we are able to calculate not only the band head ($J = \Omega$) but also the excited states of the parent nucleus by putting $J = \Omega + 1, \Omega + 2, \ldots$ in (21). We will refer to the calculations based on (21) as "dynamically corrected Nilsson-orbit model" or "dynamically corrected adiabatic description" (ADI-D). The standard Nilsson-orbit description will be referred to as the adiabatic approach (ADI).

NUMERICAL RESULTS

The numerical tests have been performed for the nucleus ^{141}Ho, viewed as the composite system of a proton and the daughter nucleus ^{140}Dy (collective core). For the rotational bands in the daughter nucleus, we have fixed the maximum spin to $I = 12$. In the resonance Nilsson-orbit model, the maximum of the proton j value was taken to be $27/2$. As for the parameterization of the Woods-Saxon (WS) potential, we have used the Chepurnov set employed in Ref. [2].

Dynamically corrected adiabatic approach

For each value of $E_2 = E_{20}$, we determine the WS potential strength in the weak-coupling model so as to get the $J = \frac{7}{2}^-$ state at the experimental value of 1.19 MeV. Having established the single-particle potential, we carried out the ADI and ADI-D calculations. The position of the resulting resonance is displayed in Fig. 1. It is seen that the real part of the energy calculated in ADI-D is dramatically improved as compared with ADI. The difference between the ADI-D and the weak-coupling treatment is due to the off-diagonal Coriolis coupling. Since the Coriolis coupling is completely neglected in the adiabatic model, the ADI description significantly deviates from the weak-coupling result.

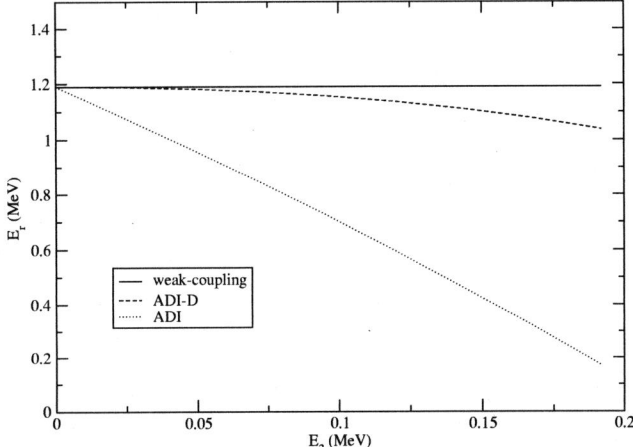

FIGURE 1. The real part of the energy of the $J = \frac{7}{2}^-$ resonance in ^{141}Ho as a function of the excitation energy E_2 of the 2^+ state of the daughter nucleus. The WS strength has been adjusted at each E_2 in the weak-coupling model to reproduce the experimental Q_p value.

Effect of the structure of the daughter nucleus

The excitation energies of the daughter nucleus are calculated using the pure rotational expression $E_{I0} = \frac{\hbar^2}{2\Theta} I(I+1)$. Instead of the moment of inertia parameter, we use the excitation energy E_2 of the 2^+ state in ^{140}Dy as a parameter of the calculations. Figure 2 shows the proton half-life $T_{1/2}$ as a function of E_2. Not surprisingly, $T_{1/2}$ is very sensitive to the position of the 2^+ state. Taking the experimental 2^+ excitation energy 0.202 MeV [15], one obtains $T_{1/2}$=40 msec. If we repeat the weak-coupling calculation in such a way that the experimental excitation energies are used and the energies of the remaining rotational states are taken from the VMI fit to the experimental data (not from the pure rotational formula), then $T_{1/2}$ is reduced to 22 msec (filled circle in the Fig. 2). This demonstrates that $T_{1/2}$ is sensitive not only to the correct position of 2^+ level but also to the placement of other rotational states of the daughter nucleus.

The excited states of the daughter nucleus correspond to open or closed channels. To check the importance of the closed channels, we carried out the following weak-coupling calculation. For the open channels (the 0^+, 2^+, 4^+, and 6^+ states of the daughter nucleus) we have used the experimental excitation energies, and for the closed channels we have taken the energies calculated by the pure rotational formula. This yields $T_{1/2}$=32 msec (shown by a square in Fig. 2). From this exercise we can conclude that the effect of the closed channels is as important as the open ones.

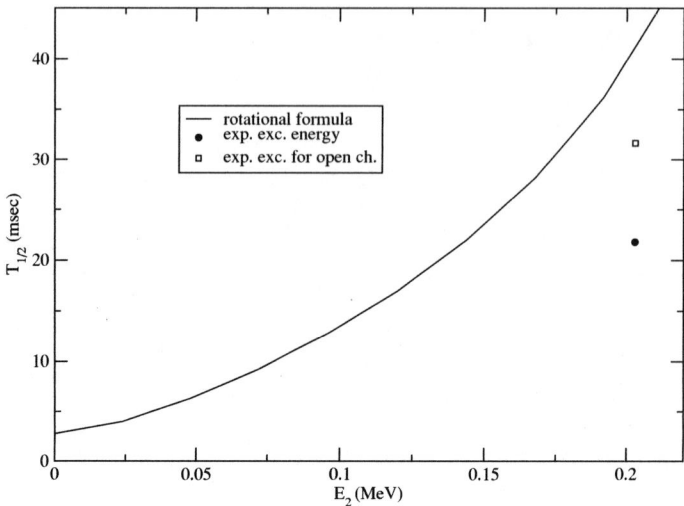

FIGURE 2. Half-life of the $J = \frac{7}{2}^{-}$ resonance in ^{141}Ho as a function of the excitation energy E_2 of the 2^+ state in ^{140}Dy. The solid curve shows the result when the excitation energies are calculated by the pure rotational formula. The result obtained by using the experimental energies of the ground-state band in ^{140}Dy is marked by a dot. The square marks the result obtained by assuming that the energies of closed-channel states are taken from the pure rotational formula.

Effect of γ-vibrations

We have seen that the structure of the daughter nucleus (e.g., the exact placement of rotational members of the ground-state band) has a large influence on $T_{1/2}$. If we assume that the daughter nucleus ^{140}Dy has other excited states, such as a low-lying γ-vibrational band, this will also have a large effect on the calculated proton-emission observables. The coupling of the γ-vibrational $K = 2$ rotational band to the ground-state band is possible if the proton-daughter interaction has a non-axial component in the body-fixed system.

Figure 3 shows results of coupling to the γ-vibrational band of the daughter nucleus. Guided by experimental systematics, we assumed the energy of the $K=2$ bandhead to be 0.881 MeV. The positions of higher-lying states of the γ-band were calculated using the pure rotational formula. For the ground-state band, we have used the experimental values of the excitation energies. The deformation parameter a_0 was set to the value of 0.244, which is consistent with earlier investigations [12, 15]. Different curves in Fig. 3 display results of calculations carried out using different number of states in the γ-band. It is seen that in our preliminary calculations we have not yet reached full convergence with respect to the number of states in the γ-band. The trend, however, is such that

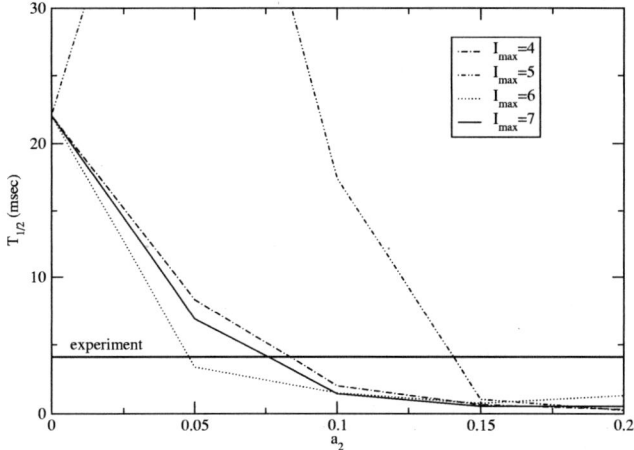

FIGURE 3. Half-life of the $J = \frac{7}{2}^-$ resonance in ^{141}Ho as a function of the deformation parameter a_2. Results of calculations carried out assuming a different number of active states in the γ-band of ^{140}Dy are shown.

triaxiality certainly helps to improve the agreement with the experimental lifetime (4.1 msec).

Our calculations have clearly demonstrated that the structure of the daughter nucleus strongly influences the results of the non-adiabatic coupled-channel model. The description of the proton radioactivity of ^{141}Ho is a very challenging task for theory. Different models have been proposed in order to describe simultaneously the half-life time and branching ratio. Our model shows that the assumption of dynamical coupling to low-lying γ-vibrational states should be treated very seriously.

ACKNOWLEDGMENTS

This research was supported by Hungarian OTKA Grant Nos. T37991 and T29003 and by the U.S. Department of Energy under Contract Nos. DE-FG02-96ER40963 (University of Tennessee) and DE-AC05-00OR22725 with UT-Battelle, LLC (Oak Ridge National Laboratory).

REFERENCES

1. Kruppa A.T., Barmore B., Nazarewicz W., and Vertse T., *Phys. Rev. Lett.* **84**, 4549 (2000).
2. Barmore B., Kruppa A.T., Nazarewicz W., and Vertse T., *Phys. Rev. C* **62**, 054315 (2000).
3. Esbensen H., and Davids C.N., *Phys. Rev. C* **63**, 014315 (2001).
4. Davids C.N. and Esbensen H., *Phys. Rev. C* **64**, 034317 (2001).
5. Hagino K., *Phys. Rev. C* **64**, 041304 (2001).
6. Maglione E., Ferreira L.S., and Liotta R.J., *Phys. Rev. Lett.* **81**, 538 (1998).
7. Davids C.N., et al., *Phys. Rev. Lett.* **80**, 1849 (1998).

8. Maglione E., Ferreira L.S., and Liotta R.J., *Phys. Rev. C* **59**, R589 (1999).
9. Rykaczewski K., et al., *Phys. Rev. C* **60**, 011301 (1999).
10. Sonzogni A.A, et al., *Phys. Rev. Lett.* **83**, 1116 (1999).
11. Bohr A., and Mottelson B.R., *Nuclear Structure Vol. II*, Benjamin, New York, 1975.
12. Seweryniak D., et al., *Phys. Rev. Lett.* **86**, 1458 (2001).
13. Yan J., Brentano von V.P., and Gelberg A., *Phys. Rev. C* **48**, 1046 (1993).
14. Kortelahti M.O., et al., *Phys. Rev. C* **42**, 1267 (1990).
15. Królas W., et al., *Phys. Rev. C* **65**, 031303(R) (2002).
16. Tamura T., *Rev. Mod. Phys.* **37**, 679 (1965).
17. Fröman P.O., *Mat. Fys. Skr. Dan. Vid. Selsk.* **1**, No 3 (1957).

Effects Of Nuclear Reconstruction For Proton Decay Of Spherical Nuclei

Kadmensky S.G., Ivankov Yu.V.

Voronezh State University, University square 1,
394693, Voronezh, Russia

Abstract. The generalization of the unified theory of proton decay on the basis of methods of unified theory nucleus and the open Fermi-systems theory has been done taking into account the reconstruction of the proton-emitting nucleus frame in the proton decay process. The new integral formulae based on the methods of projection operators and orthogonal projecting for the proton partial width amplitudes have been constructed. These formulae have been used for the analysis of proton widths for case of even-odd spherical proton-emitting nuclei. It has been shown that the deviations of correct proton widths taking into account the differences between shell-model proton energies and proton separation energies from proton widths calculated on the basis of well-depth procedure are small.

INTRODUCTION

Proton decay is practically the only source of information about properties of nuclei lying beyond the proton drip line, which determines the boundary of the existence of nuclei in nature [1]. The unified theory of the proton decay taking into account the manyparticle character of the proton decay was constructed in refs. [2,3] for spherical and in refs. [4,5] for deformed parent nuclei on the basis of the integral formula for decay width that was derived at first time in refs. [6,7]. Later, different modifications of this theory were built for the description of proton decay of spherical [8–9] and deformed [10–13] nuclei. At the present time the unified theory gives the possibility to interpret the results of a considerable amount of experimental works in the field of proton decay from ground and low-lying single-particle and multiparticle isomeric states of nuclei with $50 < Z < 84$. It has been demonstrated that this theory has given the unique information about the structure and shapes of the proton decaying nuclei by the new experimental results connected with the observation of proton emission from intruder states and highly deformed proton emitters and of fine structure of proton spectra of both spherical [14] and deformed [15] nuclei.

At the present time all calculations of the proton partial widths for the spherical and deformed proton-emitting nuclei are based on the well-depth procedure in which the shell model energy of emitted proton E_{clj}^{sh} in parent nucleus is supposed to be equal to experimentally found proton separation energy Q_{clj}. In the work [2] it has been shown in the simplified scheme by the method of analytical continuation that taking into

CP681, *Proton-Emitting Nuclei: Second International Symposium; PROCON 2003*,
edited by E. Maglione and F. Soramel
© 2003 American Institute of Physics 0-7354-0150-0/03/$20.00

account the differences between E_{clj}^{sh} and Q_{clj} has been led to the small changing of the proton width values for spherical proton emitting nuclei. In the work [16] the new integral formula for proton partial width amplitude based on the method of unified theory of nucleus [17] and the open Fermi-systems theory [18–20] taking into account the reconstruction of proton-emitting nucleus frame in the proton-decay process have been constructed. The aims of the present work are the generalization based on the orthogonal projection method [21] of this formula for the proton partial width amplitude of the spherical and deformed nuclei and correct responsivences of the differences of E_{clj}^{sh} and Q_{clj} for the proton widths of spherical proton-emitting nuclei.

Let us investigate the deep-subbarrier proton decay of a parent nucleus, having an atomic weight A and a charge Z and occuring in an isolated quasistationary state described by the wave function $\Psi_{\sigma_i}^{J_i M_i}$ connected with the real part of the energy E_i^0 and characterized by spin J_i, it's projection M and other quantum numbers σ_i. The wave function $\Psi_{\sigma_i}^{J_i M_i}$ can be represented as the sum [17]:

$$\Psi_{\sigma_i}^{J_i M_i} = \left(\Psi_{\sigma_i}^{J_i M_i}\right)_{sh} + \left(\Psi_{\sigma_i}^{J_i M_i}\right)_{cl} \qquad (1)$$

where the function $\left(\Psi_{\sigma_i}^{J_i M_i}\right)_{sh}$ build on the basis of the multiparticle shell-model and gives a possibility to calculate the proton spectroscopical factor taking into account the superfluid and normal nucleon-nucleon correlations and the influence of the collective modes of nuclear movement. The function $\left(\Psi_{\sigma_i}^{J_i M_i}\right)_{cl}$ describes the cluster region of the parent nuclei where the decay fragments are completely formed and has the form,

$$\left(\Psi_{\sigma_i}^{J_i M_i}\right)_{cl} = \left\langle G^{J_i M_i}(x,r;x',r')Q(H-E)P\left|\Psi_{\sigma_i}^{J_i M_i}\right)_{sh}\right\rangle, \qquad (2)$$

where H and E are the total Hamiltonian and energy of parent nucleus, P and $Q = 1 - P$ are the projection operators:

$$P = \sum_n \left|\left(\Psi_n^{J_i M_i}\right)_{sh}\right\rangle\left\langle\left(\Psi_n^{J_i M_i}\right)_{sh}\right|, \qquad (3)$$

The multiparticle shell model wave functions $\left|\left(\Psi_n^{J_i M_i}\right)_{sh}\right\rangle$ form the basis for construction of the parent nucleus stationary wave functions $\left(\Psi_{\sigma_i}^{J_i M_i}\right)_{sh}$ in internal region, where nucleus has a compact form.

In the formula (2) $G^{J_i M_i}(x,r;x',r')$ is the multiparticle cluster Green-function which is defined by the equation [17–18]:

$$Q(H-E)QG^{J_i M_i}(x,r;x',r') = \delta(x-x')(r-r'), \qquad (4)$$

where r is the module of relative coordinate $\vec{r}$ of the emitted proton and daughter nucleus, x is total set of parent nucleus coordinates with the exception of coordinate r. Using the orthogonal projection method [21] the operators $Q(H-E)P$ and $Q(H-E)Q$ can be replaced by the operators $(H-E)$ and $(\tilde{H}-E)$ correspondingly, where $\tilde{H} = H + \chi P = H_0 + \tilde{V}$, $\tilde{V} = V + \chi P$, $H_0 = H - V$, V is interaction potential

72

between emitted proton and daughter nucleus, $\chi \to \infty$. The solution of equation (4) can be represented as [22, 23]

$$G^{J_iM_i}(x,r;x',r') = \sum_\alpha \int \frac{dn_\alpha}{dE_\alpha} dE_\alpha \frac{\Psi_\alpha^{J_iM_i+}(x,r)\Psi_\alpha^{*J_iM_i-}(x',r')}{E_\alpha - E + i\delta},$$ (5)

where $\dfrac{dn_\alpha}{dE_\alpha}$ is energy density of α-states, $\Psi_\alpha^{J_iM_i\pm}(x,r)$ is solution of Shrödinger equation:

$$\left(\hat{H} - E_\alpha\right)\Psi_\alpha^{J_iM_i\pm}(x,r) = 0,$$ (6)

with asymptotic boundary conditions including outgoing (+) and ingoing (–) spherical waves. The wave functions $\Psi_\alpha^{J_iM_i\pm}(x,r)$ are orthogonal to the wave functions $(\Psi_{\sigma_i}^{J_iM_i\pm})_{sh}$ and describe the potential scattering of proton by daughter nucleus. The function $\Psi_\alpha^{J_iM_i\pm}$ is normalized by δ - function of energy and can be represented as [22, 23]:

$$\Psi_\alpha^{J_iM_i\pm}(x,r) = \sum_{\alpha'} U_{\alpha'}^{J_iM_i}(x)\frac{f_{\alpha'\alpha}^{J_i\pm}(r)}{r},$$ (7)

where $U_\alpha^{J_iM_i}(x)$ is the channel function:

$$U_\alpha^{J_iM_i}(x) = \left\{(\Psi_{\sigma_f}^{J_fM_f})_{sh}\, i^l\left\{\chi_{\frac{1}{2}m}(\sigma_p)Y_{lm_l}(\Omega_r)\right\}_{jm_j}\right\}_{J_iM_i},$$ (8)

with the multiparticle shell-model wave function of daughter nucleus $(\Psi_{\sigma_f}^{J_fM_f})_{sh}$ and $\alpha = cjl$, $c = \sigma_f J_f$. The radial form-factors $f_{\alpha'\alpha}^{J_i\pm}(r)$ are solution of the set of coupling equations:

$$\left(\frac{d^2}{dr^2} - \frac{l'(l'+1)}{r^2} + k_{c'}^2\right)f_{\alpha'\alpha}^{J_i\pm} - \frac{2m}{\hbar^2}\sum_{\alpha''}\mathcal{V}_{\alpha'\alpha''}^{J_i}(r)f_{\alpha''\alpha}^{J_i\pm}(r) = 0,$$ (9)

where $k_c^2 = \dfrac{2m}{\hbar^2}Q_c$; $\mathcal{V}_{\alpha'\alpha'}^{J_i}(r)$ is matrix element of type $\left\langle U_{\alpha'}^{J_iM_i}\left|\mathcal{V}\right|U_{\alpha'}^{J_iM_i}\right\rangle$ and m is the reduced mass, with boundary conditions:

$$f_{\alpha'\alpha}^{J_i\pm}(0) = 0\ ;\quad f_{\alpha'\alpha}^{J_i\pm}(r) \underset{r\to\infty}{\to} \left\{e^{\mp i(k_c'r - \frac{l'\pi}{2})}\delta_{\alpha'\alpha} \mp S_{\alpha'\alpha}^{J_i\pm}e^{\pm i(k_c'r - \frac{l'\pi}{2})}\right\},$$ (10)

In (10) $S_{\alpha'\alpha}^{J_i}$ is matrix element of S-matrix $S_{\alpha'\alpha}^{J_i-} = \left(S_{\alpha\alpha'}^{J_i+}\right)^*$.

Then the cluster wave function of parent nucleus $\left(\Psi_{\sigma_i}^{J_iM_i}\right)_{cl}$ in asymptotic region $r \to \infty$ can be represented as:

$$\left(\Psi_{\sigma_i}^{J_iM_i}\right)_{cl} \to \sum_\alpha U_\alpha^{J_iM_i}\frac{e^{i(k_cr - \frac{l\pi}{2})}}{r}\sqrt{\frac{\Gamma_\alpha^{J_i}}{\hbar v_c}},$$ (11)

where the partial proton width amplitude is defined as:

73

$$\sqrt{\Gamma_\alpha^{J_i}} = \sqrt{2\pi} \sum_{\alpha'} \left\langle U_{\alpha'}^{J_i M_i} \frac{f_{\alpha'\alpha}^{J_i-}(r)}{r} \middle| H - E \middle| \left(\Psi_{\sigma_i}^{J_i M_i} \right)_{sh} \right\rangle, \quad (12)$$

The formula (12) takes into account the coupling of different proton channels and describes not only subbarrier proton decays but and underbarrier proton decays of spherical and deformed nuclei. This formula gives the possibility to solute the problem of reconstruction of the parent nucleus frame in proton decay process, including the problem of influence of differences between E_{clj}^{sh} and Q_{clj} to proton partial widths.

THE PROTON DECAY OF GROUND AND ISOMERIC STATES

If it has been neglected by the coupling between different proton decay channels, the form-factors $f_{clj}(r)$ is the solution of the integro-differential equation:

$$\left(\frac{d^2}{dr^2} - \frac{l(l+1)}{r^2} + k_c^2 - \frac{2m}{\hbar^2} V_{clj,clj}^{J_i}(r) \right) f_{clj}(r) - \frac{2m}{\hbar^2} \chi \left| \varphi_{clj}^{sh}(r) \right\rangle \left\langle \varphi_{clj}^{sh} \middle| f_{clj} \right\rangle = 0, \quad (13)$$

with boundary conditions:

$$f_{clj}(0) = 0 \; ; f_{clj}(r) \xrightarrow[r \to \infty]{} \sqrt{\frac{2}{\pi \hbar v_c}} \sin\left(k_c r - \frac{l\pi}{2} + \delta_{clj} \right) e^{i\delta_{clj}} ; \quad (14)$$

where δ_{clj} is the scattering phase. Because of the orthogonality of the form-factor $f_{clj}(r)$ to shell-model single proton radial wave functions $\varphi_{clj}^{sh}(r)$ the phase δ_{clj} has the potential character and has no single particle resonance behavior. Then the amplitude of proton partial width (12) for deep subbarier proton decay of odd-even spherical parent nucleus being in ground or isomeric states to ground state of even-even daughter nucleus transits to the form

$$\sqrt{\Gamma_{clj}^{J_i}} = \sqrt{2\pi} Z_{clj}^{1/2} \left\langle \frac{f_{clj}}{r} U_{clj}^{J_i} \middle| H - E \middle| \frac{\varphi_{clj}^{sh}}{r} U_{clj}^{J_i} \right\rangle, \quad (15)$$

where Z_{clj} is the proton spectroscopic factor. The single proton radial shell-model wave function $\varphi_{clj}^{sh}(r)$ is the solution of Schrödinger equation:

$$\left[\frac{d^2}{dr^2} - \frac{l(l+1)}{r^2} + \frac{2m}{\hbar^2} V_{clj,clj}(r) + \tilde{k}_c^2 \right] \varphi_{clj}^{sh}(r) = 0, \quad (16)$$

with boundary conditions for $E_{clj}^{sh} > 0$:

$$\varphi_{clj}^{sh}(0) = 0 ; \varphi_{clj}^{sh}(r) \xrightarrow[r \to R_1]{} G_{cl}(\tilde{k}_c r) \sqrt{\frac{\left(\Gamma_{clj}(\tilde{k}_c) \right)_{sh}}{\hbar v_c}} ; \varphi_{clj}^{sh}(r) = 0 , \; r \geq R_1, \quad (17)$$

where the point R_1 lies in subbarrier region, $G_{cl}(\tilde{k}_c r)$ is irregular radial Coulomb function with $\tilde{k}_c = \frac{2m}{\hbar^2} E_{clj}^{sh}$, $\left(\Gamma_{clj} \right)_{sh}$ is the proton width of the single particle quasistationary state $\varphi_{clj}^{sh}(r)$.

For deep subbarrier proton decay taking into account the hermiticity of the potential V formula can be transformed to the form:

$$\sqrt{\Gamma_{clj}^{J_i}} = Z_{clj}^{1/2} \sqrt{\left(\Gamma_{clj}(\tilde{k}_c)\right)_{sh}} \left[G_{cl}(\tilde{k}_c R_1) \frac{d}{dr} f_{clj}(k_c R_1) - f_{clj}(k_c R_1) \frac{d}{dr} G_{cl}(\tilde{k}_c R_1) \right] \times$$

$$\times \frac{1}{\sqrt{k_c \tilde{k}_c}}. \tag{18}$$

If the proton separation energy Q_{clj} is equal to shell-model single proton energy E_{clj}^{sh} then $f_{clj}(k_c R_1) \approx F_{cl}(k_c R_1)$, where $F_{cl}(k_c r)$ is radial regular Coulomb function and $\Gamma_{clj}^{J_i} = Z_{clj}\left(\Gamma_{clj}(\tilde{k}_c)\right)_{sh}$. The detail analysis has demonstrated that values of proton partial widths (18) practically don't change for different values of radius R_1 in subbarrier region, where nuclear interaction of emitted proton and daughter nucleus is very small.

In the table for different proton decaying nuclei, represented as spherical nuclei, the experimental proton separation energy Q_{nlj} [1] and the shell-model single proton energy E_{nlj}^{sh} calculated for set of parameters of proton shell model potential from works [24] have been shown. It has been seen that differences between E_{clj}^{sh} and Q_{clj} can reach very big values $\leq 2.1 \, \text{MeV}$.

TABLE 1. The experimental proton separation energy Q_{nlj}, the shell-model single proton energy E_{nlj}^{sh}, the single proton partial widths Γ_{wd} and the proton partial widths Γ_{clj} calculated for proton decaying spherical nuclei.

Nuclei	$n\,l\,j$	Q_p, MeV	E_{nlj}, MeV	Γ_{wd}, MeV	Γ_{clj}, MeV	$\Gamma_{wd} / \Gamma_{clj}$
Sb_{51}^{105}	2d 5/2	0,49	2.33	0.243E-22	0.231E-22	1.05
I_{53}^{109}	3s 1/2	0.83	3.61	0.516E-15	0.433E-15	1.19
Cs_{55}^{113}	2d 3/2	0.98	3.10	0.840E-15	0.791E-15	1.06
La_{57}^{117}	2d 3/2	0.82	2.91	0.123E-17	0.117E-17	1.05
Tm_{69}^{147}	1h 11/2	1.07	1.92	0.298E-21	0.296E-21	1.01
Tm_{69}^{147m}	3s 3/2	1.14	0.98	0.290E-17	0.291E-17	1.00
Lu_{71}^{151}	1h 11/2	1.26	1.78	0.128E-19	0.133E-19	1.00
Lu_{71}^{151m}	2d 3/2	1.33	0.11	0.892E-16	0.939E-16	0.95
Ta_{73}^{155}	1h 11/2	1.79	1.65	0.132E-15	0.132E-15	1.00
Ta_{73}^{157}	3s ½	0.95	0.37	0.235E-20	0.245E-20	0.96
Re_{75}^{161}	3s ½	1.21	0.37	0.278E-17	0.292E-17	0.95

Re_{75}^{161m}	1h 11/2	1.34	0.87	0.913E-20	0.918E-20	0.99
Ir_{77}^{165m}	1h 11/2	1.73	0.78	0.782E-17	0.790E-17	0.99
Ir_{77}^{167m}	1h 11/2	1.26	0.15	0.404E-21	0.408E-21	0.99

CONCLUSION

The integral formulae for partial proton widths found in the present work give very good basis for calculations of the shape reconstruction effects for proton decay deformed nuclei. The analogous formulae can be used successfully for description of binary and ternary nuclear fission and of two-proton radioactivity. The work was supported by grant INTAS (project N: 99-0229).

REFERENCES

1. Woods, P.J., and Davids, C.N., *Annu. Rev. Nucl. Part. Sci.* **47**, 541 (1997).
2. Bugrov, V.P., Kadmensky, S.G., Furman, W.I. et al., *Phys of At. Nucl.* **41**, 717 (1985).
3. Bugrov, V.P., Bunakov, V.E., Kadmensky, S.G. et al., *Phys. of At. Nucl.* **42**, 34 (1985).
4. Bugrov, V.P., Kadmensky, S.G., *Phys of At. Nucl.* **49**, 967 (1989).
5. Kadmensky, S.G., and Bugrov, V.P., *Phys of At. Nucl.* **59**, 399 (1996).
6. Kadmensky, S.G., and Kalechitz, V.E., *Phys of At. Nucl.* **12**, 37 (1970).
7. Kadmensky, S.G., and Khlebostroev, V.G., *Phys of At. Nucl.* **18**, 505 (1973).
8. Aberg, S., Semmes, P.B., Nazarevich, W., *Phys. Rev.* **B56**, 1762 (1997).
9. Davids, C.N., and Esbensen, H., *Phys. Rev.* **C56**, 054302 (2000).
10. Maglione, E., Liotta, R.J., Vertse, T., *Nucl. Phys.* **A584**, 13 (1995).
11. Maglione E., Ferreira, L.S., Liotta, R.J., *Phys. Rev. Lett.* **81**, 538 (1998); *Phys. Rev.* C **59**, 589 (1999).
12. Batchelder, J.C., Bingham, C.R. et al., *in Proceedings of ENAM-98*, N. York: Woodbury, 1998, p.264.
13. Kadmensky, S.G., *Phys of At. Nucl.* **63**, 551 (2000).
14. Ginter, T.N. et al, in *Proceedings of Intern. Symposium of Proton Emitting Nuclei*, AIP, p.518, p.83 (1999).
15. Sonzogni, A.A., Davids, C.N. et al. Preprint PHY-9348-H1-99 (Argonne Nat. Lab., 1999).
16. Kadmensky, S.G., in *Proceedings of Intern. Symposium of Exotic Nuclei at the Proton Drip Line* (Unicam, Italy, 2002) p. 133
17. Wildermuth, K., Tang, Y.C., *A Unified Theory of the Nucleus,* Vieweg: Braunschweig, 1977.
18. Kadmensky, S.G., *Phys of At. Nucl.* **62**, 201 (1999).
19. Kadmensky, S.G., *Phys of At. Nucl.* **64**, N3 (2001).
20. Kadmensky, S.G. and Sonzogni, A.A., *Phys. Rev.* **C 62**, 054601 (2000).
21. Kukulin, V.I., Neudachin, V.G., Smirnov, Yu.F., *Particles and Nuclei* **10**, 1236 (1979).
22. Kadmensky, S.G., *Phys. of At. Nucl.* **66**, N10, (2003).
23. Golberger, M., and Watson, K.,*Collision Theory*, N.York: J. Willey and Sons, 1964.
24. Chepurnov, V.A., *Phys. of At. Nucl.* **6**, 955 (1967)

Proton decay of near-spherical nuclei

S. V. Lukyanov*, E. Maglione† and L. S. Ferreira*

*Centro de Física Interações Fundamentais, and Departamento de Física, Insituto Superior
Tecnico, Av. Rovisco Pais, P1049-001 Lisbon, Portugal
†Dipartimento di Fisica "G. Galilei", Via Marzolo 8, I-35131 Padova, Italy and Insituto Nazionale
di Fisica Nucleare, Padova, Italy

Abstract. Fine structure and proton decay widths of odd-A spherical proton emitters has been described [1] taking into account the vibrational excitations of the daughter nucleus. In the present work, we show that, if the residual pairing interaction is also included, two phonon states are required in order to reproduce the experimental results.

INTRODUCTION

Proton emissions from a $h_{11/2}$ orbital has been observed in the last few years [2, 3] in experiments with exotic nuclei beyond the proton drip line. The experimental features found for most of these nuclei were explained considering them as spherical [4]. Nevertheless the simple spherical approach does not provide a full description of the proton emission process because it allows decay only to the ground state [5]. Fine structure was observed in the case of ^{145}Tm showing decay to the lowest 2^+ excited state of the daughter nucleus [6]. It is reasonable to expect a correlation between the outgoing proton and the vibrational spectra, that should modify the spectroscopic factor. In a spherical system, the odd proton is considered to be coupled only to the 0^+ ground state of the daughter nucleus, thus preventing these decays. However, if the nucleus is considered as having a time–averaged spherical shape, but the wave function contains components due to particle vibration coupling, fine structure decay can be automatically described. This was shown, in the case of odd-A proton emitters, by the authors of Ref. [1] that obtained good agreement with ^{145}Tm experiment. Similar calculations were also performed in Ref. [7].

The pairing residual interaction has been taken into account in these calculations only multiplying the final widths by a BCS spectroscopic factor. In the present study, it is taken into account exactly in the structure of the decaying nucleus. In this way, one also solves the ambiguity of which state one has to choose as the decaying state, since it will be the one with lowest energy given rigorously by the model.

THE MODEL

Following the calculations of Refs.[1, 7] we consider the decaying nucleus as a spherical core with a vibrational coupling to the single proton. Thus we expand the total wave

CP681, *Proton-Emitting Nuclei: Second International Symposium; PROCON 2003*,
edited by E. Maglione and F. Soramel

function of the system for a given total spin (J, M) as,

$$|JM> = \sum_{jL} C_{jL}^{n_\lambda} |jn_\lambda, LJM>, \tag{1}$$

where $C_{jL}^{n_\lambda}$ are the normalization amplitudes, and $|jn_\lambda, LJM>$ the channel-spin wave function obtained by coupling the single-particle wave function $|j>$ of the proton, to the core wave function $|n_\lambda, L>$ with n_λ phonons and angular momentum $L = 0, ..., \lambda n_\lambda$. The quantity λ designates the angular momentum of the phonon, that will be taken as $\lambda = 2$. We will consider the 0^+ ground state and 1-phonon and 2-phonon excited states of the vibrational core nucleus. Whereas the 1-phonon state has only angular momentum equal to 2, the 2-phonon state can have the values $0, 2$ and 4.

The functions of Eq. 1 are eigenfuctions of the total Hamiltonian, consisting of a single-particle Hamiltonian H_{sp} with eigenvalue ϵ_j, the intrinsic vibrational Hamiltonian of the daughter nucleus $\hat{H}_{vib}$ with eigenvalues $\epsilon_{n_\lambda} = n_\lambda E_{2^+}$ according to the number of phonons present. The energy E_{2^+} is the 1-phonon excitation energy in the daughter nucleus. The vibration coupling interaction, is taken as [1],

$$\delta V_{vib} = F_{vib}(r) \sum_\mu \alpha_{2\mu} Y_{2\mu}^*(\hat{r}), \tag{2}$$

with a radial form factor with the spin-orbit interaction,

$$F_{vib}(r) = -R_N \frac{dV_N(r)}{dr} - R_c \frac{4\pi Z_D e^2}{5} \int dr' r'^2 \frac{d\rho_D(r')}{dr'} \frac{r_<^2}{r_>^3}$$
$$- R_N V_{so} [j(j+1) - l(l-1) - 3/4] \frac{1}{r} \frac{d^2}{dr^2} f\left(\frac{r - R_N}{a_N}\right). \tag{3}$$

The quantities $R_N = r_0 A_D^{1/3}$ and $R_c = r_c A_D^{1/3}$ are the radii associated with the nuclear interaction and the charge density of the core nucleus, respectively, and the functions $r_< = \min(r, r')$ and $r_> = \max(r, r')$. The nuclear mean field $V_N(r) = V_N^{(0)} f[(r - R_N)/a_N]$ and charge density $\rho_D(r) = \rho_D^{(0)} f[(r - R_c)/a_c]$ of the daughter nucleus are described in terms of the Fermi function $f(x) = 1/(1 + \exp(x))$, and normalized to unity. The vibrational coupling, Eq. 2 only has matrix elements different from zero for states that differ by one phonon.

After projection of the Schrödinger equation on the channel-spin wave function $|jn_\lambda, LJM>$, we obtain the set of equations,

$$\left(\epsilon_{j'} + \epsilon_{n_{\lambda'}} - \epsilon\right) C_{j'L'}^{n_{\lambda'}} + \sum_{jL, n_\lambda} C_{jL}^{n_\lambda} \delta\epsilon_{j'L', jL} = 0, \tag{4}$$

where $\delta\epsilon_{j'n_{\lambda'}L', jn_\lambda L} = <j'n_{\lambda'}, L'JM|\delta\hat{V}_{vib}|jn_\lambda, LJM>$.

The effect of the residual pairing interaction can be included making the usual Bogoliubov transformation from single particles to quasiparticles with the energy $\tilde{\epsilon}_k = [(\epsilon_k - \epsilon_F)^2 + \Delta^2]^{1/2}$, depending on the pairing gap Δ, and on the Fermi energy ϵ_F. Thus in Eqs. (4) quasi particle energies $\tilde{\epsilon}_k$ replace the particle ones ϵ_k. Under this transformation the energy matrix $\epsilon_k \delta_{k'k} + \delta\epsilon_{k'k}$ changes and it has to be multiplied by a factor

$f_{uv} = u_{j'}u_j + v_{j'}v_j$, where v^2 and u^2 are the probability of having the single particle state occupied or, empty, respectively. Therefore Eqs. (4) become

$$\sum_{kn_\lambda} \left(\tilde{\epsilon}_k \delta_{k'k} + f_{uv} \delta \epsilon_{k'k} \right) C_k^{n_\lambda} = \epsilon C_{k'}^{n_{\lambda'}}. \tag{5}$$

where a simplified notation, $k = (jn_\lambda L)$ and $\epsilon_k = \epsilon_j + \epsilon_{n_\lambda}$, was used. Since the coupling interaction, only relates states which differ by one phonon, the off-diagonal part of the energy matrix element in Eqs. (5) has two contributions $\delta \epsilon_{k'k} = \delta \epsilon_{k'k}^{(1)} + \delta \epsilon_{k'k}^{(2)}$ corresponding to the coupling between the states with no phonons and only one, and the coupling between states with one phonon and the ones with two, respectively. The first term, can be calculated as [8]

$$\delta \epsilon_{k'k}^{(1)} = \delta_{n'_\lambda 1} \sqrt{\frac{(2\lambda + 1)}{4\pi}} \alpha_2^{(0)} < j \tfrac{1}{2} 20 | j' \tfrac{1}{2} >< l'j'|F_{vib}(r)|lj > + h.c., \tag{6}$$

where $\alpha_2^{(0)} = < 2\mu | \alpha_{2\mu} | 00 >$ is the vibrational transition matrix element with $\alpha_{2\mu}$, and $< j'|F(r)|j >$ the radial coupling matrix element. The coupling between states with one and states with two phonons is analogously,

$$\delta \epsilon_{k'k}^{(2)} = \delta_{n_{\lambda'} 2} (-1)^{J+j'} \sqrt{\frac{(2\lambda + 1)(2L + 1)(2j + 1)}{2\pi}} \alpha_2^{(0)} \times$$

$$\left\{ \begin{matrix} L & J & j' \\ j & 2 & 2 \end{matrix} \right\} < j \tfrac{1}{2} 20 | j' \tfrac{1}{2} >< j'|F(r)|j > + h.c. \tag{7}$$

The total width for decay is a sum of partial widths $\Gamma = \sum_{n_\lambda j} \Gamma_{jL}^{n_\lambda}$ for all possible j of the emitted proton restricted by angular momentum and parity conservation [9] given by,

$$\Gamma_{jL}^{n_\lambda} = \frac{\hbar^2 \kappa}{\mu} |N_j|^2 C_{jL}^{n_\lambda 2} u_j^2. \tag{8}$$

where μ is the reduced mass, κ the proton wave number, u_j^2 the probability that the single particle level in the daughter nucleus is empty obtained in the BCS approach, and N_j the asymptotic normalization of the single particle wave function of the outgoing proton. The decay width to the ground and excited states, is obtained by setting n_λ equal to zero, or one and two, respectively.

RESULTS AND DISCUSSION

Let us consider odd-A spherical emitters with $J = 11/2^-$. Then, in addition to the $j = 11/2^- \otimes 0^+$ component, which can only decay to the daughter ground state, the parent wave function will contain additional components corresponding to the single-particle orbits associated with the 1-phonon and 2-phonon excited states. We first calculate the decay widths taking only into account the particle-vibrational coupling as it was done in [1], using the same potential parameters for the nuclear and Coulomb parts, and

TABLE 1. Results of particle-vibrational coupling calculations for spherical proton emitters with initial spin $J_i^\pi = 11/2^-$.

Nuclide	E_{2^+} (keV)	Γ_{calc} (10^{-16} MeV)	$\Gamma_{calc}[+so]$ (10^{-16} MeV)	Γ_{exp} (10^{-16} MeV)	Γ_{calc} **from Ref. [1]** (10^{-16} MeV)
^{145}Tm	326	2.41 ($J_f = 0^+$)	2.51 ($J_f = 0^+$)	1.34	1.58 ($J_f = 0^+$)
^{145}Tm	326	0.15 ($J_f = 2^+$)	0.16 ($J_f = 2^+$)	0.18	0.22 ($J_f = 2^+$)
^{147}Tm	510	1.58×10^{-6}	1.64×10^{-6}	1.18×10^{-6}	9.36×10^{-7}
^{151}Lu	600	5.59×10^{-5}	5.59×10^{-5}	3.65×10^{-5}	3.34×10^{-5}
^{161}Re	610	1.81×10^{-5}	1.81×10^{-5}	1.37×10^{-5}	1.52×10^{-5}
^{165}Ir	548	8.82×10^{-3}	9.20×10^{-3}	1.31×10^{-2}	1.01×10^{-2}
^{167}Ir	431	4.27×10^{-7}	4.44×10^{-7}	6.08×10^{-7}	5.45×10^{-7}

$\alpha_2^{(0)} = 218/A\sqrt{E_{2^+}}$, where A is the core mass number. The depth of the nuclear potential $V_N^{(0)}$ is adjusted so that to obtain a resonant solution for each case. The results are given in Table 1, and compared with the experimental data Γ_{exp}, and the results of [1], with and without the spin-orbit interaction, using a depth $V_{so} = 20$ MeV·fm^2.

The decay widths of Table 1, were calculated for the state which has the largest percentage of zero phonons, and the pairing residual interaction only enters trough the BCS spectroscopic factor of Eq. 8. This is the reason of the differences observed with the results of Ref. [1] which considered another state, and used the degenerate $h11/2$ with $s1/2$ and $d3/2$ states in the calculation of the BCS factor. Therefore, this factor is larger than the one used in our calculation. As one can see from the table, the agreement with experiment is not as good as the one obtained in Ref. [1], including the corresponding BCS factors.

The calculated decay widths show some differences from the experimental data, especially for lighter nuclei. The branching ratio of 5.9% (6.0% with the spin-orbit interaction) obtained for decay to the 2^+ in ^{145}Tm differs substantially from the experimental value of 11.8% [6]. As it was observed in Ref. [1] the spin-orbit interaction does not change significantly the decay widths only increasing them slightly. The branching ratios obtained for other nuclei in the all cases are very small.

Taking into account the residual pairing interaction and introducing in the description coupling with 1-phonon excitations makes the description more appropriate on the whole range of nuclei. The results are presented in Table 2. The spin-orbit interaction in this case, again slightly increases the decay widths. However, in the case of ^{145}Tm, the decay width is almost twice the experimental value, and the branching ratio of 4.5% (4.4% with spin-orbit interaction) corresponds to 40% of the measured value.

Since the coupling to the 1 phonon state is very strong, one might wonder what happens putting in the basis the 2 phonon states. Taking them into account, the results are shown in Table 3. As it is seen, excluding only the last two nuclei, better understanding of the experimental data is achieved. Thus, after including in the description the coupling with 2-phonon excitations, the branching ratio for ^{145}Tm is 7.9% (8.3% with spin-orbit interaction) and the agreement with experiment is better.

Some of the differences obtained for heavier nuclei, can be attributed to the potential used, that has a very low $h11/2$ state with respect to the $s1/2$ and $d3/2$, thus leading to very small u^2 BCS factors, and consequently to very small decay widths for the heavier

TABLE 2. The same calculations as in Table 1 but taking into account the residual pairing interaction in the hamiltonian and the coupling with 1-phonon states.

Nuclide	E_{2^+} (keV)	Γ_{calc} (10^{-16} MeV)	$\Gamma_{calc}[+so]$ (10^{-16} MeV)	Γ_{exp} (10^{-16} MeV)
^{145}Tm	326	2.56 ($J_f = 0^+$)	2.60 ($J_f = 0^+$)	1.34
^{145}Tm	326	0.12 ($J_f = 2^+$)	0.12 ($J_f = 2^+$)	0.18
^{147}Tm	510	1.64×10^{-6}	1.68×10^{-6}	1.18×10^{-6}
^{151}Lu	600	5.75×10^{-5}	5.92×10^{-5}	3.65×10^{-5}
^{161}Re	610	1.84×10^{-5}	1.90×10^{-5}	1.37×10^{-5}
^{165}Ir	548	8.91×10^{-3}	9.19×10^{-3}	1.31×10^{-2}
^{167}Ir	431	4.33×10^{-7}	4.43×10^{-7}	6.08×10^{-7}

TABLE 3. The same calculations as in Table 2 but taking into account also coupling with 2-phonon states.

Nuclide	E_{2^+} (keV)	Γ_{calc} (10^{-16} MeV)	$\Gamma_{calc}[+so]$ (10^{-16} MeV)	Γ_{exp} (10^{-16} MeV)
^{145}Tm	326	1.17 ($J_f = 0^+$)	1.21 ($J_f = 0^+$)	1.34
^{145}Tm	326	0.10 ($J_f = 2^+$)	0.11 ($J_f = 2^+$)	0.18
^{147}Tm	510	8.44×10^{-7}	9.08×10^{-7}	1.18×10^{-6}
^{151}Lu	600	3.21×10^{-5}	3.52×10^{-5}	3.65×10^{-5}
^{161}Re	610	1.07×10^{-5}	1.19×10^{-5}	1.37×10^{-5}
^{165}Ir	548	4.99×10^{-3}	5.49×10^{-3}	1.31×10^{-2}
^{167}Ir	431	2.20×10^{-7}	2.36×10^{-7}	6.08×10^{-7}

nuclei.

In order to observe the importance of two-phonon states in the base, we present in Table 4 the structure for the wave function of ^{145}Tm calculated with the spin-orbit interaction. Considering only the coupling to the 1-phonon excitations, the wave function is dominated by the $h_{11/2-}$ orbital coupled to the 0^+ ground state $\sim 61\%$. If one includes in the description 2-phonon excitations, this component drops to 28% while the component of the proton wave function coupled to the 2^+ state was found to be maximum and consists approximately of 48%.

TABLE 4. The wave function structure for ^{145}Tm.

n_λ	$C_k^{n_\lambda 2}$ 0-phonon	$C_k^{n_\lambda 2}$ 1-phonon	$C_k^{n_\lambda 2}$ 2-phonon	$C_k^{n_\lambda 2}$ from Ref. [1]
0	1	0.61	0.28	0.33
1	0	0.39	0.48	0.67
2	0	0.00	0.24	0.00

CONCLUSIONS

We have studied the influence of the residual pairing interaction in the particle-vibration model of near-spherical proton emitters. The results have shown that to reproduce simultaneously the fine structure of ^{145}Tm and the decay widths for the whole range of nuclei one must include in the description with the residual pairing interaction not only 1-phonon states but also 2-phonon ones. Including 2-phonon states in the description of the core, the structure of the wave function changes and the 1-phonon components are found to be the dominant ones, $\sim$48%. As previously found [1], the spin-orbit term does not play a significant role in the description of the decay widths.

ACKNOWLEDGMENTS

This work was supported by the Grant FJ08 and Project POCTI-36575/99 of Fundação para a Ciência e a Tecnologia, Ministério da Ciência e da Tecnologia, Portugal.

REFERENCES

1. C.N. Davids and H. Esbensen, Phys. Rev. C **64**, 034317 (2001).
2. C. N. Davids et al., Phys. Rev. C **55**, 2255 (1997).
3. C. R. Bingham et al., Phys. Rev. C **59**, R2984 (1999).
4. S. Åberg, P. B. Semmes and W. Nazarewicz, Phys. Rev. C **56**, 1762 (1997); **58**, 3011 (1997).
5. K. P. Rykaczewski et al., Nucl. Phys. A**682**, 270c (2001).
6. M. Karny, *et al*, Phys. Rev. Lett. **90**, 012502-1 (2003).
7. K. Hagino, Phys. Rev. C **64**, 041304 (2001).
8. A. Bohr and B. R. Mottelson, *Nuclear Structure* Vol. II. (World Scientific, Singapore, 1998).
9. E. Maglione and L.S. Ferreira, Phys. Rev. C **61**, 047307 (2000).

Properties of Proton Emitters in Relativistic Mean Field Calculations

P. Ring*, D. Vretenar† and G. A. Lalazissis**

*Physics Department T30, Technische Universität München,
D-85748 Graching, Germany
†Physics Department, Faculty of Science, University of Zagreb,
10 000 Zagreb, Croatia
**Department of Theoretical Physics, Aristotle University of Thessaloniki,
Thessaloniki GR-54006, Greece

Abstract. An effective relativistic field theory is used to investigate nuclei along the proton drip line together with proton emission in heavy deformed nuclei. Nucleons are described as Dirac particles moving in self-consistent relativistic scalar and time-like vector fields. Pairing correlations are treated within the relativistic Hartree-Bogoliubov (RHB) approximation using a finite range Gogny interaction. This method provides a fully microscopic and universal description of nuclei in the vicinity and far from the valley of stability. It allows the determination of the proton drip line and the investigation of proton emitters, in particular the calculation of the proton separation energies, the quantum numbers of the corresponding proton levels and their spectroscopic factors. We find good agreement for experimentally known proton emitters in medium heavy nuclei and we predict new possible proton emitters below Z=50.

INTRODUCTION

In recent years experiments with radioactive beams made it possible to investigate nuclei far from stability with a large neutron or proton excess. Because of the strongly repulsive Coulomb force the number of isotopes with proton excess is much more limited than the number of isotopes with neutron excess. Therefore, in contrast to the neutron drip line, which today is experimentally known only up to Z=17, the location of the proton drip line is today well investigated for isotopes with odd proton numbers known up to Z=91. The region of proton rich nuclei is of special interest because for several reasons: a) Proton radioactivity provides a new and very precise experimental tool to investigate details of nuclear structure at the edge nuclear stability, b) the N=Z-line, which is interesting for a number of theoretical reasons comes close to the proton drip line around Z=50 and passes it afterwards and c) the mass region $60 < A < 100$ is important for the process of nucleosynthesis during explosive hydrogen burning. The path of the rp-process lies between the valley of stability and the proton drip line. It is a very complicated function of the physical conditions governing the explosion [1]. The input for nuclear reaction network calculations includes nuclear masses, i.e. proton separation energies of the neutron deficient isotopes, proton capture rates, inverse photodisintegration rates, β-decay and electron capture rates [2] The structure and decays modes of nuclei beyond the proton drip-line represent one of the most active areas

CP681, *Proton-Emitting Nuclei: Second International Symposium; PROCON 2003*,
edited by E. Maglione and F. Soramel
© 2003 American Institute of Physics 0-7354-0150-0/03/$20.00

of experimental and theoretical studies of exotic nuclei with extreme isospin values.

For a realistic description of nuclei far from stability, which is applicable over the full range charge numbers one needs a fully self-consistent theory with a proper treatment of the spin orbit splitting, the basis of nuclear shell structure. We therefore use Relativistic Hartree Bogoliubov (RHB) theory [3, 4]. This theory is an effective field theory based on average fields with a definite Lorentz structure.In addition a density dependence is taken into account by a non-linear coupling of the corresponding meson fields. Pairing effects are described in this method by the effective finite range pairing force of Gogny. In general, the calculated properties have been found in excellent agreement with available experimental data, and with the predictions of the macroscopic-microscopic mass model [5].

RELATIVISTIC HARTREE BOGOLIUBOV THEORY.

In the framework of relativistic mean field theory nucleons are described as point particles that move independently in mean fields which originate from complicated processes between the nucleons having its origin originally in QCD. Since there is now way to solve QCD at low energies and to calculate these processes in detail one is left with an effective field theory, where the fields are parametrized and represented in a relativistic way by the exchange of effective mesons. The theory is fully Lorentz invariant. Conditions of causality and Lorentz invariance impose that the interaction is mediated by the exchange of point-like effective mesons, which couple to the nucleons at local vertices. The single-nucleon dynamics is described by the Dirac equation

$$\left\{-i\alpha \cdot \nabla + \beta(m+g_\sigma\sigma)+g_\omega\omega^0+g_\rho\tau_3\rho_3^0+e\frac{(1-\tau_3)}{2}A^0\right\}\psi_i = \varepsilon_i\psi_i. \tag{1}$$

σ, ω, and ρ are the meson fields, and A denotes the electromagnetic potential. g_σ g_ω, and g_ρ are the corresponding coupling constants for the mesons to the nucleon. This is an effective field theory. The meson do not correspond to existing particles on the energy shell and the corresponding fields are not quantized. Rather these are effective fields and their names σ, ω, and ρ characterize only the quantum numbers of spin, parity and isospin imposed by the Lorentz structure of the corresponding fields.

In detail the sources of the meson fields are defined by the nucleon densities and currents. The ground state of a nucleus is described by the stationary self-consistent solution of the coupled system of the Dirac (1)and Klein-Gordon equations

$$\left[-\Delta+m_\sigma^2\right]\sigma(\mathbf{r}) = -g_\sigma\rho_s(\mathbf{r})-g_2\sigma^2(\mathbf{r})-g_3\sigma^3(\mathbf{r}), \tag{2}$$

$$\left[-\Delta+m_\omega^2\right]\omega^0(\mathbf{r}) = g_\omega\rho_v(\mathbf{r}), \tag{3}$$

$$\left[-\Delta+m_\rho^2\right]\rho^0(\mathbf{r}) = g_\rho\rho_3(\mathbf{r}), \tag{4}$$

$$-\Delta A^0(\mathbf{r}) = e\rho_p(\mathbf{r}), \tag{5}$$

for the sigma meson, omega meson, rho meson and photon field, respectively. Due to charge conservation, only the 3rd-component of the isovector rho meson contributes.

The source terms in equations (2) to (5) are sums of bilinear products of baryon amplitudes, and they are calculated in the *no-sea* approximation, i.e. the Dirac sea of negative energy states is not included in the calculation of nucleonic densities and currents.

Due to time reversal invariance, there are no currents in the static solution for an even-even system, and therefore the spatial vector components ω, ρ_3 and $\mathbf{A}$ of the vector meson fields vanish.

The quartic potential, which has first been introduced by Boguta and Bodmer [6]

$$U(\sigma) = \frac{1}{2}m_\sigma^2\sigma^2 + \frac{1}{3}g_2\sigma^3 + \frac{1}{4}g_3\sigma^4 \tag{6}$$

introduces an effective density dependence. The non-linear self-interaction of the σ field is essential for a quantitative description of surface properties of finite nuclei.

In addition to the self-consistent mean-field potential, pairing correlations have to be included in order to describe ground-state properties of open-shell nuclei. For nuclei close to the β-stability line, pairing can been taken into account in the relativistic mean-field model in the form of a simple BCS approximation [7]. For more exotic nuclei further away from the stability line, however, the BCS model presents only a poor approximation. In particular, in order to correctly reproduce density distributions in nuclei close to the drip lines , mean-field and pairing correlations have to be described in a unified framework, i.e. in the Hartree-Fock-Bogoliubov model or the relativistic Hartree-Bogoliubov (RHB) model. In the unified framework the ground state of a nucleus $|\Phi\rangle$ is represented by the product of independent single-quasiparticle states. These states are eigenvectors of the generalized single-nucleon Hamiltonian which contains two average potentials: the self-consistent mean-field $\hat{h}_D$ which encloses all the long range particle-hole (*ph*) correlations through the effective meson fields, and a pairing field $\hat{\Delta}$ which sums up the particle-particle (*pp*) correlations. In the Hartree approximation for the self-consistent mean field, the relativistic Hartree-Bogoliubov equations read

$$\begin{pmatrix} \hat{h}_D - m - \lambda & \hat{\Delta} \\ -\hat{\Delta}^* & -\hat{h}_D + m + \lambda \end{pmatrix} \begin{pmatrix} U_k(\mathbf{r}) \\ V_k(\mathbf{r}) \end{pmatrix} = E_k \begin{pmatrix} U_k(\mathbf{r}) \\ V_k(\mathbf{r}) \end{pmatrix}. \tag{7}$$

where $\hat{h}_D$ is the single-nucleon Dirac Hamiltonian (1), and m is the nucleon mass. The chemical potential λ has to be determined by the particle number subsidiary condition in order that the expectation value of the particle number operator in the ground state equals the number of nucleons. The column vectors denote the quasi-particle spinors and E_k are the quasi-particle energies. The pairing field $\hat{\Delta}$ in (7) is defined as an integral operator with the kernel

$$\Delta_{ab}(\mathbf{r},\mathbf{r}') = \frac{1}{2}\sum_{c,d}V_{abcd}(\mathbf{r},\mathbf{r}')\sum_{k>0}U_{ck}^*(\mathbf{r})V_{dk}(\mathbf{r}'), \tag{8}$$

where a,b,c,d denote quantum numbers that specify the Dirac indices of the spinors, $V_{abcd}(\mathbf{r},\mathbf{r}')$ are matrix elements of a general two-body pairing interaction. The sum over $k > 0$ means a short hand notation for the no-sea approximation.

The RHB equations are solved self-consistenfly, with potentials determined in the mean-field approximation from solutions of Klein-Gordon equations for the meson

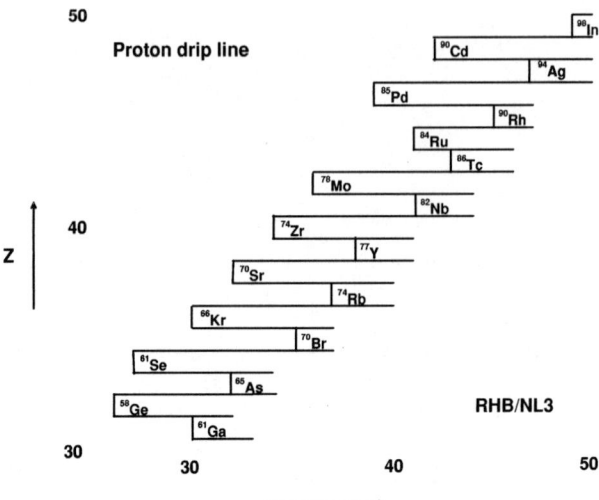

FIGURE 1. The proton drip line in the region $31 \leq Z \leq 49$. On this section of the chart of the nuclides the last bound isotopes for each element are indicated. Nuclei to the left are predicted to be proton unstable by the present RHB (NL3+D1S) calculation.

fields. The sources are given by the densities

$$\rho_s(\mathbf{r}) = \sum_{p>0} \bar{V}_p(\mathbf{r})V_p(\mathbf{r}) + \sum_{n>0} \bar{V}_n(\mathbf{r})V_n(\mathbf{r}), \tag{9}$$

$$\rho_v(\mathbf{r}) = \sum_{p>0} V_p^\dagger(\mathbf{r})V_p(\mathbf{r}) + \sum_{n>0} V_n^\dagger(\mathbf{r})V_n(\mathbf{r}), \tag{10}$$

$$\rho_3(\mathbf{r}) = \sum_{p>0} V_p^\dagger(\mathbf{r})V_p(\mathbf{r}) - \sum_{n>0} V_n^\dagger(\mathbf{r})V_n(\mathbf{r}), \tag{11}$$

$$\rho_c(\mathbf{r}) = \sum_{p>0} V_p^\dagger(\mathbf{r})V_p(\mathbf{r}). \tag{12}$$

where the indices p and n run over the corresponding quasi-particle states for protons and neutrons. The current version of the model [8] describes axially symmetric deformed shapes. The Dirac-Hartree-Bogoliubov equations and the equations for the meson fields are solved by expanding the nucleon spinors $U_k(\mathbf{r})$ and $V_k(\mathbf{r})$, and the meson fields in terms of the eigenfunctions of a deformed axially symmetric oscillator potential as discussed in detail in Ref. [7]. A simple blocking procedure is used in the calculation of odd-proton and/or odd-neutron systems. The blocking calculations are performed without breaking the time-reversal symmetry.

The input parameters of the RHB model are the coupling constants and the masses for the effective mean-field Lagrangian, and the effective interaction in the pairing channel. In most applications we have used the NL3 effective interaction [9] for the RMF Lagrangian. Properties calculated with NL3 indicate that this is one of the best effective interaction so far, both for nuclei at and away from the line of β-stability. For

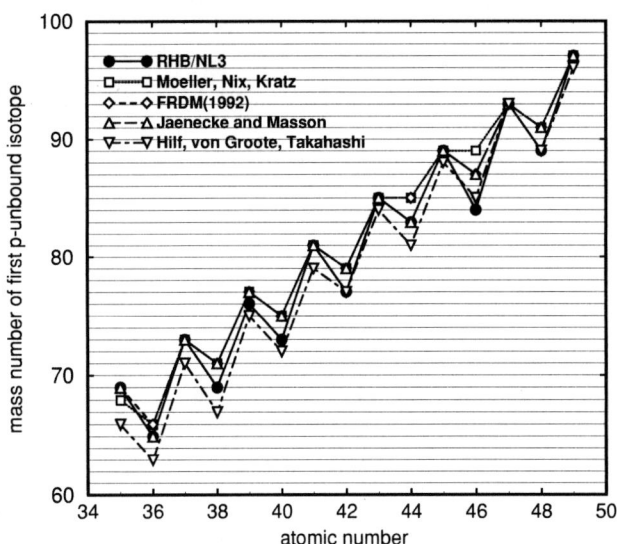

FIGURE 2. The proton drip line in the region $35 \leq Z \leq 49$. The mass number of the first proton unbound nucleus along each isotopic chain is plotted as a function of the atomic number. The results of the present RHB (NL3+D1S) calculation are compared with the predictions of various mass models.

the pairing field we employ the pairing part of the Gogny interaction

$$V^{pp}(1,2) = \sum_{i=1,2} e^{-((\mathbf{r}_1 - \mathbf{r}_2)/\mu_i)^2} (W_i + B_i P^{\sigma} - H_i P^{\tau} - M_i P^{\sigma} P^{\tau}), \qquad (13)$$

with the set D1S [10] for the parameters μ_i, W_i, B_i, H_i and M_i ($i = 1,2$). This force has been very carefully adjusted to the pairing properties of finite nuclei all over the periodic table. In particular, the basic advantage of the Gogny force is the finite range, which automatically guarantees a proper cut-off in momentum space.

THE PROTON DRIP LINE FROM Z=31 TO Z=49

The relativistic Hartree-Bogoliubov model based on the effective interactions NL3 and D1S has been used with great succes to investigate the proton dripline in several studies of proton-rich nuclei [11, 12, 8, 13]. We analyzed the structure of proton drip line nuclei in the $60 < A < 100$ mass range: the location of the proton drip-line, the ground-state quadrupole deformations and one-proton separation energies at and beyond the dripline, the deformed single-particle orbitals occupied by the odd valence proton in odd-Z nuclei, and the corresponding spectroscopic factors.

In Fig. 1 we display the section of the chart of the nuclides along the proton drip line in the region $31 \leq Z \leq 49$. The calculation predicts the last bound isotopes for

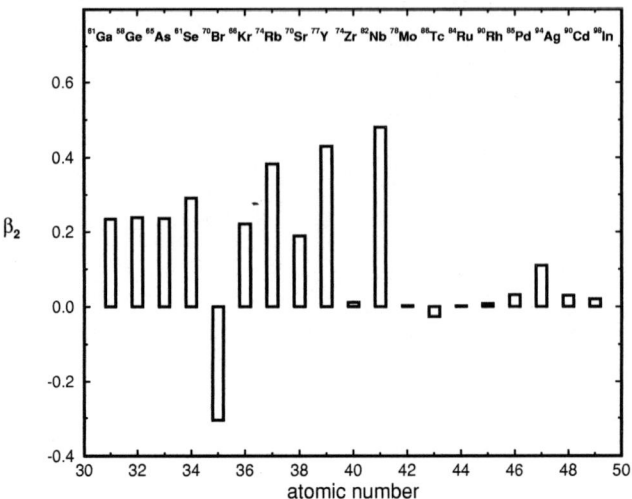

FIGURE 3. Calculated ground state quadrupole deformations for the nuclei at the proton drip line $31 \leq Z \leq 49$.

each element. Nuclei to the left are proton unstable. For odd-Z nuclei the proton drip line can be compared with available experimental data and with the predictions of other, well known and frequently used mass models [14, 15, 16, 5]. The comparison is illustrated in Fig. 2, where the mass number of the first proton unbound nucleus along each isotopic chain is plotted as a function of the atomic number. We notice that, with the exception of the classical mass formula by Hilf et al. [14], all models agree on the location of the proton drip line for odd-Z nuclei (see also Fig. 9 in Ref. [2]). The only significant difference is ^{77}Y which, in contrast to the mass models of Refs. [15, 16, 5], is predicted to be the heaviest proton bound $T_z = -\frac{1}{2}$ nucleus by the present RHB/NL3 calculation. For even Z-nuclei, the theoretical models differ in their predictions for the location of the proton drip line. As it is shown in Fig. 2, the differences are especially pronounced for Sr, Zr, Ru and Pd, and they reflect the different treatment of pairing correlations and deformation effects. In fact, the combined effect of pairing correlations and nuclear deformation is responsible for the large difference between the T_z values for even-Z and odd-Z nuclei at the proton drip line. The calculated ground-state quadrupole deformations of the last proton bound nuclei are shown in Fig. 3. For $Z \leq 33$ the drip line nuclei are moderately deformed, between $34 \leq Z \leq 41$ the odd-Z drip line nuclei are highly deformed, and for $Z > 41$ (protons in the $g \, 9/2$ orbital) the drip line enters a region of spherical nuclei. Between $34 \leq Z \leq 41$ one notices that the odd-Z nuclei are much more deformed than their even-Z neighbors on the drip line. The reason is that pairing correlations are strongly reduced in odd-Z nuclei, and as a result the nucleus is driven toward larger deformations. Much stronger pairing in even-Z nuclei results

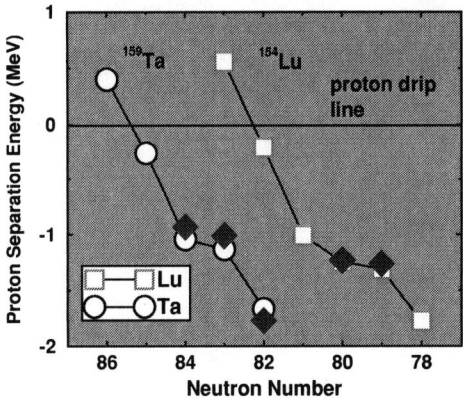

FIGURE 4. Proton separation energies for Lu and Ta isotopes at and beyond the drip-line. Results of self-consistent RHB calculations are compared with experimental transition energies for ground-state proton emission in ^{150}Lu, Filled diamonds denote the negative values of the transition energies E_p.

in almost spherical shapes, which in turn shift the drip line to extremely low values $T_z \approx -3$. One could also say that the strong reduction of pairing in odd-Z nuclei causes the drip line to lie at $T_z = -\frac{1}{2}$ or $T_z = 0$. We have verified that the blocking of odd proton orbitals is essential for the correct description of the drip line in odd-Z nuclei. Without blocking, the calculated drip line in Fig. 1 is shifted to the left, to the position of the drip line of even-Z nuclei.

PROTON RADIOACTIVITY IN DEFORMED AND TRANSITIONAL NUCLEI.

In medium heavy and heavy nuclei the Coulomb wall is so high, that protons in levels above the continuum limit cannot escape directly, rather they have to tunnel through a rather large barrier. This tunneling process leads to a delayed emission of protons. Usually this proton decay can be observed in a certain energy window. For lower Q_p values the total half-life will be completely dominated by β^+ decay; higher transition energies result in extremely short proton-emission half-lives which cannot be observed directly. Experimentally one can determine amongst other quantities the separation energy of the emitted proton and its quantum numbers, the spectroscopic factor and its life time. With the assumption that the deformation does not change drastically during the emission process, RHB-theory can be used for a fully self-consistent and universal description of such a process. In order to do this we first calculate the wave function of the parent nucleus with odd proton number. This is achieved by the self-consistent solution of the RHB-equations in the parent nucleus. The lst proton is kept in the quasi-

TABLE 1. Lu ground-state proton emitters. Results of the RHB calculation for the one-proton separation energies S_p, quadrupole deformations β_2, and the deformed single-particle orbitals occupied by the odd valence proton, are compared with predictions of the macroscopic-microscopic mass model, and with the experimental transition energies. All energies are in units of MeV; the RHB spectroscopic factors are displayed in the sixth column.

	N	S_p	β_2	p-orbital	u^2	Ω_p^π [5]	S_p [5]	β_2 [16]	E_p exp.
^{149}Lu	78	-1.77	-0.158	$7/2^-[523]$	0.60	$5/2^-$	-1.51	-0.175	
^{150}Lu	79	-1.31	-0.153	$7/2^-[523]$	0.61	$5/2^-$	-1.00	-0.158	1.261(4) [17]
^{151}Lu	80	-1.24	-0.151	$7/2^-[523]$	0.58	$7/2^-$	-0.99	-0.150	1.233(3) [17]

particle level with the quantum numbers under investigation. Blocking is taken into account in a self-consistent way. This leads us the wave function $|\Phi_k(Z,N)\rangle$ of the parent nucleus, the corresponding binding energy $B_k(Z,N)$ and the selfconsistent potential with its Coulomb barrier. In a second step we calculate in the same way the wave function $|\Phi_0(Z-1,N)\rangle$ and the binding energy $B(Z-1,N)$ of the daughter nucleus with even proton number.

The one-proton separation energies are then given by

$$S_p^{(k)}(Z,N) = B_k(Z,N) - B(Z-1,N). \tag{14}$$

The spectroscopic factor of the odd-proton orbital k is defined as the probability that this state is found empty in the daughter nucleus with even number of protons.It is found to be

$$u_k^2 = |\langle \Phi(Z,N)|a_k^\dagger|\Phi(Z-1,N)\rangle|^2 \tag{15}$$

where $a_k^\dagger$ describes the particle with the quantum numbers k in the canonical basis. In this basis the Hartree-Bogoliubov function $|\Phi_0(Z-1,N)\rangle$

$$|\Phi(Z-1,N)\rangle = \prod_k(u_k + v_k a_k^\dagger a_{\bar{k}}^\dagger)|-\rangle \tag{16}$$

has the form of a BCS-function. Having the precise form of the Coulomb barrier and the energy of the tunneling particle we should in principle be able tot calculate the life time for the tunneling process. In the spherical case, where one has a one-dimensional barrier this should be in particular simple. In the deformed case it is more complicated. One has to solve a coupled channel problem. Because of the long range of the Coulomb potential this is not trivial, but it has been solved for the non-relativistic case of a fixed potential of Saxon-Woods shape [18, 19, 20]. In such calculations the depth of the potential is adjusted to the proton separation energy. This is crucial, because the life time depends exponentially on the energy. It is well known, that fully self-consistent calculations cannot determine this energy with the necessary precision. Therefore it makes no sense to calculate life times based on the self-consistent solution.

Relativistic RHB calculations therefore provide complementary information to the calculations of the tunneling probability, such as the quantum numbers of the emitted proton, its rought separation energy, and its spectroscopic factors, i.e. very valuable information about the structure of these nuclei close to the proton drip line.

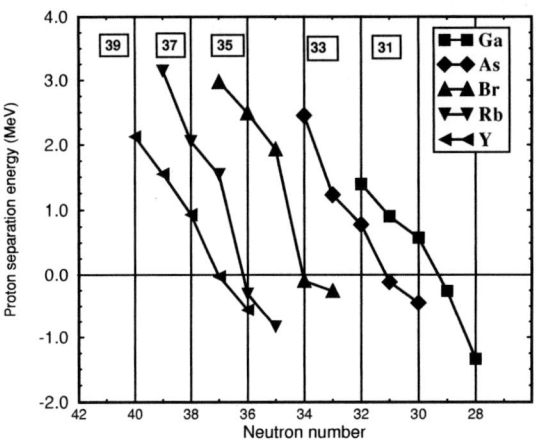

FIGURE 5. Predictions of the RHB model for the one-proton separation energies of odd-Z nuclei $31 \leq Z \leq 39$, at and beyond the drip-line.

As an example for proton emission we discuss in the following the deformed Lu-isotopes. In Fig. 4 we compare the calculated separation energies with experimental transition energies for ground-state proton emission in ^{150}Lu, ^{151}Lu [17], ^{155}Ta [21], ^{156}Ta [22], and ^{157}Ta [23]. In all five cases an excellent agreement is observed between model predictions and experimental data. In addition to ^{151}Lu, which was the first ground-state proton emitter to be discovered [24], and ^{150}Lu, the self-consistent RHB calculation predicts ground-state proton decay in ^{149}Lu. The calculated one-proton separation energy -1.77 MeV corresponds to a half-life of a few μs, if one assumes that the nucleus is spherical. Direct proton emission with a half-life of the order of few μs is just above the lower limit of observation of current experimental facilities. For the Lu ground-state proton emitters, in Table 1 the results of the RHB model calculation are compared with the predictions of the finite-range droplet (FRDM) mass model: the projection of the odd-proton angular momentum on the symmetry axis and the parity of the odd-proton state Ω_p^π [5], the one-proton separation energy [5], and the ground-state quadrupole deformation [16]. We have also included the RHB spectroscopic factors, and compared the separation energies with the experimental transition energies in ^{150}Lu and ^{151}Lu. Both theoretical models predict oblate shapes for the Lu proton emitters, and similar values for the ground-state quadrupole deformations. On the other hand, while the FRDM assigns spin and parity $5/2^-$ to the deformed single-particle orbitals occupied by the odd valence proton in all three proton emitters, the RHB model predicts the $7/2^-[523]$ Nilsson orbital to be occupied by the odd proton. We also notice that the RHB separation energies are much closer to the experimental values. The spectroscopic factors of the $7/2^-[523]$ orbital are displayed in the sixth column of Table 1.

While the relatively high potential energy barrier enables the observation of ground state proton emission from medium-heavy and heavy nuclei, no examples of ground state proton radioactivity have been discovered so far below $Z = 50$. The reason is, of course, the low Coulomb barrier. Nuclei beyond the proton drip line in this region exist only as short lived resonances.

Using an approximate formula introduced by Goldansky [25] for the calculation of half-lives we find that for the half live interval $T = 10 - 10^{-4}$ sec, the corresponding energy range of the emitted protons is: $0.2 - 0.3$ MeV (for Z=30) and $0.35 - 0.5$ MeV (for Z=40).

In Fig. 5 the one-proton separation energies are displayed for the odd-Z nuclei $31 \leq Z \leq 39$, as function of the number of neutrons. The energy window extends beyond the proton drip line, in order to include those nuclei for which a direct observation of ground-state proton emission is in principle possible on the basis of the calculated separation energies.

ACKNOWLEDGMENTS

This work has been supported by the Bundesministium für Forschung und Bildung under the contract No. TM 979 and by the Alexander von Humboldt Foundation.

REFERENCES

1. A. E. Champagne and M. Wiescher, Annu. Rev. Nucl. Part. Sci. **42**, 39 (1998).
2. H. Schatz et al., Phys. Rep. **294**, 167 (1998).
3. P. Ring, Progr. Part. Nucl. Phys. **37**, 193 (1996).
4. T. Gonzalez-Llarena, J. L. Egido, G. A. Lalazissis, and P. Ring, Phys. Lett. **B379**, 13 (1996).
5. P. Möller, J. R. Nix, and K. L. Kratz, At. Data Nucl. Data Tables **66**, 131 (1997).
6. J. Boguta and A. R. Bodmer, Nucl. Phys. **A292**, 413 (1977).
7. Y. K. Gambhir, P. Ring, and A. Thimet, Ann. Phys. (N.Y.) **198**, 132 (1990).
8. G. A. Lalazissis, D. Vretenar, and P. Ring, Nucl. Phys. **A650**, 133 (1999).
9. G. A. Lalazissis, J. König, and P. Ring, Phys. Rev. **C55**, 540 (1997).
10. J. F. Berger, M. Girod, and D. Gogny, Nucl. Phys. **A428**, 32 (1984).
11. D. Vretenar, G. A. Lalazissis, and P. Ring, Phys. Rev. **C57**, 3071 (1998).
12. D. Vretenar, G. A. Lalazissis, and P. Ring, Phys. Rev. Lett. **82**, 4595 (1999).
13. G. A. Lalazissis, D. Vretenar, and P. Ring, Phys. Rev. **C60**, 051302 (1999).
14. E. R. Hilf, H. von Groote, and K. Takahashi, Proc. 3rd Int. Conf. Nuclei far from Stability, Cargese, CERN, Geneva **76-13**, 142 (1976).
15. J. Jänecke and P. Masson, At. Data Nucl. Data Tables **39**, 265 (1988).
16. P. Möller, J. R. Nix, W. D. Myers, and W. J. Swiatecki, At. Data Nucl. Data Tables **59**, 185 (1995).
17. P. J. Sellin et al., Phys. Rev. **C47**, 1933 (1993).
18. E. Maglione, L. S. Ferreira, and R. J. Liotta, Phys. Rev. Lett. **81**, 538 (1998).
19. E. Maglione, L. S. Ferreira, and R. J. Liotta, Phys. Rev. **C59**, R589 (1999).
20. K. Rykaczewski et al., Phys. Rev. **C60**, 011301 (1999).
21. J. Uusitalo et al., Phys. Rev. **C59**, R2975 (1999).
22. R. D. Page et al., Phys. Rev. Lett **68**, 1287 (1992).
23. R. J. Irvine et al., Phys. Rev. **C55**, R1621 (1997).
24. S. Hofman, W. Reisdorf, G. Münzenberg, F. P. Hessberger, J. R. H. Schneider, and P. Armbruster, Z. Phys. **A305**, 125 (1982).
25. V. I. Goldansky, Nucl. Phys. **19**, 482 (1960).

N=Z and proton-rich nuclei in the Hartree-Fock mean field model with a separable nucleon-nucleon interaction

J. Rikovska Stone*†, K. Schofield*, P. D. Stevenson‡*, M. R. Strayer** and
W. B. Walters†

*Department of Physics, Oxford University, Parks Road, Oxford OX1 3PU, UK
†Department of Chemistry, University of Maryland, College Park, MD 20742, USA
‡Department of Physics, University of Surrey, Guildford, GU2 7XH, UK
**Physics Division, Oak Ridge National Laboratory, Oak Ridge, TN 37831-6373, USA

Abstract. A new model for the effective two-body nucleon-nucleon interaction has been recently successful in calculation of ground state properties of spherical, doubly closed shell nuclei from ^{16}O to ^{208}Pb [1] and nuclear matter and neutron star properties [2]. The application of the density dependent finite range separable monopole (SMO) interaction has been now extended to axially symmetrical deformed nuclei [3]. In the present paper we report on HF+BCS calculation of ground-state properties of even-even spherical and deformed $N = Z$ and proton-rich nuclei between the $N = Z$ and the proton-drip line for $28 \leq Z \leq 50$. The SMO results shown include total energy surfaces and deduced shapes, single-particle energies, two-dimensional distribution of nucleon densities, charge radii and two-proton separation energies. A comparison is made with results obtained using the SkO Skyrme interaction and some other theoretical models as well as with experimental data, where available.

HARTREE-FOCK CALCULATION USING THE MONOPOLE SEPARABLE NUCLEON-NUCLEON INTERACTION FOR AXIALLY DEFORMED NUCLEI

The separable monopole interaction

This interaction consists of two main parts, attractive (a) and repulsive (r) which have the same mathematical form for the density and isospin dependence and differ only in the values of adjustable parameters. Having attractive and repulsive terms with different ranges is a common feature of density-dependent effective interactions and it is these terms which are responsible for most of the binding energy in finite nuclei and for the volume energy in nuclear matter. To describe better surface properties of nuclei, a third term is added, depending on the second derivative of density.

Each term in the expression for the interaction is separable in the space coordinates of individual nucleons (and in isospin where applicable). This is the essential new property of this effective nucleon-nucleon interaction which makes it suitable for building a perturbation series in powers of the strength of the SMO interaction in order to include correlations in finite nuclei, beyond the mean field, with convergent results.

CP681, *Proton-Emitting Nuclei: Second International Symposium; PROCON 2003*,
edited by E. Maglione and F. Soramel

In coordinate space, the SMO is written as [1]

$$
\begin{aligned}
V(\vec{r}_1, \vec{r}_2) &= W_a f_a \rho^{\beta_a}(\vec{r}_1) \rho^{\beta_a}(\vec{r}_2)(1 + a_a(t_1^+ t_2^- + t_1^- t_2^+) + 4b_a t_{1z} t_{2z}) \\
&+ W_r f_r \rho^{\beta_r}(\vec{r}_1) \rho^{\beta_r}(\vec{r}_2)(1 + a_r(t_1^+ t_2^- + t_1^- t_2^+) + 4b_r t_{1z} t_{2z}) \\
&+ k \nabla_1^2 \rho(\vec{r}_1) \nabla_2^2 \rho(\vec{r}_2),
\end{aligned}
\tag{1}
$$

where the function f_ξ is defined as

$$
f_\xi = \left[\int d^3\vec{r} \rho^{\alpha_\xi}(\vec{r}) \right]^{-1},
\tag{2}
$$

for subscripts $\xi = a, r$. $\rho(\vec{r}_i)$ are nuclear densities, t are isospin operators and W_ξ, α_ξ, β_ξ, a_ξ, b_ξ and k are adjustable parameters.

In addition, we adopt the standard form of the spin-orbit interaction as used in mean-field calculations with the Skyrme interaction [17]:

$$
V_{ls} = c(\nabla \rho \cdot \sigma)(\sigma \cdot \nabla)
\tag{3}
$$

which provides an energy

$$
E_{ls} = -c \int d^3\vec{r} \rho \nabla \mathscr{J}
\tag{4}
$$

and a contribution to the mean field of

$$
U_{ls} = c \nabla \mathscr{J},
\tag{5}
$$

where $\nabla \mathscr{J}$ is the spin-orbit current. The constant c in the above equations stands for the strength of the spin-orbit potential and corresponds to $t_4/2$ in the notation of Skyrme interactions. The Coulomb interaction which also has to be included in the model, is treated in the standard way and both direct and the exchange terms (in the Slater approximation) contribute to the total energy. The set of parameters SMO2 [2] was used in the present work. Both direct and exchange terms of the interaction are included into the Hartree-Fock equations.

Pairing had to be added to the model for correct treatment of open-shell nuclei. We follow the method of Bender et al. [15] who use the HF+BCS approximation and parametrise the effective δ-force pairing interaction in terms of a local pairing energy functional

$$
\mathscr{E} = \frac{1}{4} \sum_{q \in p,n} \int d^3\vec{r} \chi_q^*(\vec{r}) \chi_q(\vec{r}) G_q(\vec{r}),
\tag{6}
$$

where $\chi_q(\vec{r})$, the local part of the pair density matrix, is expressed in terms of single-particle Hartree-Fock wave functions and pairing amplitudes u and v. The pairing strength G_q is dependent on nucleon density as

$$
G_q(\vec{r}) = V_{0,q} \left[1 - \left(\frac{\rho(\vec{r})}{\rho_0} \right) \right]
\tag{7}
$$

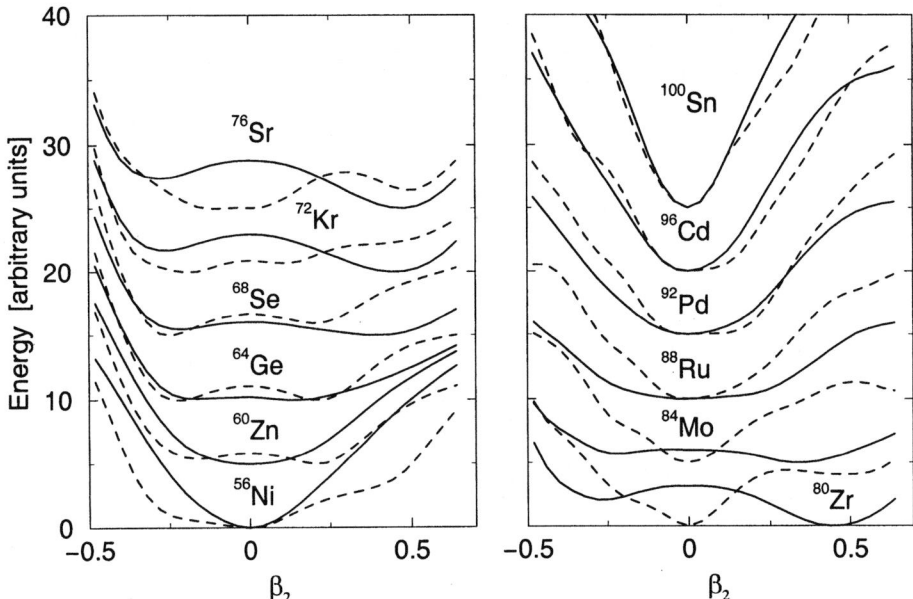

FIGURE 1. Calculated ground state total energy as a function of mass quadrupole deformation for $28 \leq Z \leq 50$ N=Z nuclei. The solid (dashed) line shows results for the SMO (SkO Skyrme) interaction.

where $V_{0,q}$ are constants fitted to empirical pairing gaps of selected nuclei and $\rho_0 = 0.16$ fm^{-3} is the nuclear saturation density. This method treats pairing as a valence-particle effect and supresses to a large extent contributions from unbound continuum states close to the Fermi surface. This effect is in principle smaller for finite range forces [19]. Instead of a rather cumbersome implementation of these forces, the finite range effect can be simulated by introducing smooth energy-dependent cut-off weights (form-factors) F_a in evaluation of the local pair density $\chi_q(\vec{r})$ [15, 16]. The parameters of F_a are the width of the energy range of pairing active states and the average level density in the vicinity of the Fermi surface. These are adjusted to include approximately one additional shell above the Fermi energy.

The computer code developed for calculation of properties of axially deformed nuclei in Hartree-Fock + BCS model with a Skyrme effective interaction, used recently for calculation of shape-coexistence effects in light and medium neutron rich nuclei [17], has been adapted for the separable monopole interaction (Eq. 1). In this code the calculation is performed on a grid in cylindrical coordinates r, z, ϕ, with imposed symmetry with respect to the z-axis.

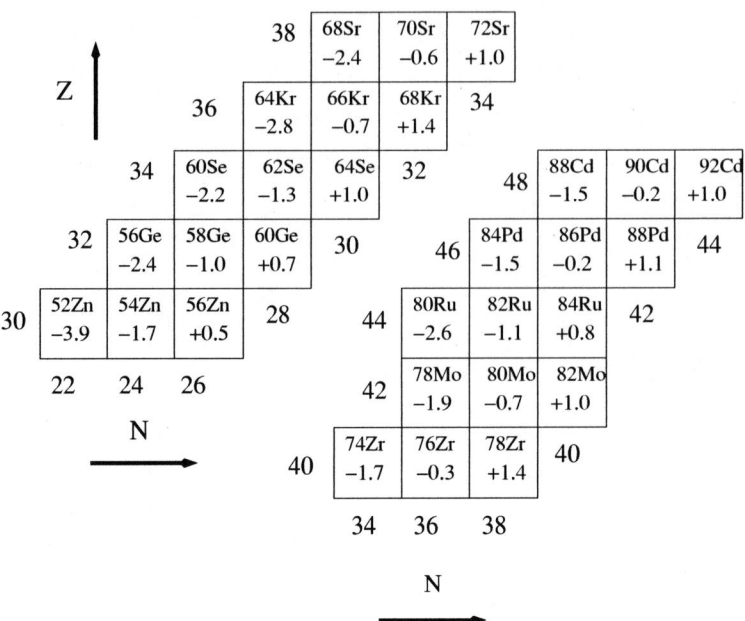

FIGURE 2. S_{2p}=B(A,Z)-B(A+2,Z+2) for $30 \leq Z \leq 48$ N=Z even-even nuclei. Values (in MeV) are given for three isotopes of each element. These span the stability limit allowing for some uncertainty in calculated total binding energies B.

Calculated properties

After adding a quadrupole constraint $\hat{Q} \propto r^2 Y_{20}$ to the HF field, a systematic survey has been performed of ground state properties of N=Z even-even nuclei in the region $28 \leq Z \leq 50$ and neutron-deficient nuclei between the N=Z line and the two-proton drip line towards the limits of nuclear existence. The calculated deformed shapes are characterized by a dimensionless quadrupole mass deformation parameter

$$\beta_2 = \sqrt{\frac{\pi}{5}} \frac{<r^2 Y_{20}>}{A <r^2>} \tag{8}$$

where $<r^2>$ is the mean square radius of the mass distribution. Deformation parameters β_2^{π} (β_2^{ν}) of the proton (neutron) distributions were defined in an analoguous way. We do not consider β_4 deformation parameters in the present work (although their values have been calculated) as they are much less clearly connected to experimental data. Mean square radii are calculated in the usual way for the total, proton and neutron distributions [17]. The correction for the finite size of the proton is included in the radius of the charge distribution as

$$r_{ch}^2 = r_p^2 + \underbrace{r_{ch,p}^2}_{0.743\,\text{fm}^2} - \frac{N}{Z} \underbrace{r_{ch,n}^2}_{0.119\,\text{fm}^2} \quad , \tag{9}$$

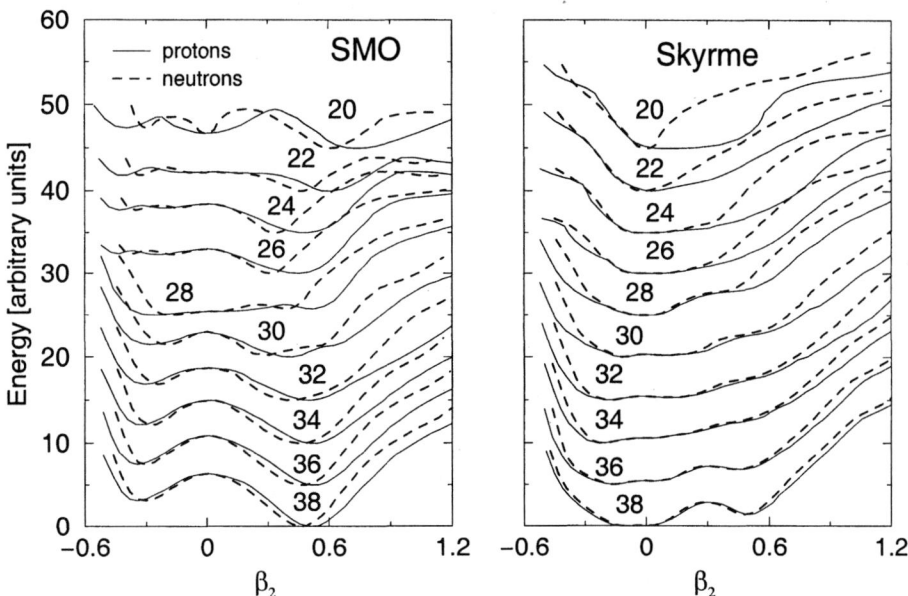

FIGURE 3. Total energy surfaces for protons (solid line) and neutrons (dashed line) in even-even strontium nuclei as calculated with the SMO (left) and SkO Skyrme (right) interactions. The numbers indicate the number of neutrons and are placed above the minimum of each surface to guide the eye.

where $r_{ch,p/n}$ are the nucleon radii.

The SkO Skyrme interaction [17] has been selected to demonstrate the similarities and differences between results obtained with the separable SMO model and a Skyrme model. This modern parametrisation has been used rather recently to study shape coexistence within the self-consistent Hartree-Fock method in N = 20, 28, 40 and 56 regions and belongs to the class of Skyrme interaction which behave well in calculation of nuclear matter and neutron star properties [18]. The total energy as a function of quadrupole deformation β_2 for ^{56}Ni, ^{60}Zn, ^{64}Ge, ^{68}Se, ^{72}Kr, ^{76}Sr, ^{80}Zr, ^{84}Mo, ^{88}Ru, ^{92}Pd, ^{96}Pd and ^{100}Sn in the SMO and SkO Skyrme models are shown in Fig. 1. The SMO model predicts a gradual departure from sphericity with increasing softness and prolate-oblate shape coexistence which starts at Z=32, peaks at large prolate deformation of ^{76}Sr and ^{80}Zr and decreases to zero at Z=44. The SkO Skyrme model predicts softer shape of ^{56}Ni, prolate-oblate shape coexistence in ^{60}Zn and ^{64}Ge with tendency to oblate shapes which develops in ^{68}Se and ^{72}Kr. For ^{76}Sr and heavier even-even N=Z nuclei up to ^{100}Sn the Skyrme model predicts well defined spherical shapes. As expected, our calculation shows that the proton and neutron matter separately follows the same pattern of shapes as the mass distribution. The calculated values of deformation parameter β_2, corresponding to the minimum total energy, are summarized in Table. 1 in comparison with results of some other model calculations. For nuclei exhibiting shape coexistence, deformation parameters are quoted for both the prolate and oblate shapes together with the absolute energy difference between the coexisting minima. We notice that if there

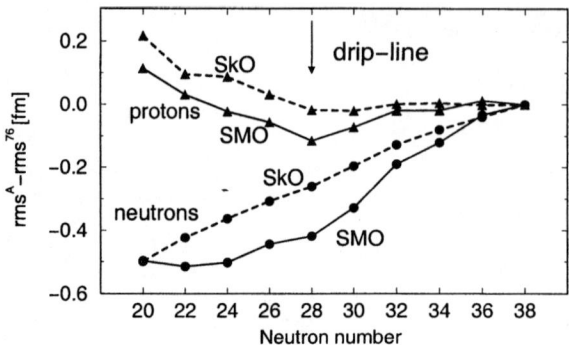

FIGURE 4. Diferences in proton (solid line) and neutron (dashed line) between r.m.s. for a given mass number A and ^{76}Sr for light Sr nuclei. Results for the SMO model (filled triangles) are compared with the SkO Skyrme model (filled circles).

TABLE 1. Quadrupole deformation parameters of N=Z nuclei as calculated in different nuclear models.

A	Element	SMO	SkO [17]	MSk7 [6]	RMF [5]	FRDM [7]
60	Zn	0.0	-0.16	0.12	0.17	0.18
			0.20			
			(0.439)			
64	Ge	0.12	-0.24	-0.19	0.22	0.22
		-0.16	0.20			
		(0.130)	(0.008)			
68	Se	0.40	-0.24	-0.25	-0.28	0.29
		-0.20	0.20			
		(0.518)	(0.032)			
72	Kr	0.44	-0.20	-0.20	-0.36	-0.34
		-0.24	0.12			
		(1.744)	(0.778)			
76	Sr	0.48	-0.12	0.42	0.41	0.42
		-0.28	0.48			
		(2.398)	(1.435)			
80	Zr	0.44	0.0	0.40	0.44	0.43
		-0.28				
		(2.104)				
84	Mo	0.36	0.0	0.0	-0.25	0.053
		-0.24				
		(0.709)				

is a discrepancy in extracted values of β_2 between the SMO model and some of the other calculations, usually the SMO indicates two coexisting minima and one of them will be in agreement with the rest. This may indicate that there is delicate balance between calculated coexisting minima in all models. It would be useful if the deformation corresponding to the two lowest energy minima in all models was always given.

The present model is not adapted for a proper calculation of odd-A nuclei as yet and

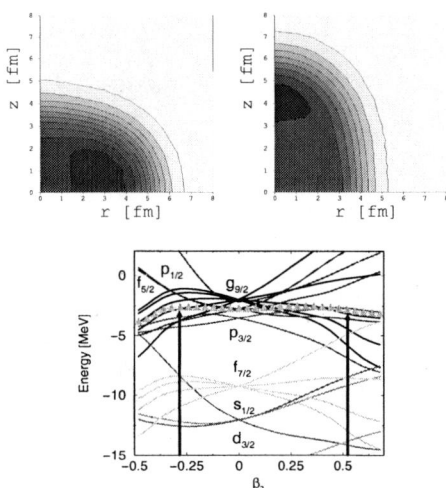

FIGURE 5. Proton density distribution and single particle states, corresponding to the oblate and prolate minimum of the total energy surface in ^{76}Sr. In this and the following figures, the maximum density is ~ 0.08 fm^{-3} and the regions between two adjacent contours corresponds to a decrease in density ~ 0.01 fm^{-3} as going away from the center. The area from the last contour towards the figure frame has zero density. The Fermi surface is depicted by filled triangles. The arrows connect single-particle energies at deformations corresponding to the two minima.

cannot be used to calculate one-proton separation energy, but allows predictions of the two-proton drip line. The results are shown in Fig. 2 and are in a good agreement with predictions of the other current models.

It is interesting to study the predicted development of nuclear shapes beyond the 2p drip-line. The nature of the proton-drip line is different from the neutron drip-line because in proton rich nuclei the proton decay occurs via tunneling of protons through the Coulomb potential barrier rather then leaving the nucleus freely when it becames unbound. Mean field calculations of this kind are technically demanding as a special care must be given to ensure that calculation of such exotic systems is meaningful. Standard nuclear models are designed for more ordinary systems with parameters fitted to properties of doubly closed shell nuclei. We present total energy surfaces for $^{58-76}$Sr nuclei with the SMO (left) and SkO Skyrme model (right) in Fig. 3. There is a difference between the predictions of the two models. SMO model predicts development of well defined deformed shapes with decreasing neutron number. On the contrary, the SkO Skyrme model indicates spherical shapes of proton-rich Sr isotopes well beyond the proton drip-line. An important observable closely connected to nuclear shapes is the mean-square radius of nucleon density distributions in atomic nuclei. We compare in Fig. 4 the increase of the charge radius with increasing neutron defficiency taking the radius of ^{76}Sr as a base line. The two models, the SMO and SkO Skyrme model, predict very similar trends for both charge and neutron radii. No experimental data on radii exist in this region to date.

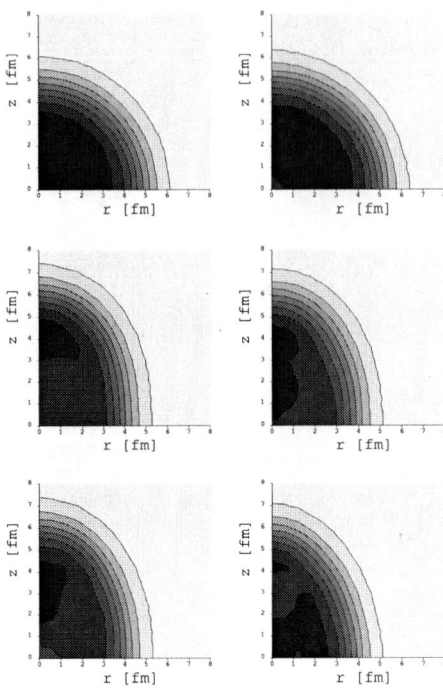

FIGURE 6. Proton (left) and neutron (right) density distributions, calculated with the SMO separable interaction for [88]Sr (top), [74]Sr (middle) and [70]Sr (bottom) ground states

Whilst bulk deformation parameters and r.m.s.radii show limited sensitivity to details of nuclear models, the nucleon density distribution itself, in particular its spatial angular distribution is more interesting. In Fig.5 we show the calculated projection of the proton density distribution onto the rz plane (in cylindrical coordinates r, z, ϕ) for oblate and prolate minima of the total energy surface of [76]Sr. Taking z as the symmetry axis, individual single-particle orbitals will be spread from the origin depending upon the value of the orbital angular momentum **l**. The resulting total distribution depends on the occupied single particle states close to the Fermi surface. This is illustrated in the lower part of Fig.5. Both for the oblate and prolate minimum of the total energy surface, the dominant contribution to the maximum density comes mainly from particles occupying high l orbitals $f_{5/2}$ and $g_{9/2}$. The occupation of the orbitals is of course smeared out due to the pairing interaction.

The distinctive sensitivity of the calculated density distribution to the single-particle spectrum makes it a powerful tool in distinguishing between models for the effective nucleon-nucleon interaction. Ground state proton and neutron density distributions differ considerably between the SMO model and the Skyrme SkO [17] model, as illustrated in Figs. 6,7. Data are shown for the stable spherical nucleus [88]Sr (N=50), the N=Z [76]Sr (N=38) nucleus, and the drip-line nucleus [68]Sr (N=20). Interesting spatial distributions

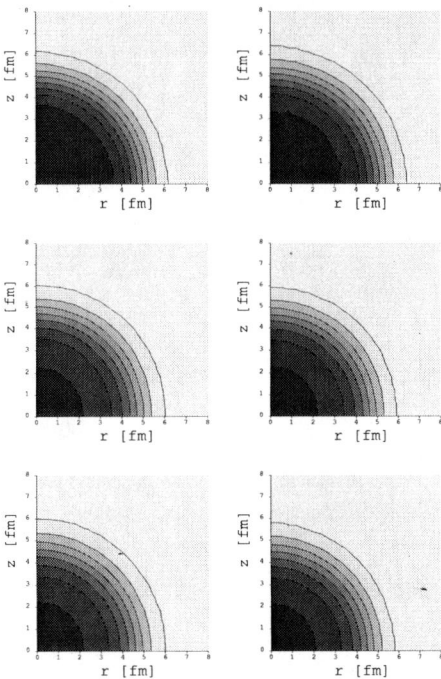

FIGURE 7. The same as Fig. 6 but for the SkO Skyrme model.

for protons and neutrons can be seen for ^{68}Sr, indicating overlapping orbitals, in the SMO model. The Skyrme model predicts development of spherical shapes at the drip-line and beyond, in variance with the SMO model which indicates increasing quadrupole deformation with decreasing neutron number.

CONCLUSIONS

We have shown that the newly developed SMO interaction describes N=Z and proton rich nuclei in a Hartree-Fock+BCS model giving predictions which are very similar to those of other mean-field models based on very different physics. This is true for observables obtained from one-body densities like nuclear shapes and mean radii of the charge and neutron distributions. These observables are therefore not useful criteria for selection between different models. On the contrary, the angular distribution of the proton density, which is in principle measurable directly, is very sensitive to details of the single particle spectra calculated using different nucleon-nucleon interactions. It may be that real progress in eliminating inadequate theories will require a new generation of difficult but possible experiments. Angular distribution of charge density in nuclei oriented in space, probed by a well tuned electron beam, is one of the challenges.

ACKNOWLEDGMENTS

This research has been sponsored by the Division of Nuclear Physics of the U.S. DOE under contract DE-AC05-00OR 22725 managed by UT–Battelle, LLC, grant no. DE-FG02-94ER40834 and the UK EPSRC.

REFERENCES

1. Stevenson, P., Strayer, M. R. and Rikovska Stone, J.,*Phys. Rev.*, **C63**, 054309 (2001)
2. Rikovska Stone, J., Stevenson, P. D. and Strayer, M. R.,*Phys. Rev.*, **C65**, 064312 (2002)
3. Stevenson, P. D., Rikovska Stone, J. and Strayer, M. R.,*Phys. Lett.*, **B545**,291 (2002)
4. Lalazissis, G. A. and Sharma, M. M., *Nucl. Phys.*, **A586**, 201 (1995)
5. Lalazissis, G. A., Raman, S. and Ring, P., *ADNDT*, **71**, 1 (1999)
6. Goriely, S., Tondeur, F. and Pearson, J. M., *ADNDT*, **77**, 311 (2001)
7. Moller, P., Nix, J. R., Meyers, W. D. and Swiatecki, W. J., *ADNDT*, **59**, 185 (1995)
8. Petrovici, A., Schmid, K. W. and Faessler, A.,*Nucl Phys.*, **A605**, 290 (1996)
9. Sarriguren, P., Moyay de Guerra, E. and Escuderos, A., *Nucl. Phys.*, **A658**, 13 (1999)
10. Engel, J., Langanke, K.-H. and Vogel, P., *Phys. Lett.*, **B389**, 211 (1996)
11. Yamagami, M., Matsuyanagi, K. and Matsuo, M., *Nucl. Phys.*, **A693**, 579 (2001)
12. Lister, C. J. et al., *Phys. Rev.*, **C42**, R1191 (1990)
13. Bucurescu, D., et al., *Phys. Rev.*, **C56**, 2497 (1997)
14. Marginean, N., et al., *Phys. Rev.*, **C63**, 031303(R) (2001)
15. Bender, M., Rutz, K.,Reinhard, P.-G. and Maruhn, J. A., *Eur. Phys. J.*, **A8**,59 (2000)
16. Krieger, S. J., Bonche, P., Flocard, H., Quentin, P. and Weiss, M. S., *Nucl. Phys.*, **A517**, 275 (1990)
17. Reinhard, P.-G., Dean, D. J., Nazarewicz, W., Dobaczewski, J., Maruhn, J. A. and Strayer, M. R., *Phys. Rev.*, **C60**, 014316 (1999)
18. Rikovska Stone, J., Miller, J. C., Koncewicz, R., Stevenson, P. D. and Strayer, M. R., submitted to *Phys. Rev. C*
19. Dobaczewski, J., Nazarewicz, W., Werner, T. R., Berger, J. F., Chinn, C. R. and Decharge, J., *Phys. Rev.*, **C53**, 2809 (1996)
20. Beiner, M., Flocard, H., Van Giai, N. and Quentin, P., *Nucl. Phys.*, **A236**, 269 (1974)
21. Van Giai, N. and Sagawa, H., *Phys. Lett.*, **106B**, **379** (1981)

TWO PROTON EMISSION

Evidence for the 2p decay of ^{45}Fe from GSI

M. Pfützner*, E. Badura†, C. Bingham**, B. Blank‡, M. Chartier§,
H. Geissel†, J. Giovinazzo‡, L.V. Grigorenko†, R. Grzywacz*,
M. Hellström†, Z. Janas*, J. Kurcewicz*, A.S. Lalleman‡, C. Mazzocchi†,
I. Mukha†, G. Münzenberg†, C. Plettner†, E. Roeckl†, K.P. Rykaczewski*¶,
K. Schmidt‖, R.S. Simon†, M. Stanoiu†† and J.-C. Thomas‡

*Institute of Experimental Physics, Warsaw University, PL-00-681 Warszawa, Poland
†GSI, Planckstrasse 1, D-64291 Darmstadt, Germany
**Dept. of Physics and Astronomy, University of Tennessee, Knoxville 37996, USA
‡CEN Bordeaux-Gradignan, F-33175 Gradignan Cedex, France
§Oliver Lodge Laboratory, Dept. of Physics, University of Liverpool, Liverpool, L69 3BX, UK
¶Physics Division, ORNL, Oak Ridge, TN 37831-6371, USA
‖Dept. of Physics and Astronomy, University of Edinburgh, Edinburgh EH9 3JZ, UK
††GANIL, BP 5027, F-14021 Caen Cedex, France

Abstract. In an experiment performed at the GSI Fragment Separator the decay of ^{45}Fe has been investigated. Five ^{45}Fe atoms were implanted in a silicon telescope mounted inside a high-efficiency NaI detector. The implantation events were correlated with radioactive decay signals. One of them was consistent with the β decay of ^{45}Fe accompanied by the emission of a 10 MeV proton. In each of the four other decay events, a signal corresponding to the emission of 1.1 ± 0.1 MeV particle(s) was recorded with no coincident signals in the NaI detector. This observation represents the first evidence for the two-proton ground state decay of ^{45}Fe. The time distribution of the observed decay events corresponds to a half-life of $3.2^{+2.6}_{-1.0}$ ms.

INTRODUCTION

Two-proton $2p$ radioactivity, a process predicted by Goldansky in 1960 [1], may occur in a nucleus located beyond the proton drip line when due to the pairing interaction the emission of a single proton is energetically forbidden. It requires that the energy difference between the ground states of the initial and the $1p$ daughter nuclei is larger than the sum of their widths. Such a condition is expected to be fulfilled for medium-mass, even-Z nuclei where due to Coulomb barrier the relevant states are narrow. Prior to our work, the emission of two protons from the nuclear ground state was observed only in two cases, i.e. ^{6}Be [2] and ^{12}O [3]. In both of them, however, the initial as well as the ground state of the intermediate nucleus are so broad that the decay may proceed sequentially through the tails of the respective widths. Thus, these disintegration modes do not exhibit the characteristics of the $2p$ decay in the sense mentioned above.

In order to identify possible candidates for the $2p$ decay, accurate mass predictions for nuclei beyond the drip line are necessary. One of the most exact methods is the application of the isobaric multiplet mass equation (IMME) combined with the experimentally measured mass of the neutron-rich member of the multiplet. For this purpose,

CP681, *Proton-Emitting Nuclei: Second International Symposium; PROCON 2003*,
edited by E. Maglione and F. Soramel
© 2003 American Institute of Physics 0-7354-0150-0/03/$20.00

coefficients of the IMME can be calculated within the shell model or deduced from the Coulomb-energy systematics. Both approaches were undertaken [4, 5, 6] yielding a few candidate nuclides. One of them is ^{45}Fe predicted to decay either by $2p$ emission with a Q-value of approximately 1 MeV or by EC/β^+ decay with a Q_{EC} value of about 19 MeV.

^{45}Fe was identified for the first time at GSI Darmstadt among fragmentation products of a 600 MeV/nucleon ^{58}Ni beam, analyzed with the Fragment Separator (FRS) [7]. Three ions of this isotope were observed at the final focus of the FRS but no spectroscopic properties were determined. In a subsequent experiment, performed at the GANIL LISE3 separator, a study of decay properties of nuclei around ^{45}Fe was attempted [8]. Apart from the first observation of ^{48}Ni, decay events correlated with the implantation of ^{45}Fe were detected. However, due to an unfortunate definition of the trigger of the data acquisition, it could not be established whether these events represent the β decay of ^{45}Fe or that of its $2p$ daughter.

To clarify the issue of the dominant decay mode of ^{45}Fe, a further experiment was undertaken at the FRS in July 2001. It yielded the first evidence for the $2p$ decay of ^{45}Fe. Here, we report briefly on this experiment. The detailed description of the results as well as of the applied detection system were published in Refs. [9, 10].

EXPERIMENTAL DETAILS

The proton-rich nuclei of interest were produced by projectile fragmentation of a 650 MeV/nucleon ^{58}Ni beam, delivered by the synchrotron SIS, impinging on a 4 g/cm^2 thick beryllium target. The average beam intensity was approximately 4×10^9 ions per spill. Each spill lasted 2 s, the spill repetition period being 7.6 s. The FRS was operated in the achromatic mode with two aluminum degraders, one homogeneous of 3.2 g/cm^2 thickness mounted between the first and the second dipole magnets, and the second wedge-shaped of 3.6 g/cm^2 average thickness, located at the intermediate focal plane.

Ions transmitted to the final focus were identified by the standard ΔE-TOF-$B\rho$ method. Energy loss (ΔE) information was provided by by the ionization chamber MU-SIC. The time-of-flight (TOF) of ions was measured with help of three scintillation detectors mounted at the intermediate, third, and final focal planes of the FRS. The same detectors delivered the horizontal position of the ions, necessary to determine their magnetic rigidity $B\rho$. The combination of these observables yielded the atomic number Z and the mass-to-charge ratio A/Z for each ion (all of them were fully stripped of atomic electrons).

After passing the identification detectors at the final focus, ions were implanted into a telescope of eight silicon (Si) detectors, each 300 μm thick and 60 mm in diameter. The telescope was mounted inside a NaI(Tl) "barrel" composed of six 30 cm long crystals (see figure 1). Its main purpose was to discriminate against β^+ decay events which are accompanied by β-delayed γ rays, including two 511 keV annihilation quanta. The whole detection set-up was calibrated by implanting the known β-delayed proton (βp) emitters ^{49}Fe and ^{50}Co. It was found that the total efficiency for detecting any γ ray accompanying the βp event in the energy range 0.9–4 MeV is 93 %.

Signals from all Si detectors were processed by standard low-gain charge-sensitive

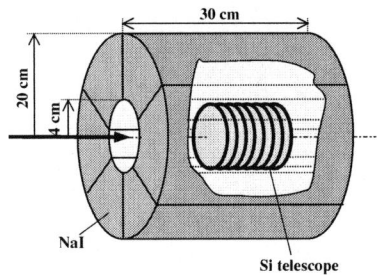

FIGURE 1. Schematic drawing of the implantation set-up.

preamplifiers in order to record the stopping of the ions. Additionally, all detectors but the first one were connected to high-gain preamplifiers with a fast-reset function. They were switched-off for 2 μs by an external logical signal generated each time an ion was passing through the first Si detector. Thus, already 2 μs after the implantation of a heavy ion, accompanied by the release of up to 1 GeV energy, the preamplifiers were sensitive to low-energy decay signals.

Signals from all detectors were analyzed by two independent acquisition systems running in parallel. The first one used standard analogue NIM modules for signal processing and CAMAC ADC and TDC modules for digitizing them. It was triggered by heavy ions reaching the final FRS focus. All data accumulated within a few microseconds after the trigger were read out and stored as an event on a magnetic tape. This system suffered from a large dead-time and served mainly for the monitoring and auxiliary purposes. The second system was based on digital DGF-4C modules [11]. Their function consisted of digitizing each input with 40 MHz frequency, amplitude determination and time stamping by the on-board digital signal processor and subsequent storage in an output memory buffer. This system was triggered only when an ion with a low A/Z ratio entered the telescope. All incoming signals were accepted for an interval of 10 ms following the trigger. Only after this time period, the DGF buffers were read out and stored on a magnetic tape. Thus, the full "history" of all events following the trigger was recorded for 10 ms practically dead-time free. A more detailed description of the DGF-based acquisition system is given in Ref. [10].

RESULTS AND DISCUSSION

The measurement optimized for the transmission of ^{45}Fe lasted 5.6 days. During this time the DGF-based acquisition system was triggered about 2000 times. The identification plot corresponding to those 2000 ions is shown in figure 2. The six events of ^{45}Fe can be seen. Other ions which initiated acquisition are longer lived species lying closer to the stability line. One of the most abundant among them, and with the shortest half-life, is ^{44}V ($T_{1/2} = 150$ ms).

The 10 ms "history" periods recorded after the arrival of each ^{45}Fe ion were carefully examined. The results are compiled in Table 1. One ion (event number 4) was stopped

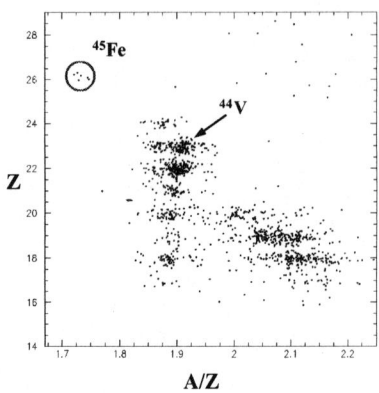

FIGURE 2. Identification plot of those ions which triggered the DGF-based acquisition system. Six events of ^{45}Fe and a group of ^{44}V ions are indicated.

in the detector no. 5 while the corresponding preamplifier suffered from a temporary malfunction, so that no decay signals could be recorded. The implantation of the fifth ion was followed by a release of about 10 MeV energy accompanied by a coincident γ-ray in the NaI detector. Such an event is consistent with the βp scenario. In the four other cases, however, a different pattern was observed : the energy of about 1 MeV was released in the Si detector where the ion was stopped, with no other signal registered in coincidence. Such pattern is expected if ^{45}Fe decays by the $2p$ emission. The decay event no. 6 occurred after the sensitivity period of the DGF system and was recorded only by the standard acquisition system.

In the fifth column of Table 1 all ions are listed which entered the telescope after the triggering ^{45}Fe ion and prior to the decay signal. They represent either stable or long-lived species and all but one were stopped in a different Si detector than the ^{45}Fe ion of reference. In case of the second event, a contaminant ion could possibly be stopped in the same detector as ^{45}Fe but it was identified as a stable isotope (^{38}Ar). These ions could not be responsible for the observed decay signals. Moreover, the inspection of 10 ms periods following all triggering ions except ^{45}Fe revealed that only for two ions of ^{44}V (out of a total number of 60 identified) a decay signal was recorded. This is consistent with the number of decays expected within 10 ms for an activity with a half-life of 150 ms, i.e. that of ^{44}V.

During the measurement less exotic contaminant ions, with half-lives between seconds and hours, were implanted into the telescope at a constant rate of about 200 per second. There is a possibility that some of them may decay within a 10 ms observation period started by a different ion. To estimate this probability, a singles spectrum of decay-like events detected in the telescope during the entire experiment was constructed. It contained all events which were not accompanied by any other signal including the NaI detector. Only 5 such events were found in the region above 700 keV, 4 representing signals correlated with the implantation of ^{45}Fe and one which could not be correlated to any ion. This latter event is assigned to the random background originating from long-lived activities. The probability to detect such a random event during a 10 ms observation

TABLE 1. Event number, implantation detector, decay energy and decay time recorded for each of the six ^{45}Fe ions observed. All ions which entered the telescope after the implantation of ^{45}Fe and before its decay are listed in the rightmost column. For each of them the arrival time and the detector number in which the ion was stopped are given.

Event	Detector	E [keV]	T [ms]	Comment
1	4	1000 ± 120	0.644	
2	3	990 ± 130	5.276	1.7 ms : ^{38}Ar in det. 2/3 1.9 ms : ^{37}Ar in det. 5
3	5	10010 ± 100	3.395	0.1 ms ^{39}K in det. 3 1.3 ms : ^{39}Ar in det. 1
4	5	-	-	DC problem in the preamp. 5
5	2	1150 ± 100	1.196	0.5 ms : Ar or Cl in det. 6 0.8 ms : ^{40}Ca in det. 4
6	2	1200 ± 100	12.617	0.7 ms : ^{41}Ca in det. 1 1.2 ms : ^{36}Cl in det. 7

period is of the order of 10^{-3}. It is thus very unlikely that all four signals correlated with ^{45}Fe represent this sort of background.

Therefore, we are led to conclude that four events observed after implanting ^{45}Fe do represent the decay of this nucleus. The maximum-likelihood analysis of the decay times yields half-life of $3.2^{+2.6}_{-1.0}$ ms. The experimental decay energies suggest that ^{45}Fe decays predominantly (4 events out of 5) by the emission of charged particle(s) with a total energy of 1.1 ± 0.1 MeV.

Turning to the interpretation of this dominant decay mode, we first note that anticoincidence with γ rays practically rules out the β-decay scenario. The escape probability of γ rays in all four events equals $(1 - 0.93)^4 = 2 \times 10^{-5}$. Next, we consider the hypothesis that ^{45}Fe emits a single proton. The WKB approximation [12] for a proton emitted with the energy of 1.1 MeV with orbital angular momentum of 1 and 3, and assuming a spectroscopic factor of 1, yields the half-lives of 4×10^{-18} s and 1×10^{-15} s, respectively, which are utterly incompatible with the measured value. The half-life for α emission estimated from systematics [13] for an α-particle energy of 1.1 MeV would be of the order of 10^{10} s. Actually, the α separation energy, predicted by the IMME approach [4] for ^{45}Fe, is equal to 8 MeV. Thus, the α decay has also to be excluded. On the other hand, the measured energy is in excellent agreement with predictions for the $2p$ decay which are (1.154 ± 0.094) MeV [4], (1.279 ± 0.181) MeV [5], and (1.218 ± 0.049) MeV [6]. Additionally, the measured half-life is compatible with a value of 13 ms predicted by a rigorous three-body model developed by Grigorenko et al. [14, 15].

Finally, we conclude that our experimental findings represent the evidence that the main decay channel of ^{45}Fe proceeds by the emission of two protons from the ground state.

109

SUMMARY

The decay of ^{45}Fe was studied in an experiment performed at GSI Darmstadt. Six atoms of this nuclide, produced by the fragmentation of ^{58}Ni beam at 650 MeV/nucleon, were implanted into a telescope of Si detectors in the final focus of the FRS. For five atoms, decay signals were recorded. The half-life of ^{45}Fe was determined to be $3.2^{+2.6}_{-1.0}$ ms. In four decay events, an energy release of 1.1 ± 0.1 MeV was observed. These findings, together with the non-observation of coincident γ rays, provide the first evidence for the two-proton ground-state radioactivity of ^{45}Fe. This is the first case of such a decay mode established for a narrow nuclear ground state.

Further experiments are clearly needed to confirm this result, in particular by a separate detection of the two emitted protons. Energy and angular correlation of the two protons are necessary to clarify the mechanism of the *2p* radioactivity. The important question to be answered is whether a two-body process involving the diproton state plays a significant role in the *2p* decay.

It should be noted that our results were confirmed by the analysis of data obtained at GANIL [16, 17].

ACKNOWLEDGMENTS

This work was partially supported by the EC under contract HPRI-CT-1999-50017 and by the U.S. DOE through contract DE-FG02-96ER40983 (University of Tennessee). ORNL is managed by UT-Battelle, LLC, for the U.S. DOE under contract DE-AC05-00OR22725.

REFERENCES

1. V.I. Goldansky, Nucl. Phys. **19**, 482 (1960).
2. O.V. Bochkarev et al., Sov. J. Nucl. Phys. **55**, 955 (1992).
3. R.A. Kryger et al., Phys. Rev. Lett. **74**, 860 (1995).
4. B.A. Brown, Phys. Rev. C **43**, R1513 (1991).
5. E. Ormand, Phys. Rev. C **53**, 214 (1996).
6. B.J. Cole, Phys. Rev. C **54**, 1240 (1996).
7. B. Blank et al., Phys. Rev. Lett. **77**, 2893 (1996).
8. B. Blank et al., Phys. Rev. Lett. **84**, 1116 (2000).
9. M. Pfützner et al., Eur. Phys. J. **A 14**, 279 (2002).
10. M. Pfützner et al., Nucl. Instr. and Meth. in Phys. Res. **A 493**, 155 (2002).
11. B. Hubbard-Nelson, M. Momayezi and W.K. Warburton, Nucl. Instr. and Meth. in Phys. Res. **A 422**, 411 (1999); see also http://www.xia.com .
12. S. Hoffmann, in *Particle Emission from Nuclei*, ed. M. Ivascu and D.M. Poenaru, CRC Press Bocaraton, 1989, p. 25.
13. B.A. Brown, Phys. Rev. C **46**, 811 (1992).
14. L.V. Grigorenko et al., Phys. Rev. C **64**, 054002 (2001).
15. L.V. Grigorenko, I.G. Mukha, M.V. Zhukov, Nucl. Phys. **A 714**, 425 (2003).
16. J. Giovinazzo et al., Phys. Rev. Lett. **89**, 102501 (2002).
17. J. Giovinazzo et al., contribution to this volume.

Two-proton radioactivity: the case of ^{45}Fe

J. Giovinazzo*, B. Blank*, C. Borcea†, B.A. Brown**, M. Chartier¹*,
S. Czajkowski*, A. Fleury*, R. Grzywacz²‡, M. Lewitowicz§,
M.J. Lopez Jimenez³*, V. Maslov⁴§, F. de Oliveira Santos§, M. Pfützner‡,
M.S. Pravikoff*, M. Stanoiu§ and J.-C. Thomas*

*CEN Bordeaux-Gradignan, Le Haut Vigneau, BP 120, F-33175 Gradignan cedex, France
†IAP, Bucharest-Magurele, P.O. Box MG6, Romania
**NSCL, Michigan State University, East Lansing, Michigan 48824-1321
‡Institute of Experimental Physics, University of Warsaw, PL-00-681 Warsaw, Poland
§Grand Accélérateur National d'Ions Lourds, BP 5027, F-14076 Caen cedex, France

Abstract. According to most mass models, several of the best candidates for two-proton radioactivity lie in the $A = 50$ mass region, at the proton drip-line. Recent experiments using the fragmentation of a ^{58}Ni beam allowed us to reach this region and observe some of these candidates, like ^{45}Fe and ^{48}Ni. During an experiment at the GANIL / LISE3 facility helded in 2000, we could identify 22 events of implantation of ^{45}Fe nuclei. The energy distribution of the corresponding decay events shows a clear peak at (1.14 ± 0.04) MeV that could be assigned to the direct 2-proton emission from the ground state. This energy is in very good agreement with several theoretical predictions, and no coincident β particle has been observed experimentally. A half-life of $(4.7^{+3.4}_{-1.4})$ ms has been measured for the decay of ^{45}Fe.

INTRODUCTION

For nuclei located just beyond the drip-lines, the nuclear forces are not able any more to bind the last nucleons. Nevertheless, in the case of proton-rich nuclei, the last proton for an odd-Z nucleus or the last two protons for an even-Z nucleus can be kept inside the unstable nucleus due to the proton electric charge. Such particles will then be emitted by a tunnel effect through the Coulomb barrier. The one-proton radioactivity, for odd-Z nuclei, has been observed for the first time in 1981 [1, 2]. Nowadays, about 30 cases of proton radioactivity have been observed in the drip-line region from $Z = 53$ to $Z = 83$.

The two-proton radioactivity has been predicted since 1960 [3] for nuclei beyond the drip-line with even Z. Indeed, due to the pairing interaction, for candidates to 2-proton radioactivity, the one-proton emission channel is energetically forbidden. Therefore, the sequential decay of two protons is not accessible, and the nucleus can only decay via the simultaneous 2p emission.

---

¹ Present address: Oliver Lodge Laboratory, University of Liverpool, L69 7ZE, United Kingdom
² Present address: Physics Division, ORNL, Oak Ridge, Tennessee 37831-6371
³ Present address: CEA/DPTA/SPN, F-91680 Bruyères-le-Châtel, France
⁴ Present address: Flerov Laboratory of Nuclear Reactions, 141980 Dubna, Russia

CP681, *Proton-Emitting Nuclei: Second International Symposium; PROCON 2003*,
edited by E. Maglione and F. Soramel
© 2003 American Institute of Physics 0-7354-0150-0/03/$20.00

The two-proton decay may occur either as an isotropic emission with no angular correlation between the two emitted particles, or as a correlated emission of a ^2He resonance, which is unbound and would break up into two single protons. In both cases, the protons are expected to share equally the available energy, but in the second case, an angular correlation could remain after the two protons have left the nucleus.

The isotropic emission has been observed in light nuclei like ^6Be [4] and ^{12}O [5], with very short half-lives of the order of characteristic nuclear reaction times (10^{-21} s). In the decay of these nuclei, the states involved in the process are very broad, and the emission could be interpreted in terms of a sequential emission through the tail of the intermediate state.

According to recent theoretical calculations [6, 7, 8, 9], the best candidates for ^2He radioactivity may be ^{45}Fe, ^{48}Ni and ^{54}Zn, with a Q_{2P} value ranging from 1.1 to 1.8 MeV and a half-live in the order of 10^{-4} to 10^{-2} s. While ^{54}Zn has never been observed yet, ^{45}Fe [10] and ^{48}Ni [11] could be identified in previous projectile fragmentation experiments at GSI (Darmstadt) and GANIL (Caen).

In a recent experiment performed at GANIL, the decay of ^{45}Fe could be observed, leading to a consistent picture of a direct two-proton emission from the ground state [12]. This result is in agreement with the data from a recent experiment at GSI [13].

EXPERIMENTAL PROCEDURE AND DATA ANALYSIS

The experiment has been performed at the GANIL/LISE3 facility. A high intensity $^{58}Ni^{26+}$ primary beam was accelerated to 75 MeV/A using the CSS1 and CSS2 cyclotrons. The projectile fragmentation in a 240 μm natural nickel target located in the SISSI device produced a wide range of isotopes. Proton rich nuclei with $Z = 22$ to 28 were selected with the Alpha and the LISE3 spectrometers: a first dipole for $B\rho$ selection, a 50 μm achromatic degrader at the intermediate focal plane coupled to the second dipole and a Wien filter for a final velocity selection.

The selected ions were stopped in the detection setup devoted to both the identification of the implanted ions and the decay measurements (see fig. 1). For the identification, we used crossed conditions on several energy and energy loss measurements in a silicon telescope, and times of flight in the LISE line. The time of flight measurements were performed either between the cyclotron high frequency signal and the first silicon detector, or between micro-channel plates located before the Wien filter and the same silicon detector. The result of the identification analysis is summarized in fig. 2, which corresponds to the full running time on the setup for very exotic nuclei, approximately 36 hours. This plot shows 22 events that have been identifed as an implantation of a ^{45}Fe nucleus. The redundance of parameters used for identification allows a significant background reduction: almost no count appear where unbound nuclei would be expected in such an identification matrix.

The detector used for implantation is a double-side silicon detector with 16×16 XY strips. Since the implantation and decay energies differ by several orders of magnitude, each strip had a low and a high gain electronic and readout chains. To build time and energy distributions of decay events, a pixel correlation is performed after the

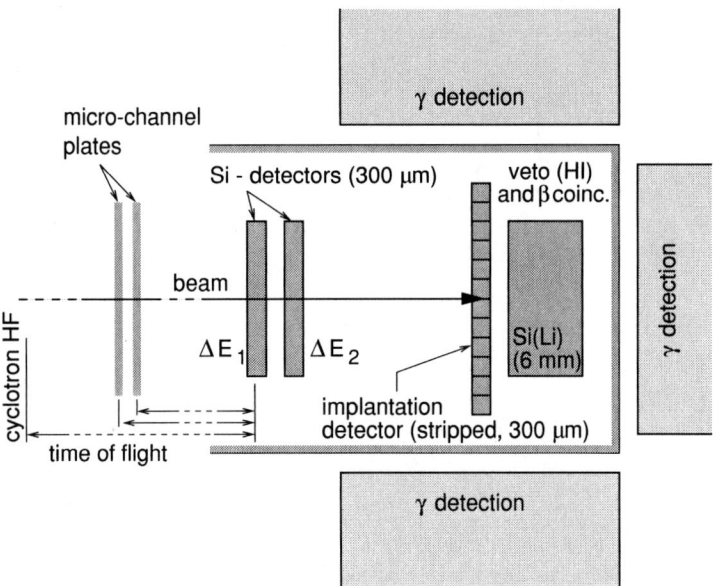

FIGURE 1. The selected ions (beam) are implanted in a double-side XY silicon strip detector. The identification is performed with two energy loss measurements ΔE_1 and ΔE_2 in 300 μm silicon detectors, the residual energy $E3$ in the implantation detector (300 μm), and several time of flight measurements between 2 micro-channel plate detector or the high frequency signal from the cyclotrons and the first silicon detector of the telescope. In addition, the 6 mm thick $Si(Li)$ detector is used as a veto for light particles going across the implanation detector. The decay of implanted isotopes occurs in the implantation detector: protons with less than few MeV energy are stopped in this detector, while β particles may escape. The $Si(Li)$ detector, located close to the implantation detector, has an efficiency about 30% to measure those β particles in coincidence with the protons. The telescope is surrounded by an array of germanium detectors to perform a detailed spectroscopy of the several nuclei accessible with the spectrometer setting. It has an efficiency in the order of few percent at 1 MeV.

implantation of each identified nucleus. This allows to reduce from a factor 10 to 100 the background in decay distributions. The nuclei of interest are implanted in the center of the strip detector. As a consequence, protons with a few MeV energy emitted either directly or after β decay from these nuclei deposit all their energy in that detector. Since most β particles escape the implantation detector, another silicon detector of the telescope is located very close to the strip detector. With this set-up, the β-proton coincidence measurement efficiency is about 30%. The silicon telescope is in addition surrounded with an array of germanium detectors for γ rays measurement, used for spectroscopic studies of proton-rich nuclei in the mass region accessible in the experiment. The γ detection efficiency was of the order of a few percents at 1 MeV.

The results for the decay analysis after the 22 implantation events of ^{45}Fe are given in figures 3 and 4. The energy distribution (fig. 3) shows a peak at (1.14 ± 0.05) MeV, with a FWHM $= 0.06$ MeV. From the time distribution (dark spectrum in fig. 4) of the

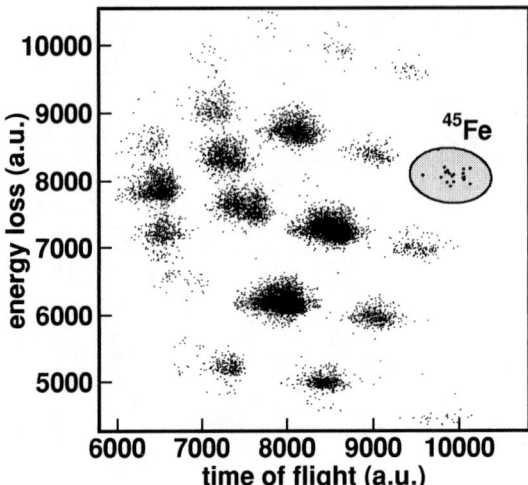

FIGURE 2. The identification matrix of nuclei selected with the spectrometer: time-of-flight versus energy loss in the first silicon detector of the implantation telescope. This plot takes into account conditions on other parameters that do not explicitly appear here, but contribute highly to the background suppression. The shadowed area shows the events for which a ^{45}Fe has been implanted and identified.

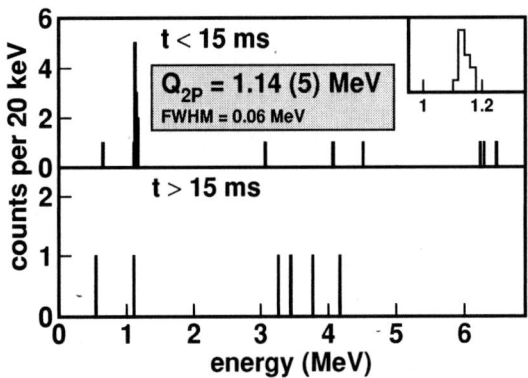

FIGURE 3. Energy distribution of decay events correlated to a ^{45}Fe implantation. A peak clearly appears for short times after implantation (upper part). This peak is identified as the two-proton decay from the ground state. Longer time events (lower part) are compatible with a subsequent decay of ^{43}Cr, the daughter of ^{45}Fe after two-proton emission.

events in this peak, a half-life of $T_{1/2} = 4.7^{+3.4}_{-1.4}\ ms$ is deduced. A fit including all decay events and taking into account the decay of the daughter nucleus leads to half-lives of $T_{1/2} = 5.7^{+2.7}_{-1.4}\ ms$ for ^{45}Fe and of $T_{1/2} = 16.7 \pm 7.0\ ms$ for ^{43}Cr, the daughter nucleus of ^{45}Fe after two-proton emission. This result is in agreement with the ^{43}Cr half-life from previous work [14].

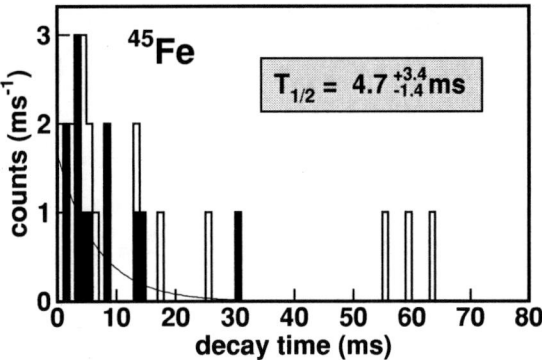

FIGURE 4. Time distribution of decay events after a ^{45}Fe implantation. The dark spectrum corresponds to the time distribution of the events in the $1.14\ MeV$ two-proton peak, for which a half-life of $4.7^{+3.4}_{-1.4}\ ms$ is estimated. If all events are taken into account, this distribution is consistent with the decay of the daughter nucleus ^{43}Cr with an estimated half-life of $(16.7 \pm 7.0)\ ms$.

For the events in the $1.14\ MeV$ peak in the decay of ^{45}Fe, no β particle is observed in coincidence, as shown in figure 5. The signal in the last detector is in the level of the noise of the amplification chain. As a comparison, the same coincidence distribution is build for the decay of ^{46}Fe, selecting the same energy conditions. This nucleus is known as a β-delayed proton emitter [14], and β particles are clearly detected above the noise level.

DISCUSSION

The $1.14\ MeV$ energy peak in the decay of ^{45}Fe is assigned to the two-proton emission from the ground state for several reasons. The energy of the peak and the observed half-lives give a picture that is fully consistent with such a decay process for both the parent and the daughter nucleus. For all events in the peak, no β particle is observed in coincidence, while the same analysis performed on ^{46}Fe decay events with the same energy conditions clearly shows the β energy signal over the noise level. The probability of the non-observation of any coincidence signal for events in the peak in the decay of ^{45}Fe would be of the order of 1% in the case of a β-proton decay. In addition, the $1.14\ MeV$ peak appears to be 30% narrower than peaks in the decay of β-delayed proton emitters. This indicates that in the case of ^{45}Fe, the peak is not broadened due to β energy pile-up in the strip detector.

The observation of only 12 counts in the peak indicates that the branching ratio for two-proton emission may not be 100%. Due to the relatively low half-life, a few events (3 or 4) may be lost in the dead-time of the acquisition system (0.3 to 0.5 ms). We estimate that the $2p$ branch has a branching ratio of around 70-80%.

The energy of the peak is in nice agreement with the theoritical predictions for Q_{2P} from Brown [6]: $(1.15 \pm 0.09)\ MeV$, from Ormand [7]: $(1.28 \pm 0.18)\ MeV$ and

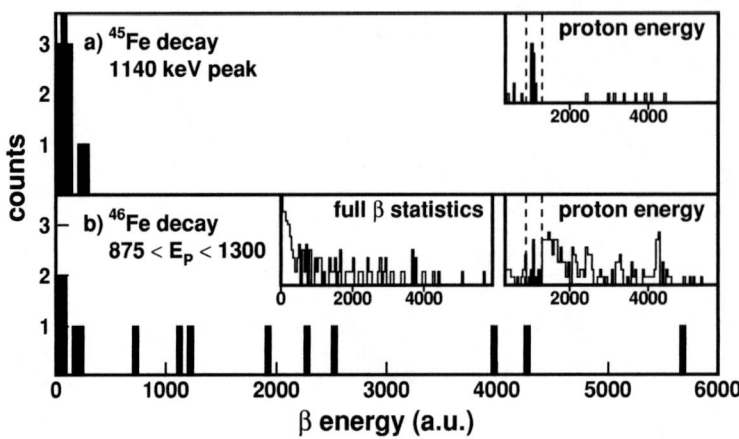

FIGURE 5. The upper distribution show the energy signal in the silicon detector located close the strip detector where implantation occurs. This signal is measured in coincidence with decay events in the 1.14 MeV peak in the decay of ^{45}Fe, shown in the proton energy distribution, between the dashed lines. No signal is observed above the noise of the electronic amplification chain. The lower part shows the same distribution in the decay of ^{46}Fe which is known to be a beta-delayed proton emitter, for decay events with a proton in the same energy window. For same condition energies, we observe a coincidence signal due to beta particles. The additional inserted figure shows the beta energy distribution for all events in the decay of ^{46}Fe. With a 30 to 35% efficiency for coincident beta detection, those distributions indicate that the probability that there is no beta particle in the decay of ^{45}Fe is in the order of 99%.

from Cole [8]: (1.22 ± 0.05) MeV. A very simple barrier penetration description for the emission of a ^2He particle gives a half-life of $T_{1/2} = 0.024^{+0.074}_{-0.017}$ ms for $Q_{2P} = (1.14 \pm 0.05)$ MeV, with a spectroscopic factor of 1. From shell-model calculations, Brown estimated a spectroscopic factor [15] of 0.2, which increases the tunneling half-life to $T_{1/2} = 0.12^{+0.38}_{-0.09}$ ms. A very recent description of 2p emission from ^{45}Fe within the framework of R-matrix theory yields half-lives between 4 ms and 41 ms [16].

Grigorenko et al. [17] developed a three body model which takes into account the proton-core interaction and the proton-proton correlation explicitly. The half-life calculation has been performed for the protons in a p- or a f-wave and for emission of two independant protons in a s-wave. The experimental values are in good agreement with the p-wave two-proton emission. But according to the mirror nucleus ^{45}K, the last two protons in ^{45}Fe are expected to be in a f state.

While the comparison of experimental data with theoretical calculations for the simultaneous emission of 2 protons (either correlated or not) is not fully satisfying, a sequential decay may be excluded. Indeed, the intermediate state in such a case would be the ground state of ^{44}Mn. The Q_{1P} from mass estimates ranges from -24 to $+10$ keV, depending on the prediction considered [6, 7, 8] and the expected half-life for the first proton emission would then be in the order of hours or more.

116

CONCLUDING REMARKS

In a projectile fragmentation experiment performed at GANIL in 2000, we have observed the two-proton emission from the ground state of ^{45}Fe. The Q_{2P} deduced from the data is (1.14 ± 0.05) MeV and the estimated half-life of ^{45}Fe is $T_{1/2} = 4.7^{+3.4}_{-1.4}$ ms. While $2p$ emission is most likely the main decay mode, a small βp branch may also exist for ^{45}Fe.

The data from an experiment performed at GSI [13] are in agreement with our results, but with lower statistics. The energy found for the two-proton decay is (1.1 ± 0.1) MeV and the half-life is $T_{1/2} = 3.8^{+2.0}_{-0.8}$ ms.

The comparison with several theoretical descriptions seems to favour a ^2He emission rather than an uncorrelated emission of the two protons, but detailed studies are still required to give a sharper conclusion. In both cases, the two particles may share equally the available energy, but in the ^2He decay, an angular correlation could also be observed experimentally. This question should be addressed in future experiments that may allow to detect individually the two protons. Increasing the current statistics may also lead to a more detailed picture of the decay, especially concerning the decay of the daughter nucleus and the branching ratio of the $2p$ decay.

REFERENCES

1. S. Hofmann *et al.*, *Z. Phys.*, **A305**, 111 (1982).
2. O. Keppler *et al.*, *Z. Phys.*, **A305**, 125 (1982).
3. V.I. Goldansky, *Nucl. Phys.*, **19**, 482 (1960).
4. O.V. Bochkarev *et al.*, *Sov. J. Nucl. Phys.*, **55**, 955 (1992).
5. A. Azhari *et al.*, *Phys. Rev.*, **C58**, 2568 (1998).
6. B.A. Brown, *Phys. Rev.*, **C43**, R1513 (1991).
7. W.E. Ormand, *Phys. Rev.*, **C53**, 214 (1996).
8. B.J. Cole, *Phys. Rev.*, **C54**, 1240 (1996).
9. L. Grigorenko *et al.*, *Phys. Rev. Lett.*, **85**, 22 (2000).
10. B. Blank *et al.*, *Phys. Rev. Lett.*, **77**, 2893 (1996).
11. B. Blank *et al.*, *Phys. Rev. Lett.*, **84**, 1116 (2000).
12. J. Giovinazzo *et al.*, *Phys. Rev. Lett.*, **89**, 102501 (2002).
13. M. Pfützner *et al.*, *Eur. Phys. Jour.*, **A14**, 279 (2002).
14. J. Giovinazzo *et al.*, *Eur. Phys. Jour.*, **A10**, 73 (2000).
15. B.A. Brown, *Phys. Rev.*, **C44**, 924 (1991).
16. B.A. Brown, F.C. Barker, *to be published* (2003).
17. L. Grigorenko *et al.*, *Phys. Rev.*, **C64**, 054002 (2001).

R-matrix Theory for Di-proton Decay

B. A. Brown * and F. C. Barker †

* Department of Physics and Astronomy, and National Superconducting Cyclotron Laboratory, Michigan State University, East Lansing, Michigan 48824-1321, USA
† Department of Theoretical Physics, Research School of Physical Sciences and Engineering, The Australian National University, Canberra ACT 0200, Australia

Abstract. *R*-matrix formulas are used to calculate di-proton decay widths of ^{18}Ne, ^{45}Fe and ^{48}Ni. This formulation allows one to relate the decay amplitude to the correlated many-body wave functions for these nuclei. The role of pairing correlations is discussed.

INTRODUCTION

The pairing interaction contribution to the nuclear binding energy gives rise to situations in proton-rich nuclei where an even Z nucleus is unbound to two-proton decay and bound to one-proton decay. These two-proton decays sometimes have narrow widths due to the Coulomb barrier. In light nuclei the intermediate states for the one-proton emission are usually very broad allowing for a combination of sequential two-proton emission through the broad intermediate state which combines with the direct two-proton emission process. The larger Coulomb barrier for heavier nuclei may allow for cases in which the sequential decay mode is relatively forbidden compared to the direct mode, and the two-proton decay width is so narrow that it may be counted as a fifth form of radioactivity (defined in [1] as a process with a lifetime of longer than 10^{-12} s), along with the previously observed fission decay, beta decay, gamma decay, alpha decay and one-proton decay modes. The recently discovered case of ^{45}Fe two-proton radioactivity is the first to be observed [2], [3] [4]. (Beta-delayed two-proton decay was first observed in [5], but in this case the time scale is set by the beta decay and the two-protons decay mainly by a fast sequential emission [6]). In addition, there are situations such as the recently discovered case of the two-proton decay of an excited state in ^{18}Ne in which two-proton decay can be observed in the presence of a much stronger one-proton decay branch.

In this review we discuss calculations for the di-proton decay in terms of a recent extension of the *R*-matrix method [7] which includes the *s*-wave $p + p$ interaction as an intermediate state. In this context we define the di-proton decay mode as that part of the two-proton decay which is related to the correlated two-proton decay through this channel. We assume that this is the dominant contribution to the direct two-proton channel. If sequential decay is allowed it is calculated separately as a decay through intermediate states. The *R*-matrix allows us to relate the amplitude for the di-proton decay to the microscopic many-body wave functions for the parent and daughter nuclei. In particular, pairing correlations are important.

CP681, *Proton-Emitting Nuclei: Second International Symposium; PROCON 2003*,
edited by E. Maglione and F. Soramel

We will concentrate here on a summary of recent results for ^{45}Fe, ^{48}Ni [8] and ^{18}Ne [9]. This R-matrix method has also been applied to ^{6}Be, ^{8}C [10] and ^{12}O [7], [11].

The work of Grigorenko et al. [12], [13] provides a more general formulation for the three-body asymptotics for two-proton decay in terms of a solution of a three-body Hamiltonian. However, in this formulation the connection to the many-body nuclear structure remains at the level of the single-particle wavefunctions, and does not include pairing correlations. It would be useful to find a way to compare our R-matrix results (a correlated di-proton decay through an intermediate state resonance of the two protons) with those of Grigorenko.

THEORETICAL FORMALISM

In the R-matrix model the di-proton decay with is given by the formula [7]

$$\Gamma^0 = 2\gamma^2 \bar{P}/(1+\gamma^2 \bar{S}'), \tag{1}$$

with

$$\bar{P} = \int_0^Q P(Q-U)\rho(U)\,dU, \tag{2}$$

$$\bar{S}' = \int_0^\infty \left[\frac{dS(E-U)}{dE}\right]_{E=Q} \rho(U)\,dU, \tag{3}$$

where Q is the available decay energy, P and S are the R-matrix penetration and shift factors [14], and ρ is the density-of-states function, which is expressed in terms of the $p+p$ s-wave phase shift by Eq.(3) of Ref. [7], with $a_2 = 2.90$ fm, $c = 0.25$ fm, $A = -0.0045$ fm^{-1}, and $B = 1.073$ fm. The half-life is $T_{1/2} = (4.56 \times 10^{-19}/\Gamma^0)$ MeV ms.

The reduced width γ^2 is related to the spectroscopic factor, S, and the dimensionless reduced width, θ_{sp}^2, by [11]

$$\gamma^2 = S\,\theta_{sp}^2\,\frac{\hbar^2}{Ma^2}, \tag{4}$$

where M is the reduced mass and a is the channel radius. This is given as in Ref. [7] by the conventional formula

$$a = 1.45(A_1^{1/3} + A_2^{1/3})\ \text{fm}, \tag{5}$$

where A_1 is the mass number of the daughter nucleus and $A_2 = 2$ for the di-proton. The single-particle dimensionless reduced width θ_{sp}^2 is given by:

$$\theta_{sp}^2 = \frac{(a/2)\,u^2(a)}{\int_0^a u^2(r)\,dr}, \tag{6}$$

where $u(r)/r$ is the single-particle radial wavefunction. θ_{sp}^2 depends upon the potential parameters for the di-proton nucleus interaction. For the potential we take the Woods-Saxon form cut off at radius $r = a$ plus a uniform-sphere Coulomb potential with radius

$R_C = r_C A_1^{1/3}$. The Woods-Saxon parameters are $R = r_0 A_1^{1/3}$ for the radius, a_0 for the diffuseness, and a well depth adjusted to reproduce the resonance energy.

The final result depends upon the channel radius a through the penetration factor $\bar{P}$, the energy derivative of the shift factor $\bar{S}'$ and the reduced width γ^2. We find that when the channel-radius a is chosen to be large enough (beyond the range of the strong interaction) the final result is insensitive to the choice of a.

APPLICATION TO ^{45}FE AND ^{48}NI

Two-proton decay of ^{45}Fe has been observed with a peak energy of (1.14 ± 0.05) MeV and a half-life of $(4.7^{+3.4}_{-1.4})$ ms [2]. An independent experiment with less statistics and a lower energy resolution gave an energy of (1.1 ± 0.1) MeV and a half-life of $(3.2^{+2.6}_{-1.0})$ ms [3]. The average half-life from the two experiments is $(3.8^{+2.0}_{-0.8})$ ms [4]. The two-proton decay branch is experimentally estimated to be $70 - 80\%$ [4]. A 75% two-proton branch would give a two-proton decay half-life of $4-8$ ms and a beta decay half-life of 12-23 ms. The lower beta decay half-life is not far from the prediction of 7 ms [15]. The energy is in good agreement with predictions of (1.15 ± 0.09) MeV [16], (1.28 ± 0.18) MeV [15], and (1.22 ± 0.05) MeV [17].

The potential parameters are taken from an analysis of low-energy deuteron scattering [18]: $r_0 = 1.17$ fm, $a_0 = 0.72$ fm, $r_C = 1.30$ fm.

To calculate the spectroscopic factor we project the shell-model wave function onto the $0s$ internal (relative) wave function for a di-proton in the pf shell by using harmonic oscillator wave functions with the general formalism of Ref. [19] to obtain:

$$S = \left(\frac{A}{A-2} \right)^\lambda G^2(pf) \, C(A,Z), \qquad (7)$$

where $G^2 = 5/16$, A is the mass of parent nucleus ($A = 45$ in our case), and

$$C(A,Z) = |< \Psi(A-2,Z-2) \,|\, \Psi_c \,|\, \Psi(A,Z) >|^2$$

is the cluster overlap for the di-proton cluster wavefunction Ψ_c in the pf shell with $L = 0$, $S = 0$ and $T = 1$ in the SU3 basis.

The spectroscopic factor depends on the model space. In [16] it was assumed that the ^{45}Fe to ^{43}Cr cluster matrix element could be approximated by that for the ^{46}Fe to ^{44}Cr matrix element. That is, the sd-shell neutron hole is an inactive spectator in the transition. We have checked this assumption by calculating the cluster overlaps for ^{45}Fe and ^{46}Fe in the $sd - pf$ model space with the full pf shell for the protons and one-neutron hole for ^{45}Fe. With the $sd - pf$ Hamiltonian from [20] the ratio $C(^{45}\text{Fe})/C(^{46}\text{Fe})$ is 0.96. Thus we confirm that the ^{45}Fe di-proton overlap is essentially the same as that for ^{46}Fe. We use the FPD6 interaction from [21] to obtain $C(^{46}\text{Fe}) = 0.480$, which gives $S = 0.197$. Thus we use $S = 0.20$ for the present calculation, which is essentially the same as the original value of 0.195 obtained in [16].

It is interesting to note that the two-proton overlaps which enter into the ^{45}Fe decay are the same as those that enter into the interpretation of the ^{46}Ca(p,t)^{44}Ca reaction on the mirror nuclei. The two-proton transition density matrix elements are -1.19, -0.28, -0.23 and -0.12 for $(0f_{7/2})^2$, $(0f_{5/2})^2$, $(1p_{3/2})^2$ and $(1p_{1/2})^2$, respectively. The amplitudes are dominated by the $0f_{7/2}$ orbit, but the coherent mixture is important. If only the $0f_{7/2}$ orbit is kept, the cluster overlap is reduced by a factor of 0.28.

With $Q = 1.14$ MeV [2], we find $\theta_{sp}^2 = 0.097$ for the di-proton in a $3s$ state (three zeros of the wave function between 0 and ∞), leading to $\gamma^2 = 8.6 \times 10^{-3}$ MeV. Also $\bar{P} = 0.638 \times 10^{-18}$ and $\bar{S}' = 0.286$ MeV^{-1}, so that the $\bar{S}'$ term in the denominator of Eq. (1) is negligible. This value of $\bar{P}$ is 3×10^{-4} times the value of $P(Q)$, the penetration factor used in the simple model [2]. The factor is small due to the integrand in Eq. (2) peaking at $U \approx 0.06$ MeV, while $\bar{P}$ is equal to $P(Q-U)$ for $U \approx 0.25$ MeV. From Eq. (1), $\Gamma^0 = 1.10 \times 10^{-20}$ MeV and $T_{1/2} = 41$ ms. The result is sensitive to the Q value. If we take the experimental upper range of 1.19 MeV we would obtain $T_{1/2} = 10$ ms, which is in reasonable agreement with experiment. Thus there is a large uncertainty in the calculated half-life due to the error in Q value. An improved experimental Q value which will be obtained in future experiments will reduce this source of error. This model allows us to make a connection to the interesting questions about pairing correlations.

With the extrapolated one-proton Q value of [16], it was estimated in [2] that the sequential decay should be negligible. Although the effective penetration factor is very small, there could be a contribution to the denominator of (1) coming from the sequential decay channel, but this is estimated to be less than 0.1, and so to have at most a 10% effect on the calculated $T_{1/2}$.

In [2] the half-life obtained with the R-matrix model (with the $p+p$ resonance) was about 10 times longer than the present result. The reason is that the channel radius of $a = 4.2$ fm used in that calculation is too small (not sufficiently outside the strong interaction potential), and it was assumed that $\theta_{sp}^2 = 1$. The half-life estimates given in [16] ignored the $p+p$ resonance (which increases the half-life by a factor of about 3000), used a small value of $a = 4.0$ fm, and also assumed $\theta_{sp}^2 = 1$. Thus the present theoretical result replaces those of [2] and [16].

^{48}Ni is also a good candidate for the observation of two-proton decay [16]. The pf-shell spectroscopic factor is 0.14. Using the extrapolated two-proton decay Q value of 1.36(13) MeV from [16] together with the same set of values for the other parameters discussed above for ^{45}Fe we obtain $T_{1/2} = 0.4$, 8 and 260 ms for $Q = 1.49$, 1.36 and 1.23 MeV, respectively.

In the present model there is some sensitivity to the potential parameters. For r_0 or a_0 increased by 0.05 fm, $T_{1/2}$ is further decreased by 20% and 18% respectively. Thus, the assumption of our use of the deuteron-scattering potentials must be checked. There is of course a sensitivity to the spectroscopic factor.

The effects of two-nucleon correlations can be compared to the information inferred from the (p,t) reactions on mirror nuclei. The (p,t) reactions are interpreted in terms of enhancement factors ε relative to a given model space. For wavefunctions which are dominated by the $0f_{7/2}$ orbit, the experimental enhancement factor relative to the full pf-shell for $L = 0$ transfer is about 2.4 [22]. (When the ^{48}Ca(p,t) data of [23] is analyzed with the FPD6 wavefunctions one obtains an enhancement factor of 2.2 for the transition to the ^{46}Ca ground state. This transition is the mirror of the ^{48}Ni di-proton decay transition.) This enhancement can be qualitatively understood in perturbation theory from the admixtures of correlated $J = 0, T = 1$ components from the major shells below and above the pf shell [22]. Qualitatively one might apply the same enhancement factor to increase the di-proton decay spectroscopic factors by 2.4 giving $T_{1/2} = 17$ ms with Q=1.14 MeV and $T_{1/2} = 4$ ms with Q=1.19 MeV which brings the result into better agreement with experiment (the new result for ^{48}Ni would be 3.3 ms with $Q = 1.36$ MeV). However, the relationship between the (p,t) cross-sections enhancements and those for di-proton decay needs to be quantified in terms of overlap functions for the two-proton removal process.

APPLICATION TO ^{18}NE

Two-proton decay has been observed for the 1^- state at 6.15 MeV in ^{18}Ne [24]. The decay of this state is dominated by single-proton decay to the low-lying bound states of ^{17}F, but there are no narrow intermediate states in ^{17}F which can contribute to the two-proton emission to the ground state of ^{16}O. The experimental width for the two-proton emission is 21±3 eV if a di-proton decay model is assumed and 57±6 eV assuming a sequential (democratic) decay model [24].

The shell-model wave functions for ^{18}Ne were obtained in a model space which includes the $0s, 0p, 1s0d$ and $1p0f$ orbits. ^{16}O is treated as an $s^4 p^{12}$ closed shell, and the lowest-lying positive parity states of ^{17}F and ^{18}Ne are taken as $s^4 p^{12} (sd)^1$ and $s^4 p^{12} (sd)^2$ configurations, respectively. The negative parity states in ^{18}Ne are treated as $1\hbar\omega$ excitations of the form $s^4 p^{11} (sd)^3$ and $s^4 p^{12} (sd)^1 (pf)^1$. With these configurations the ^{18}Ne to ^{16}O decay involves the emission of two protons in an $(sd)(pf)$ configuration. The state of interest is the second 1^- state in ^{18}Ne.

The wave functions are obtained with several Hamiltonians which have been designed for use in this type of model space, namely the MK interaction [25] and the WBP and WBT interactions [26]. The MK wave functions have been used for a number of studies ·of the analogue nucleus ^{18}O including ^{18}N (β^-) ^{18}O [27], ^{18}O (π, π') ^{18}O [28], and ^{18}O (e, e') ^{18}O [29]. The calculated energies of the lowest five 1^- states are in reasonable agreement with the energies found in the analogue nucleus ^{18}O. These low-lying 1^- states are dominated by the $s^4 p^{11} (sd)^3$ configuration. The smaller $s^4 p^{12} (sd)^1 (pf)^1$ component is the one responsible for one- and two-proton decay.

The dominant mode of decay for the 1_2^- state in ^{18}Ne is by one-proton emission to the $5/2^+$ ground state ($Q_{1p} = 2.228$ MeV) and $1/2^+$ first excited state ($Q_{1p} = 1.733$ MeV) of ^{17}F. Single-particle widths Γ^{sp} are obtained [11] from the resonance energy in a Woods-Saxon well with a depth chosen to reproduce Q_{1p}. The decay to the ground state can go by f-wave emission ($\Gamma^{sp} = 8.03$ keV) or p-wave emission ($\Gamma^{sp} = 1270$ keV). The decay to the excited state can go by p-wave emission ($\Gamma^{sp} = 627$ keV). The shell-model spectroscopic factors and the resulting decay widths $\Gamma = C^2 S \Gamma^{sp}$ are given in Table II of [9]. The widths all turn out to be about 30 keV compared with the known width of 50±5 keV [24], but the individual contributions are very different. The width obviously depends mainly on the p-wave admixtures which are small and quite variable. Thus we conclude that there is about a factor of two uncertainty in the theoretical spectroscopic factors and that the agreement with experiment is satisfactory.

Although there are no unbound states present in the energy window for the two-proton decay of the 1_2^- state in ^{18}Ne, sequential two proton decay may proceed through the tails of states outside the window. There are two possibilities, one is through the ghost associated with the $1/2^+$ bound state [30] and the other is through the low-energy tail of the broad $3/2^+$ state at 5.0 MeV which has a width of 1.5 MeV. This large width indicates that the $3/2^+$ state is dominated by the $0d_{3/2}$ single-particle component. The spectroscopic factors for the decay of the ^{18}Ne 1^- state to the $3/2^+$ state are given in Table II [9]. Barker [11] has given the R-matrix formulation for the sequential emission through a broad intermediate state. The result is a width of about 1 meV, and thus the sequential decay through the tail of the $3/2+$ state gives a negligible contribution to the width.

The ghost of the $1/2^+$ bound state turns out to be more important. Its peak is estimated to be at about 2.4 MeV in ^{17}F and it is very broad [30], so that decay through the ghost would be interpreted [24] as democratic decay. The calculated width for the sequential two-proton decay through the $1/2^+$ ghost, 9-19 eV, are comparable in size to the observed two-proton width (57±7 eV for democratic decay) which implies that this channel can be important.

For the di-proton decay the R-matrix model with $A_1 = 16$ and $A_2 = 2$. gives an effective penetration factor, the integral in Eq. (2) with $Q_{2p} = 1.628$ MeV, of 1.30×10^{-4}. If the final-state interaction between the two protons is ignored one would obtain the much larger value $P(Q_{2p})=1.31 \times 10^{-2}$. With the single-particle dimensionless reduced width of $\theta_{sp}^2=0.65$, the total width for di-proton decay is given by $\Gamma = 133\,S$ eV, where S is the spectroscopic factor.

To calculate the spectroscopic factor we project the shell-model wave function onto the $0s$ internal (relative) wave function of the $(sd)(pf)$ pair using harmonic oscillator wave functions [19]. Then,

$$ S = \left(\frac{18}{16}\right)^5 G^2(sd, pf)A^2(50) = 1.126\,A^2(50) , \tag{8} $$

where $G^2 = 5/8$ and $A(\lambda = 5, \mu = 0)$ is the amplitude of the only $(sd)(pf)$ configuration (with $L = 1$, $S = 0$) in the SU3 basis relevant to $L = 1$ di-proton emission. The cluster spectroscopic factor defined above is found immediately by calculating the two-particle (50) amplitude from the shell-model wave functions in an SU3 basis. In the more usual jj basis such as obtained in the shell-model code Oxbash, the SU3 spectroscopic amplitude is obtained by multiplying the jj two-body overlap amplitudes by the factor

$$ (i)^{\ell_1+\ell_2} \langle (20)\ell_1(30)\ell_2 \,\|\, (50)1 \rangle \begin{pmatrix} \ell_1 & 1/2 & j_1 \\ \ell_2 & 1/2 & j_2 \\ 1 & 0 & 1 \end{pmatrix} \tag{9} $$

where the $(i)^\ell$ factor accounts for the difference between the SU3 phase convention and the jj phase convention used in Oxbash.

The $(\lambda, \mu) = (50)$ intensities show considerable variation, with spectroscopic factors of 0.043, 0.024 and 0.075 for MK, WBP and WBT, respectively. When combined with $\Gamma = 133\,S$ eV, we obtain di-proton decay widths of 6, 3 and 10 eV for MK, WBP and

WBT respectively. These are somewhat smaller than the experimental width of 21 ± 3 eV obtained if a di-proton decay model is assumed [24].

SUMMARY

In summary, an R-matrix model, which includes the s-wave resonance of the two-protons, provides a basis for using di-proton decay as a quantitative spectroscopic tool. The present result for ^{45}Fe is in reasonable agreement with experiment given the rather large error due to the Q-value uncertainty. Our method provide a means of extracting unique information on the pairing correlations of protons in the nucleus. The comparison with experiment indicates an enhanced two-nucleon pairing correlation which is similar to that inferred from two-neutron transfer experiments in this mass region. A more precise Q value will be needed to make a more quantitative comparison.

We have made predictions for the two-proton decay of ^{48}Ni which should also be an excellent candidate for experimental study (it is known to have a lifetime of greater than 0.5 μs [31].) Based on the Q values obtained in [32] half-life estimates (with $S = 1$) for the heavier candidates for di-proton decay, ^{54}Zn, ^{63}Se, ^{67}Kr and ^{71}Sr are 500, 6000, 260 and 12 ms, respectively.

The two-proton decay of the 6.15 MeV 1^- state in ^{18}Ne is a more complicated process. The calculated one-proton decay width is within a factor of two of the observed width. For the two-proton decay we find that sequential decay through the ghost of the $1/2^+$ state is within a factor of three of the observed width obtained with the assumption of democratic decay [24]. The calculated width for di-proton emission is only about a factor of two smaller than that for sequential decay indicating that the observed decay may be a combination of the two processes. Given that the spectroscopic factors are small, and that the decay models involve some approximations, the agreement with experiment is satisfactory. More detailed experimental results on the two-proton decay are required to determine the fraction of the decay which can be attributed to di-proton decay.

Acknowledgements: This work was supported by NSF grant PHY-007911.

REFERENCES

1. J. Cerny and J. C. Hardy, Ann. Rev. Nucl. Part. Sci. **27**, 333 (1977).
2. J. Giovinazzo *et. al.*, Phys. Rev. Lett. **89**, 102501 (2002).
3. M. Pfützner *et. al.*, Eur. Phys. J. A **14**, 279 (2002).
4. B. Blank *et al.*, Berkeley meeting proceedings.
5. M. D. Cable, J. Honkanen, R. F. Parry, S. H. Zhou, Z. Y. Zhou and J. Cerny, Phys. Rev. Lett. **50**, 404 (1983).
6. B. A. Brown, Phys. Rev. Lett. **65**, 2753 (1990).
7. F. C. Barker, Phys. Rev. C **63**, 047303 (2001).
8. B. A. Brown and F. C. Barker, Phys. Rev. C **67**, 041304(R) (2003).
9. B. A. Brown, F. C. Barker and D. J. Millener, Phys. Rev. C **65**, 051309 (2002).
10. F. C. Barker, Phys. Rev. C **66**, 047603 (2002); C **67**, 049902(E) (2003).
11. F. C. Barker, Phys. Rev. C **59**, 535 (1999).

12. L. V. Grigorenko, R. C. Johnson, I. G. Mukha, I. J. Thompson and M. V. Zhukov, Phys. Rev. Lett. **85**, 22 (2000).
13. L. V. Grigorenko, I. G. Mukha, and M. V. Zhukov, Nucl. Phys. **A714**, 425 (2003).
14. A. M. Lane and R. G. Thomas, Rev. Mod. Phys. **30**, 257 (1958).
15. W. E. Ormand, Phys. Rev. C **53**, 214 (1996).
16. B. A. Brown, Phys. Rev. C **43**, 1513 (1991); **44**, 924 (1991).
17. B. J. Cole, Phys.Rev. C **54**, 1240 (1996).
18. W. W. Daehnick, J. D. Childs and Z. Vrcelj, Phys. Rev. C **21**, 2253 (1980).
19. N. Anyas-Weiss *et al.*, Phys. Rep. **12**, 201 (1974).
20. E. K. Warburton, J. A. Becker, and B. A. Brown, Phys. Rev. C **41**, 1147 (1990).
21. W. A. Richter, M. G. Van der Merwe, R. E. Julies and B. A. Brown, Nucl. Phys. **A523**, 325 (1991).
22. P. Decowski, W. Benenson, B. A. Brown and H. Nann, Nucl. Phys. **A302**, 186 (1978).
23. G. M. Crawley, P. S. Miller, G. J. Igo and J. Kulleck, Phys. Rev. C **9**, 574 (1973).
24. J. Gomez del Campo et al., Phys. Rev. Lett. **86**, 43 (2001).
25. D. J. Millener and D. Kurath, Nucl. Phys. **A255**, 315 (1975).
26. E. K. Warburton and B. A. Brown, Phys. Rev. C **46**, 923 (1992).
27. J. W. Olness et al., Nucl. Phys. **A373**, 13 (1982).
28. S. Chakravarti et al., Phys. Rev. C **35**, 2197 (1987).
29. D. M. Manley et al., Phys. Rev. C **43**, 2147 (1991).
30. F. C. Barker and P. B. Treacy, Nucl. Phys. **38**, 33 (1962).
31. B. Blank et al., Phys. Rev. Lett. **84**, 1116 (2000).
32. B. A. Brown, R. Clement, H. Schatz, A. Volya and W. A. Richter, Phys. Rev. C **65**, 045802 (2002).

Theory of two-proton decay and prospective candidates for experimental studies

L. V. Grigorenko*, I. G. Mukha* and M. V. Zhukov†

*Gesellschaft für Schwerionenforschung mbH, Planckstr. 1, D-64291 Darmstadt, Germany
†Department of Physics, Chalmers University of Technology, S-41296 Göteborg, Sweden

Abstract. Theoretical studies of several prospective $2p$ emitters are performed in a three-body core+p+p model. Lifetime dependencies on the decay energy are calculated for ^{19}Mg, ^{30}Ar, ^{34}Ca, ^{45}Fe, ^{48}Ni, ^{54}Zn, ^{62}Se and compared to the quasiclassical estimates. An observation of products from the in-flight decay of $2p$ emitting nuclei is discussed as an important technique supplementing a standard implantation method for a broad range of lifetimes of prospective $2p$ emitters. The sensitivity of the model to various aspects of nuclear dynamics is demonstrated on the ^{45}Fe example. Momentum correlations for emitted protons are discussed with the emphasis on ^{19}Mg and ^{45}Fe.

INTRODUCTION

The history of two-proton emission studies is as long as 40 years [1]. True two-proton emission (in terms of [1]) is a pure quantum mechanical phenomenon which has no classical analogue: there are nuclear systems near proton dripline, which can decay only via three (few) particle channel due to separation energy conditions. However, theoretical studies of the phenomenon have been limited mostly to simple quasiclassical models [2, 3, 4]. Recently a quantum mechanical theoretical method has been developed, which allows to study the phenomenon in a three-body cluster model [5, 6]. Within the method the lifetime dependencies on the decay energy and correlations among the decay fragments can be obtained. For light systems (say, up to ^{19}Mg) Coulomb displacement energies can be reliably calculated. The method has been applied to a range of the nuclear systems [7, 8, 9] providing sometimes intriguing results.

These developments coincided in time with the revival of the experimental interest to studies of the two-proton radioactivity based on modern advances in radioactive nuclear beam techniques. For example, very promising results have been recently obtained for the decay of the ^{45}Fe ground state (g.s.) [10, 11]. The experiments are being planned to study decays of ^{19}Mg, ^{48}Ni, ^{54}Zn and to improve the results for ^{45}Fe. Decent theoretical estimates are important for planning of experiments in the field because a minor uncertainty in the $2p$ separation energy can easily lead to a significant change of the lifetime, which may require a different experimental approach [8, 9].

There are still more open questions than answers in this field and the theory is also far from being checked in all possible regimes. Below we provide theoretical predictions (lifetimes and correlations) for several prospective $2p$ emitters which should encourage further experimental studies and maybe provide a basis for the further development of the theory when new experimental data become available.

CP681, *Proton-Emitting Nuclei: Second International Symposium; PROCON 2003*,
edited by E. Maglione and F. Soramel
© 2003 American Institute of Physics 0-7354-0150-0/03/$20.00

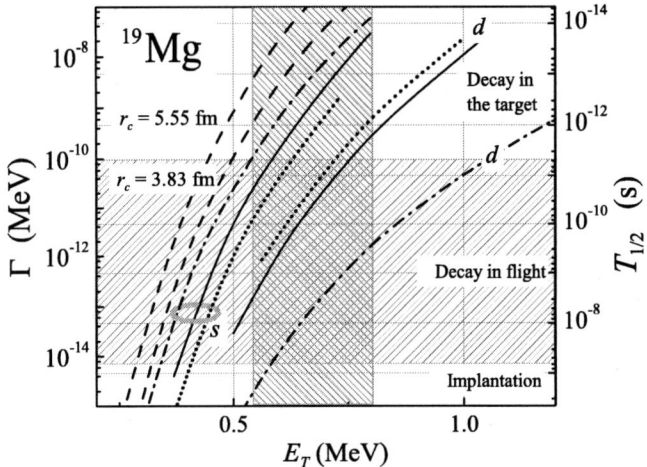

FIGURE 1. The ^{19}Mg lifetime as a function of the $2p$ decay energy E_T. The solid curves correspond to the three-body calculations with different dominating l^2 components [9]. The dashed curves correspond to the diproton estimates with different channel radii. The dash-dotted curves correspond to the simultaneous emission estimates for different l values. The dotted curves correspond to the detailed three-body calculations from [8]. In this and in the following lifetime plots the horizontal hatched bands correspond to different types of possible experiments (see [8]) and vertical bands correspond to expected decay energies (see compilation in [6]).

LIFETIMES

Theoretical lifetime calculations for particle emitters consist roughly of two ingredients: the determination of a decay energy and the determination of a lifetime at given decay energy. The decay energy can be known from experiment, but there are many exotic nuclei, for which this value is not available. The three-body model we use can provide separation energies only for the lightest systems. Here we have to rely on various theoretical calculations and systematics studies. A determination of the lifetime at given decay energy in our model has methodologically much in common with the R-matrix phenomenology as it is applied to ordinary two-body decays.

General approach

In papers [6, 9] we described what we call a "systematic l^2" model of the two-proton decay. In this model we take a realistic p-p interaction, but the width of the core-p interaction is taken from systematics and the depth is fitted so that the lowest state in this potential corresponds to the known (extrapolated) separation energy in the core-p subsystem. Varying components of the core-p interaction for different l values it is possible to vary the internal structure of the $2p$ emitter. The two-proton separation energy is adjusted using a short-range three-body force which is not large in the barrier

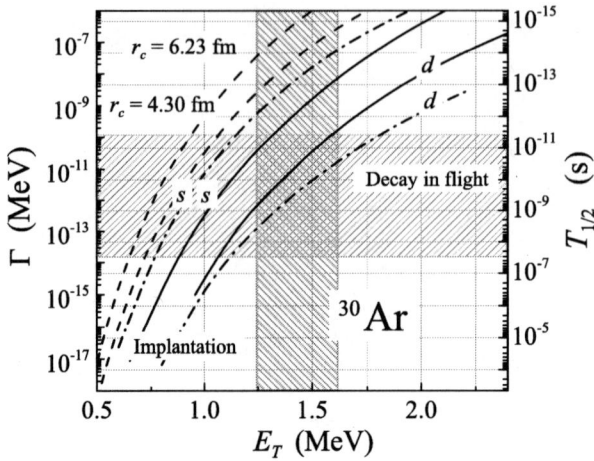

FIGURE 2. The ^{30}Ar lifetime as a function of the $2p$ decay energy [9] (see Fig. 1 caption for details).

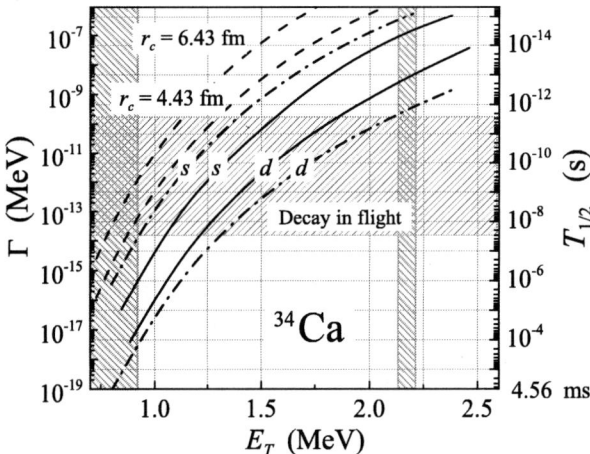

FIGURE 3. The ^{34}Ca lifetime as a function of $2p$ decay energy [9] (see Fig. 1 caption for details).

region and hence does not influence the penetration process. The model formulated in this way allows to work with poorly studied systems as it does not require much input. The model does not pretend to provide a realistic internal structure for the studied nuclei, but has enough freedom to simulate and study the effects of different dynamics. In the asymptotic region the model employs approximate boundary conditions of the three-body Coulomb problem [6] and thus it gives an opportunity to calculate consistently the penetration process.

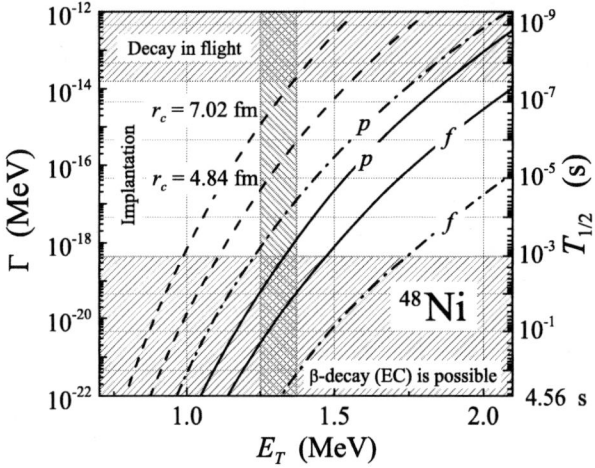

FIGURE 4. The ^{48}Ni lifetime as a function of the $2p$ decay energy (see Fig. 1 caption for details).

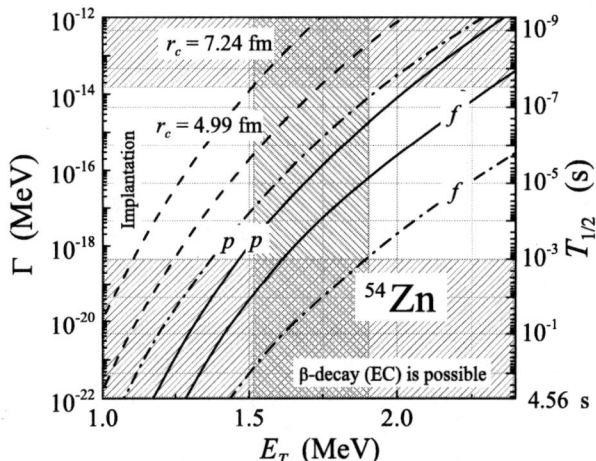

FIGURE 5. The ^{54}Zn lifetime as a function of the $2p$ decay energy (see Fig. 1 caption for details).

The lifetime dependencies on the decay energy calculated in the three-body, diproton, and simultaneous emission models (see [9] for more details) are given in Figs. 1–7. For many of these nuclei we can expect a strong (s/d or p/f) configuration mixing. The curves calculated in the assumption about total domination of one of the configurations (say s^2 of d^2) form for each nucleus a corridor in which the real lifetime curve should residue. The simultaneous emission model is also forming a corridor, but this corridor is much broader than that obtained in the three-body calculations. Finally, the diproton

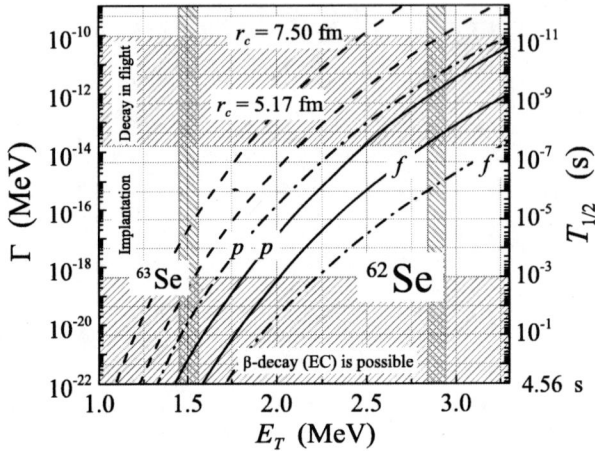

FIGURE 6. The ^{62}Se lifetime as a function of the $2p$ decay energy (see Fig. 1 caption for details).

model is providing the upper limit for the width, as shrinking of the three-body phase space to the two-body one should lead to an overestimation of the width. So, we have a sequence of models of an increasing sophistication, which presents (as we hope) a closer and closer approximation to the real situation.

We expect that ^{19}Mg, ^{30}Ar, and ^{34}Ca are characterized by a strong s/d mixing, ^{45}Fe by a significant p/f mixing (with the domination of the f^2 configuration), ^{48}Ni by the f^2 domination, ^{54}Zn and ^{62}Se by the p^2 domination. The decay of nuclei with the dominating f^2 configuration is dynamically the most complicated (and thus having the largest theoretical uncertainties) process. It suggest first of all subbarrier transition to the lower-l configurations p^2 or/and s^2, and only after that tunneling through the Coulomb barrier.

In the Figs. 1–7 the hatched areas represent lifetime regions, where different experimental techniques are required to study the $2p$ decays. The most conventional approach to study the lifetime is to implant the nucleus and to wait for a decay. For lifetimes larger than some milliseconds or hundreds of milliseconds the observation of the $2p$ decay should be suppressed by the β-decay (electron capture). For lifetimes shorter than hundreds of nanoseconds the implantation is impossible as the nuclei decay before they can be implanted. It was suggested in [12, 8] to use for the lifetime range from 5–10 ps to 10–100 ns the "decay-in-flight" technique. For this lifetime range the flight path is macroscopic. The reconstruction of fragment trajectories allows to recover the density of the decay vertexes along the trajectory and hence to deduce the lifetime. For lifetimes below some picoseconds the decay is happening directly in a target and no lifetime derivation is possible. Only when the width achieves some tens of keV, it can be defined by the invariant (missing) mass method.

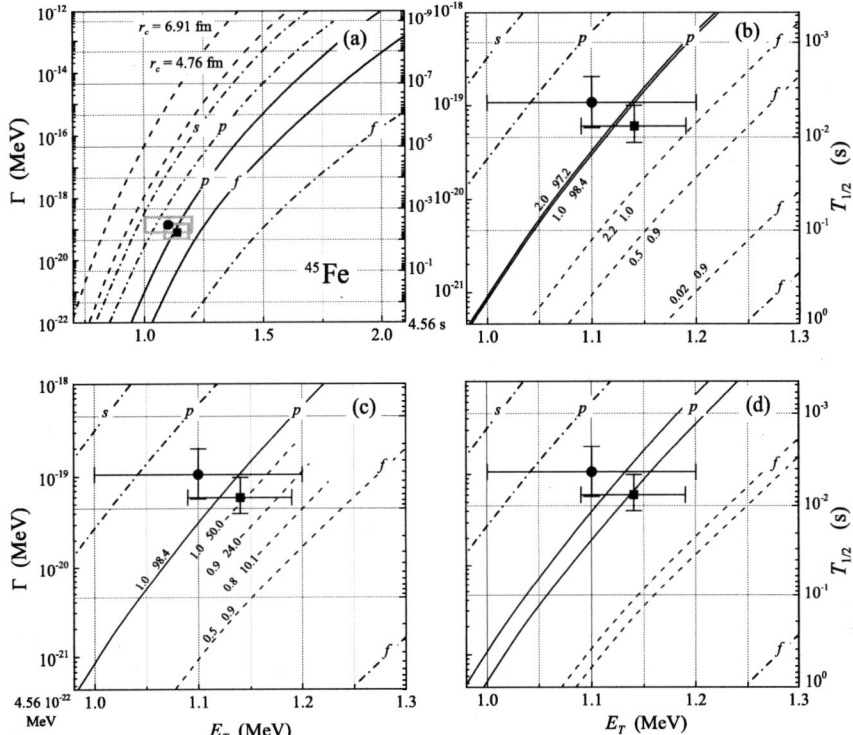

FIGURE 7. The ^{45}Fe lifetime as a function of the $2p$ decay energy. The meaning of the curves in (a) is described in the Fig. 1 caption. The sensitivity of the calculation results to different aspects of model is demonstrated in panels (b), (c), and (d). The numerical labels on the curves shows the weights of the s^2 and p^2 configurations in percent; the weight of the d^2 is ordinary negligible in these calculations, so the rest of the structure is the f^2 configuration.

Sensitivity of model on the example of ^{45}Fe

The largest uncertainty in the lifetime calculations is connected with an uncertainty of the $2p$ decay energy. 100 keV variation of the decay energy leads to 1–2 orders of magnitude variation in the lifetime. As we have seen above, the next important source of the uncertainty is the internal structure of the decaying system, which we consider in our model in an approximate way. It is important to understand the sensitivity of the calculations to these aspect of the physics. We provide the necessary illustration on the example of ^{45}Fe. This is useful for several reasons: (i) the ^{45}Fe is an important candidate for further (and more detailed) studies and (ii) the combination of factors (like, for example, non negligible p/f mixing) make this nucleus the least "predictable" of all those we have studied.

In Fig. 7a the results calculations [9] for the three-body, diproton and simultaneous emission models are shown on the same scale as in the previous plots. The circle and square show the experimental results from [10, 11]. In Figs. 7b–7d we give a more detailed analysis. In these panels the experimental data are corrected for *expected* (not yet measured!) 25% β-decay branch of ^{45}Fe with the corresponding uncertainty included in the upper width error bar. In panel (b) we show the sensitivity of the width to the variation of the s^2 configuration. While in the case of a pure p^2 structure of ^{45}Fe there is practically no sensitivity, the width in the case of a pure f^2 configuration is influenced strongly by the s^2 admixture. In panel (c) we demonstrate the dependence of the width on the p/f ratio. It is discussed below (see Fig. 10) that this ratio influence the definite momentum correlations.

One can see that within the current uncertainty of experimental data the three-body decay calculations are consistent with a broad range of possible ^{45}Fe structures, excluding only the extreme case of the f^2 configuration domination. The width curves corresponding to the realistic structure of ^{45}Fe (1–2% of s^2 and 10–15% of p^2 [13]) are currently passing closer to the error bars limits. It is, however, too early to draw some conclusions from this fact, for both theoretical (for example, the unknown exactly p/f ratio) and experimental (for example, the unknown β-decay branching) reasons.

In panel (d) we show the sensitivity of the lifetime to the variation of the core-p potential width parameter within $\pm 5\%$. The sensitivity to a possible deviation of the potential parameters from a systematic used in [9] is significant, but not sufficient to change qualitatively our previous conclusion.

CORRELATIONS

Three-particle decay is completely described by 9 parameters (say, components of three vectors). Three parameters of those describe center of mass translation. Three components corresponds to arbitrary rotation of the decay plane (three Euler angles). For narrow states the decay energy E_T can be considered as fixed. Finally, if we forget about spin degrees of freedom, the full correlation information for three-body decay can be described by two parameters. It is convenient to choose the energy distribution $\varepsilon = E_x/E_T$ between subsystems, and the angle $x = \cos\theta$ between Jacobi momenta vectors $\mathbf{k}_y$ and $\mathbf{k}_x$. Here $\mathbf{k}_y = (M_3(\mathbf{k}_1 + \mathbf{k}_2) - (M_1 + M_2)\mathbf{k}_3))/(M_1 + M_2 + M_3)$ and $\mathbf{k}_x = (M_2\mathbf{k}_1 - M_1\mathbf{k}_2)/(M_1 + M_2)$. In "T" Jacobi system particle number 3 is core; in "Y" system core is either particle 1 or 2. Total cm energy of the system is easily expressed in terms of Jacobi momenta: $E_T = E_x + E_y = k_x^2/2M_x + k_y^2/2M_y$, $M_x = M_1M_2/(M_1 + M_2)$, and $M_y = (M_1 + M_2)M_3/(M_1 + M_2 + M_3)$.

The full correlation picture is presented in Figs. 8 and 9 for selected cases of internal structure for ^{19}Mg and ^{45}Fe. Correlations for these cases have the following common features. (i) The particles are blown out from kinematical regions $\cos(\theta) = \pm 1$, $E_x/E_T \sim 0.5$ in the "T" system. This effect is more expressed in ^{45}Fe, where the Coulomb interaction is much stronger. These conditions correspond to the situation when the core and one proton are flying out with close velocity vectors. (ii) In "Y" Jacobi system the same physics is reflected in concentration of particles at $E_x/E_T \sim 0.5$. The distributions projected on the $\varepsilon = E_x/E_T$ axis form a narrow symmetric peak. These projected distributions are shown in Fig. 10 for a number of $2p$ emitters. It is easy to no-

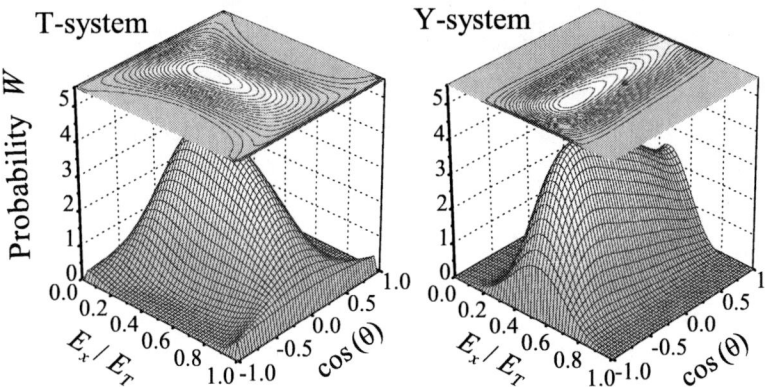

FIGURE 8. Correlations for the ^{19}Mg in "T" and "Y" Jacobi systems. Calculations from [8], case of a strong s/d mixing ($\sim 60\%$ of s-wave).

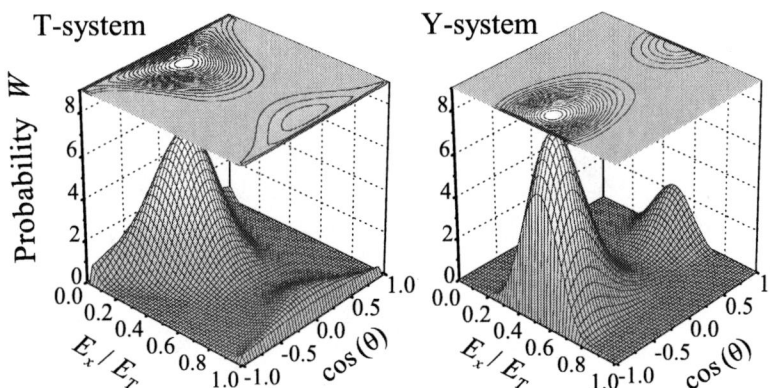

FIGURE 9. Correlations for the ^{45}Fe in "T" and "Y" Jacobi systems. Case of the f-wave domination ($\sim 75\%$ of f-wave).

tice that the heavier is the system, the narrower is this distribution. The width of this distribution is not changing with the charge monotonously because there is another factor, influencing the width of the distribution, namely the $2p$ decay energy E_T. For the shown distributions the E_T value varies from 0.7 MeV for ^{19}Mg to 2.8 MeV for ^{62}Se. The behaviour of these distributions is in an agreement with general prediction of Goldansky [1] that in the case of true two-proton emission the protons should have close energies.

The strong qualitative difference can be found between correlations in the decays of ^{19}Mg and ^{45}Fe in distributions over ε in the "T" system. This difference is also clearly reflected in the distributions over x in the "Y" system. We have found that this kind of correlation is stable in the s-d shell nuclei (for ^{12}O and ^{16}Ne see [7], and [8] for ^{19}Mg) with respect to a variation of the s/d ratio. When we change the s/d ratio drastically the variations in the distributions is rather quantitative than qualitative. This

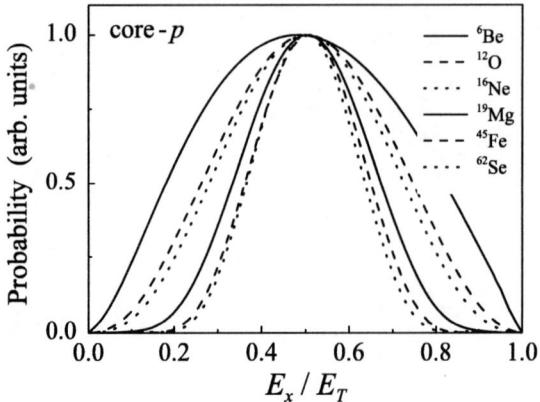

FIGURE 10. Energy distributions between the core and one of the protons for several nuclei.

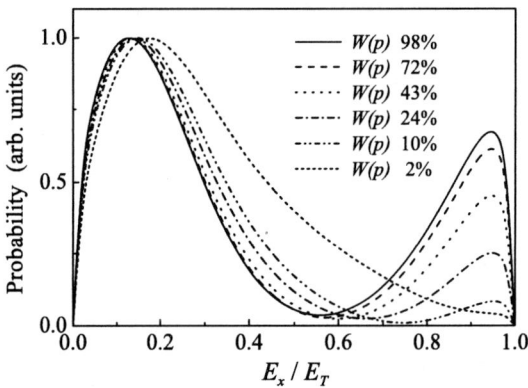

FIGURE 11. Energy distributions between two protons ("T" Jacobi system) in different assumptions about internal structure of ^{45}Fe. $W(p)$ is the weight of the p^2 configuration in the nuclear interior

appear not to be true for the p-f shell nuclei. The variation of the distribution over ε in the "T" system with variation of the p/f ratio for ^{45}Fe is given in Figure 11. In the case of the f^2 domination this distribution has a single broad peak shifted to low p-p energies (which should be connected with the p-p final state interaction). This tells that in the subbarrier region the protons first move from f^2 to s^2 configuration and only then tunnel. For that reason they "forget" about complicated correlations typical for f^2 component in the nuclear interior. In papers [5, 6] we have made a conclusion about "washing out" of the correlations typical to nuclear interior during the decay on the basis of studies of nuclei with s/d mixing and f-wave domination. It appears, however, that in the case of a considerable p/f mixing the dynamics changes abruptly. Now the protons are tunneling directly from the p^2 configuration with only a minor contribution of the s^2, which is reflected in a double-hump structure of the spectrum. Qualitatively similar double-humped spectrum has already been observed experimentally in the $2p$

decay of the p-wave nucleus ^{6}Be [14] and is well understood (see calculations in [15]). We can see in Fig. 11 how the energy distribution evolves for ^{45}Fe with the increase of the p^2 configuration, and, in Fig. 7c, how this evolution is correlated with changes in the lifetime. Interesting feature of the correlations is the fact that for very different energy correlations in the "T" system (the p-p distribution over x, Fig. 11), the core-p distribution in the "Y" system remain the same (see Fig. 10).

CONCLUSION

We have done a systematic studies of the prospective two-proton emitters: ^{19}Mg, ^{30}Ar, ^{34}Ca, ^{45}Fe, ^{48}Ni, ^{54}Zn, and ^{62}Se. The largest uncertainty of the predicted lifetimes is connected with uncertainty of the separation energies of the prospective $2p$ emitters. The next important source of the uncertainty is a structural uncertainty connected with the configuration mixing in the s-d shell and especially in the p-f shell nuclei. A detailed insight in the properties of the model is provided on the example of ^{45}Fe.

Some aspects of correlations among decay products are discussed. For ^{19}Mg and ^{45}Fe we also make detailed calculations of the possible correlations in the decay and consider their connections with other properties of the systems.

ACKNOWLEDGMENTS

We are grateful to K. Sümmerer for inspiration of this work and important ideas. The authors acknowledge the financial support from the Royal Swedish Academy of Science.

REFERENCES

1. V. I. Goldansky, Nucl. Phys. **19**, 482 (1960).
2. J. Jänecke, Nucl. Phys. **61**, 326 (1965).
3. A. I. Baz', V. I. Goldansky, V. Z. Goldberg, and Ya. B. Zeldovich, *Light and intermediate nuclei near the border of nuclear stability* ("Nauka", Moscow, 1972, in Russian).
4. W. Nazarewitz *et al.*, Phys. Rev. C **53**, 740 (1996).
5. L. V. Grigorenko, R. C. Johnson, I. G. Mukha, I. J. Thompson, and M. V. Zhukov, Phys. Rev. Lett. **85**, 22 (2000).
6. L. V. Grigorenko, R. C. Johnson, I. G. Mukha, I. J. Thompson, and M. V. Zhukov, Phys. Rev. C **64**, 054002 (2001).
7. L. V. Grigorenko, I. G. Mukha, I. J. Thompson, and M. V. Zhukov, Phys. Rev. Lett. **88**, 042502 (2002).
8. L. V. Grigorenko, I. G. Mukha, and M. V. Zhukov, Nucl. Phys. **A713**, 372 (2003).
9. L. V. Grigorenko, I. G. Mukha, and M. V. Zhukov, Nucl. Phys. **A714**, 425 (2003).
10. M. Pfützner, *et al.*, Eur. Phys. J. A **14**, 279 (2002).
11. J. Giovinazzo, *et al.*, Phys. Rev. Lett. **89**, 102501 (2002).
12. I. Mukha and G. Schrieder, Nucl. Phys. **A690**, 280c (2001).
13. A. Brown, privat communication.
14. O. V. Bochkarev *et al.*, Nucl. Phys. **A505**, 215 (1989).
15. L. V. Grigorenko, R. C. Johnson, I. G. Mukha, I. J. Thompson, and M. V. Zhukov, Eur. Phys. J. A **15**, 125 (2002).

ALPHA EMISSION

Decay of ^{114}Ba

C. Mazzocchi*, Z. Janas†, L. Batist**, V. Belleguic*, J. Döring*,
M. Gierlik†, M. Kapica*, R. Kirchner*, G.A. Lalazissis‡§, H. Mahmud¶,
E. Roeckl*, P. Ring§, K. Schmidt¶, P.J. Woods¶ and J. Żylicz†

*Gesellschaft für Schwerionenforschung mbH, 64291 Darmstadt, Germany
†Institute of Experimental Physics, University of Warsaw, 00-681 Warszawa, Poland
**St. Petersburg Nuclear Physics Institute, 188-350 Gatchina, Russia
‡Physics Department, Aristotele University of Thessaloniki, 54006 Thessaloniki, Greece
§Physics Department, Technical University Munich, 85748 Garching, Germany
¶Department of Physics and Astronomy, University of Edinburgh, Edinburgh EH9 3JZ, U.K.

Abstract. We measured for the first time the α decay of ^{114}Ba. The α decay of the daughter and granddaughter nuclei ^{110}Xe and ^{106}Te was also observed and the hitherto unknown half-life of ^{110}Xe was determined. The resulting α-decay Q-values and reduced widths as well as the Q-value for ^{12}C cluster decay of ^{114}Ba are discussed in comparison with theoretical predictions.

INTRODUCTION

Alpha-decay studies close to the doubly-magic nucleus ^{100}Sn are a rich source of nuclear structure information [1]. The existence of an α-decay island just beyond ^{100}Sn directly reflects the Z=50, N=50 shell-closure. The measurement of α-decay energies (E_α) directly yields Q_α values which are related to the N=Z=50 shell strength and allows one to test the prediction capability of mass models far from stability.

The nuclei close to ^{100}Sn have the neutrons and protons in identical $d_{5/2}$, $d_{3/2}$ and $g_{7/2}$ orbitals. This might give rise to the so-called "superallowed" α decay [2], whose main signature would be large reduced α-decay widths. Until now no convincing proof of the existence of this phenomenon was obtained. One experimental approach to find evidence for "superallowed" α decay is the study of α-decay chains above ^{100}Sn, the most interesting of these would certainly be the one starting at ^{112}Ba and ending in ^{100}Sn. Unfortunately, ^{112}Ba which has not been observed yet, is very difficult to produce and most probably lies beyond the proton drip-line [3, 4, 5]. Another interesting chain is that starting at ^{114}Ba and terminating in ^{102}Sn.

Moreover, theoretical predictions [6, 7] have suggested a new island of cluster radioactivity (phenomenon known to occur beyond ^{208}Pb) beyond the doubly-magic nucleus ^{100}Sn. The most promising candidate in this region was identified to be ^{12}C emission from ^{114}Ba [8, 9, 10].

The decay of ^{114}Ba was already investigated in previous experiments [11, 12, 13], but only lower limits for α and ^{12}C partial half-lives were established [12]. In this contribution we report on a reinvestigation of the α decay of ^{114}Ba [14] and on its relevance to ^{12}C emission.

CP681, *Proton-Emitting Nuclei: Second International Symposium; PROCON 2003*,
edited by E. Maglione and F. Soramel
© 2003 American Institute of Physics 0-7354-0150-0/03/$20.00

EXPERIMENTAL TECHNIQUE

The experiment was performed at the GSI On-line Mass Separator. The [114]Ba atoms were produced in fusion-evaporation reactions with a [58]Ni beam impinging on a [58]Ni target. The reaction products were stopped in a tantalum catcher inside a high-temperature cavity ion-source and released from the catcher as thermalized atoms. Chemical selectivity for barium isotopes with respect to the contaminants was achieved by using on-line fluorination, all contaminants including cesium being reduced to levels well below 10^{-5} [15]. The ions were then accelerated to 55 keV and mass separated. The resulting [114]Ba[19]F$^+$ beam was alternatively implanted in two thin carbon foils, each one viewed by a telescope. The telescope was composed of two silicon surface barrier detectors, one thin for energy-loss (ΔE) and one thick for residual-energy (E) measurements, see [14] for details. The total measuring time amounted to 56 hours.

RESULTS AND DISCUSSION

Since α particles with an energy up to 4.8 MeV are stopped in the ΔE detectors [14], they can be identified by recording the energy spectrum from the ΔE counter in anticoincidence with the E detector. In Figure 1, the section of the ΔE energy spectrum between 2500 and 5000 keV is displayed as obtained from both telescopes. The data were taken in anticoincidence with the related E detectors and corrected for background. Events below 2400 keV are assigned to low-energy β-delayed protons which are stopped in the ΔE detector (see [14] for details). The three peaks that can be identified in Figure 1 are assigned, in increasing energy order, to the α decay of [114]Ba and of the daughter and granddaughter nuclei [110]Xe and [106]Te, which are known α emitters [16, 17]. The resulting Q_α([114]Ba) value is (3540±40) keV.

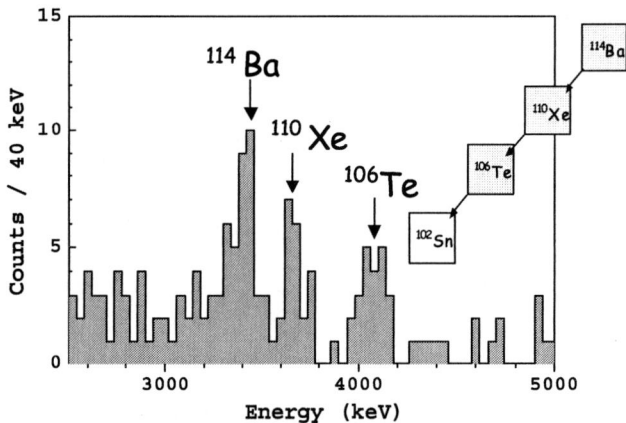

FIGURE 1. Section of the α-particle energy spectrum accumulated at mass 133=114+19. The peaks are assigned to the α decays of [114]Ba, [110]Xe and [106]Te, as indicated by arrows.

By summing the Q_α values of ^{114}Ba, ^{110}Xe [18] and ^{106}Te [17], and correcting for the binding energies of ^{12}C and of three α particles, it is possible to deduce an experimental Q value for ^{12}C emission (Q_{12C}) of (19.00 ± 0.04) MeV, which is just below the upper limit of 19.3 MeV from a previous empirical estimate [13]. This result is important when trying to obtain experimentally relevant predictions from cluster-emission calculations, as the partial half-life for ^{12}C emission depends dramatically on the adopted Q_{12C} value. This can be observed in Figure 2, where the partial half-life for ^{12}C emission is displayed as a function of Q_{12C}. The predictions from the most widely used theoretical models are reported and compared with the experimental data. As can be seen from Figure 2, the two models [10, 19] which agree with the experiment, predict for a Q_{12C} value of 19.00 MeV partial half-lives which are 4 to 7 orders of magnitude above the experimentally established limit ($1.2\cdot10^4$ [13]). This implies branching ratios for ^{12}C emission that are 4 to 7 orders of magnitude smaller than the experimental upper limit of $3.4\cdot10^{-5}$ [13], which illustrates the difficulty of a search for cluster emission of ^{114}Ba.

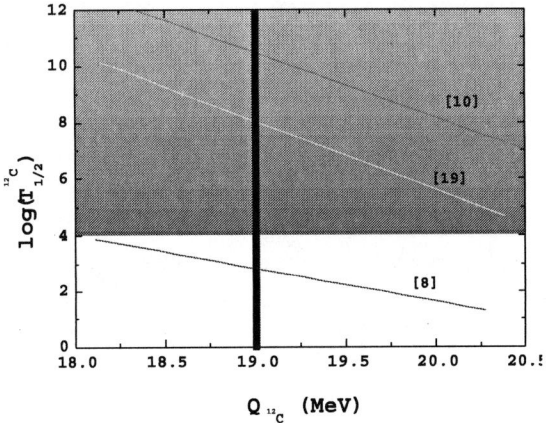

FIGURE 2. Partial half-life for ^{12}C emission ($T_{1/2}^{12C}$) as a function of Q_{12C}. The thick vertical line represents the experimental Q_{12C} with uncertainties, while the grey area indicates the "allowed" region above the lower experimental $T_{1/2}^{12C}$ limit. Theoretical predictions with the respective references are given as solid lines.

The unknown half-life of ^{110}Xe was determined on the basis of time correlations between ^{114}Ba and ^{110}Xe α-decay events. This yielded a value of $T_{1/2}=160^{+290}_{-60}$ ms. The intensities of the α lines in the ^{114}Ba$-^{110}$Xe$-^{106}$Te$-^{102}$Sn decay chain, corrected for the recoil escapes, allowed us to deduce the correspondent α-decay branching ratios. The experimental information on the nuclei which form the above-mentioned α-decay chain are summarized in Table 1.

By using the new E_α, b_α and half-life results, we determined the reduced α-decay widths for ^{114}Ba, ^{110}Xe and ^{106}Te relative to that of ^{212}Po (W_α), and thus extended the W_α systematics towards its low-mass end [1] (see Table 1). The reduced width of the ^{114}Ba α-decay of 16^{+12}_{-7} is somewhat larger than, but within the respective uncertainties consistent with the W_α values of $1.6^{+3.0}_{-1.0}$ and $5.1^{+2.6}_{-0.9}$ found for ^{110}Xe and ^{106}Te, respectively. Hence, the large uncertainties of the W_α values prevent us from drawing any firm con-

TABLE 1. Experimental α-decay properties of ^{114}Ba, ^{110}Xe and ^{106}Te. Except for the cases indicated by literature references, the data stem from this work.

Nucleus	$T_{1/2}$ (ms)	b_α (%)	E_α (keV)	Q_α (keV)	W_α
^{114}Ba	430^{+300}_{-150} [11]	0.9 ± 0.3	3410 ± 40	3540 ± 40	16^{+12}_{-7}
^{110}Xe	160^{+290}_{-60}	64 ± 35	3745 ± 14 [18]	3885 ± 14 [18]	$1.6^{+3.0}_{-1.0}$
^{106}Te	$0.060^{+0.030}_{-0.010}$ [16]	100	4128 ± 9 [17]	4290 ± 9 [17]	$5.1^{+2.6}_{-0.9}$

clusion concerning superallowed α decay.

SUMMARY AND OUTLOOK

By using the GSI On-line Mass Separator, we reinvestigated the decay of ^{114}Ba and searched for its unknown α-decay branch. We were able to identify for the first time the triple α-decay chain, linking the neutron-deficient isotopes ^{114}Ba and ^{102}Sn through ^{110}Xe and ^{106}Te. In addition to the α-decay energy and branching ratio of ^{114}Ba, the hitherto unknown half-life of ^{110}Xe was determined. The experimental Q_α and Q_{12C} values of ^{114}Ba allowed us to improve estimates of the cluster decay half-life of this isotope. The large uncertainties of the W_α values, however, prohibites any firm conclusion concerning superallowed α-decay. An experiment is planned to reinvestigate, under improved experimental conditions, all the decays along the ^{114}Ba$-^{110}$Xe$-^{106}$Te$-^{102}$Sn chain, which should allow us to clarify this uncertainty.

ACKNOWLEDGMENTS

The authors would like to acknowledge the valuable contributions of K. Burkard and W. Hüller to developing and operating the GSI On-Line Mass Separator. G.A.L. wishes to thank GSI for financial support. This work was supported in part by the Polish Committee of Scientific Research, in particular under Grant No. KBN 2 P03B 086 17, by the Programme for Scientific Technical Collaboration (WTZ) under Projects No. POL 99/009 and No. RUS 98/672, and by the European Community under Contracts No. ERBFMGECT950083 and No. HPRI-CT-1999-00001.

REFERENCES

1. E. Roeckl, Alpha decay, in *Nuclear Decay Modes*, ed. D.N. Poenaru, IoP Publishing, 1996, p. 237.
2. R.D. Macfarlane and A. Siivola, Phys. Rev. Lett. 14 (1965) 114.
3. G. Audi et al., Nucl. Phys. A 624 (1997) 1.
4. P. Möller et al., At. Data and Nucl. Data Tables 59 (1995) 185.
5. S. Liran and N. Zeldes, At. Data and Nucl. Data Tables 17 (1976) 431.
6. W. Greiner et al., in *Treatise on Heavy Ion Science*, ed. D.A. Bromley, Plenum, NY 1989, vol. 8, p. 641.
7. D.N. Poenaru et al., At. Data and Nucl. Data Tables 48 (1991) 231.

8. S.G. Kadmenski et al., Izv. Akad. Nauk. Rossii 57 (1993) 12.
9. R. Blendowske and H. Walliser, Phys. Rev. Lett. 61 (1988) 1930.
10. D.N. Poenaru et al., Phys. Rev. C 47 (1993) 2030.
11. Z. Janas et al., Nucl. Phys. A 627 (1997) 119.
12. A. Guglielmetti et al., Phys. Rev. C 52 (1995) 740.
13. A. Guglielmetti et al., Phys. Rev. C 56 (1997) R2912.
14. C. Mazzocchi et al., Phys. Lett. B 532 (2002) 29
15. R. Kirchner, Nucl. Instr. and Meth. Phys. Res. B 126 (1997) 135.
16. D. Schardt et al., Nucl. Phys. A 368 (1981) 153.
17. R.D. Page et al., Phys. Rev. C 49 (1994) 3312.
18. G. Audi and A.H. Wapstra, Nucl. Phys. A 565 (1993) 66.
19. S. Kumar and R.K. Gupta, Phys. Rev. C 49 (1994) 1922.

Evidence for the Identification of ^{178}Pb

J. C. Batchelder[1], K. S. Toth[2], M. W. Rowe[3], T. N. Ginter[3],
K. E. Gregorich[3], V. E. Ninov[3], F. Q. Guo[4], J. Powell[3],
X.-J.Xu[3,5], and Joseph Cerny[3,4]

1. UNIRIB, Oak Ridge Associated Universities, Oak Ridge, TN 37831, USA
2. Physics Division, Oak Ridge National Laboratory, Oak Ridge, TN 37831, USA
3. Nuclear Science Division, Lawrence Berkeley National Laboratory, Berkeley, CA 94720, USA
4. Department of Chemistry, University of California, Berkeley, CA 94720
5. Institute of Modern Physics, Lanzhou, China

Abstract. With the use of the Berkeley Gas-filled Separator, the α decay of ^{178}Pb was identified in ^{78}Kr irradiations of ^{102}Pd. Following their separation from the incident beam particles, reaction products were implanted in a position-sensitive silicon detector. Two events with E_α of 7602 and 7629 keV were observed and assigned to ^{178}Pb because they were position-correlated with known descendants in the ^{174}Hg α-decay chain. The half-life of ^{178}Pb was calculated with the maximum likelihood method to be 122^{+290}_{-50} μs. A mass excess of 3608(39) keV and a proton separation energy of 362(48) keV were deduced for ^{178}Pb based on the known masses of ^{174}Hg and ^{177}Tl.

INTRODUCTION

With the advent and further development of recoil separators, a large number of short-lived α-particle and proton emitters in the Z = 82 mass region have been identified and their decay properties studied. These investigations have yielded structure and mass information for nuclei near and beyond the proton dripline and, thus, have provided data to help distinguish among theoretical predictions of various nuclear models.

The most neutron-deficient lead nucleus known previously, ^{180}Pb, is located 24 mass units away from the lightest stable lead isotope. It was first identified [1] in ^{40}Ca bombardments of ^{144}Sm and its α-decay characteristics were confirmed [2] in a subsequent study wherein ^{90}Zr was irradiated with ^{92}Mo ions. The present investigation deals with our successful identification of the next proton-rich even-even lead isotope, ^{178}Pb.

Experimental

A 1.05-mg/cm^2 ^{102}Pd (enrichment of ~90%) foil was bombarded with 355-MeV ^{78}Kr ions accelerated in the Lawrence Berkeley National Laboratory 88-Inch Cyclotron. The beam energy at the mid-point of the target was calculated to be 340 MeV. The beam was pulsed so that decay spectra could be measured more cleanly

CP681, *Proton-Emitting Nuclei: Second International Symposium; PROCON 2003*,
edited by E. Maglione and F. Soramel
2003 American Institute of Physics 0-7354-0150-0

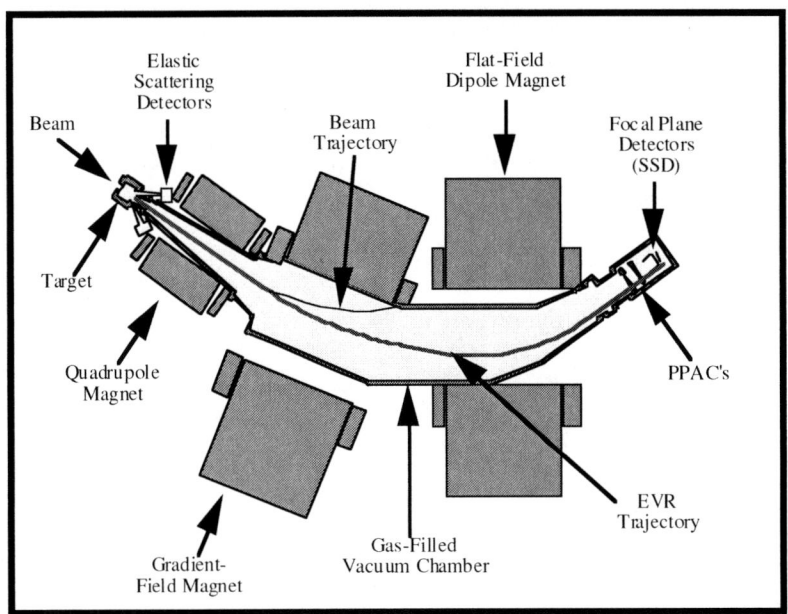

FIGURE 1. A schematic diagram showing the major components of the Berkeley Gas-Filled Separator.

during the beam-off period, though decays were also measured when the beam was on. The beam-on and beam-off duty cycle was 2 ms each.

Recoils were separated from the incident ions and transfer reaction products with the Berkeley Gas-filled Separator [3, 4] that was filled with helium gas at a pressure of 1.3 torr. A diagram of the separator is shown in Fig 1. In the focal plane region of the separator, the recoils passed through two parallel plate avalanche counters (PPACs) and were then implanted into a 16-strip 300 □m-thick silicon detector whose active area was 120-mm wide by 60-mm high. The strips were position-sensitive in the vertical direction with a resolution along a given strip of ~0.7 mm. The division of the silicon detector into 16 strips provided coarser position sensitivity along the horizontal direction. Two additional Si-strip detectors, identical in size to the focal plane detector, placed perpendicular to the focal plane detector along its upper and lower edges, were used to detect α particles escaping through the front of the detector. Behind this focal plane detector, was a another identical Si-strip detector. This permitted light, high-energy ions that were not stopped in the focal plane detector to be identified and rejected. These ions resulted from collisions between the beam and the helium atoms in the BGS. The Si-strip detector was calibrated with the energies of known α-emitters produced in the bombardment and its energy resolution for the observed α-decay peaks was ~35 keV (full-width-at-half-maximum).

Ions implanted in the Si-strip focal plane detector in coincidence with signals from the PPACs were identified as recoils, while those events anti-coincident with the PPACs were labeled as decay events. Recoils were differentiated from other beam related events via a time-of-flight measurement between the two PPACs and the Si strip detector. For each event in the detector, the time (from a continuously running clock), energy, and event type (recoil or decay) were recorded. Half-life information was determined from time differences between recoils and decays at the same detector position. Subsequent known α-decays could be correlated with the first decay event allowing for identification of the parent nucleus. The experiment ran for approximately 50 hours during which time the intensity of the ^{78}Kr^{18+} ions on target was ~35 pnA.

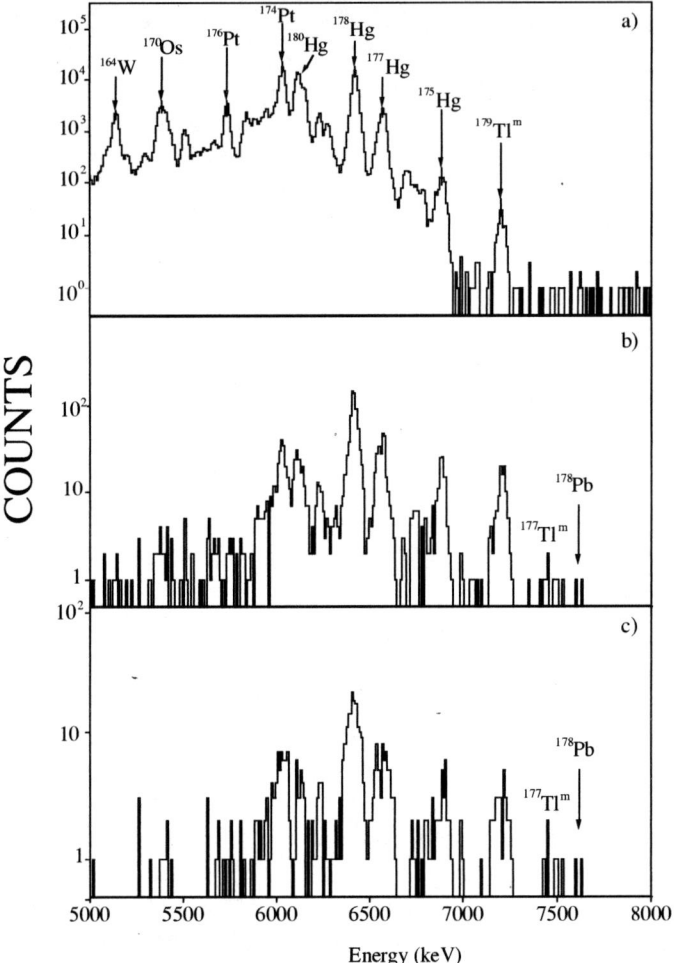

FIGURE 2. Data obtained in ^{78}Kr bombardments of ^{102}Pd. Part (a) shows the total spectrum recorded during intervals when the beam was not striking the target; parts (b) and (c) have restrictions that the decays occurred within 2 ms and 500 μs, respectively, of recoil implantation.

Results and Discussion

Figure 2(a) shows the total spectrum (from 5 MeV to 8 MeV) accumulated while the beam was off the target. Radioactivities with $Z \leq 80$ dominate the spectrum due to increased charged-particle emission from the extremely proton-rich compound nucleus ^{180}Pb. Above 7 MeV, one clearly sees 1.8-ms ^{179}Tl^m [5] whose $E_\alpha = 7201(20)$ keV. Figures 2(b) and 2(c) have an additional requirement set on the data, *i.e.*, the α particles had to have been recorded within 2 ms and 500 μs of recoil implantation, respectively. In addition to ^{179}Tl^m, we are now able to observe the recently identified 7487(13)-keV α particles of ^{177}Tl [6] and two events slightly above 7.6 MeV just where one expects to see ^{178}Pb α decays based on the extrapolation of known decay energies for ^{180,182,184}Pb [2].

Energies and times following recoil implantation for the two ^{178}Pb candidates are: 7602 keV, 202 μs; and, 7629 keV, 147 μs. Because they are position-correlated with known α emitters in the decay chain starting with ^{174}Hg [7,8], we assign them to the α decay of ^{178}Pb. The observed energy (in keV) and time correlations are as follows:

$$(1) \quad ^{178}\text{Pb} \; \frac{7602 \text{ keV}}{202 \text{ } \mu s} \rightarrow \; ^{174}\text{Hg} \; \frac{7065 \text{ keV}}{1.1 \text{ } ms} \rightarrow \; ^{170}\text{Pt} \; \frac{6570 \text{ keV}}{7.8 \text{ } ms} \rightarrow \; ^{166}\text{Os} \; \frac{escape}{not \text{ seen}} \rightarrow$$

$$^{162}\text{W} \; \frac{5506 \text{ keV}}{3.3 \text{ } s} \rightarrow \; ^{158}\text{Hf} \; \frac{5292 \text{ keV}}{0.53 \text{ } s} \rightarrow \; ^{154}\text{Yb},$$

and,

$$(2) \quad ^{178}\text{Pb} \; \frac{7629 \text{ keV}}{147 \text{ } \mu s} \rightarrow \; ^{174}\text{Hg} \; \frac{escape}{not \text{ seen}} \rightarrow \; ^{170}\text{Pt} \; \frac{6565 \text{ keV}}{86 \text{ } ms} \rightarrow escapes$$

The average of the two observed α-decay energies 7615(30) keV corresponds to a ^{178}Pb Q_α value of 7791(30) keV. When this energy is combined with the experimentally determined ^{174}Hg mass excess of 6607(34) keV [7], one arrives at a mass excess of 3609(45) keV for ^{178}Pb. For comparison, we note that predicted mass excesses of seven formulae listed in [9] range from 3.14 MeV to 5.21 MeV with the one of Möller *et al.*, 3.53 MeV, being closest to our experimental value; a newer set of calculations by Möller *et al.* [10] predicts a slightly different value, namely, 3.43 MeV. The ^{177}Tl mass excess has recently been determined by Poli *et al.* [6] to be -3318(28) keV. This mass excess and our value for ^{178}Pb show that the last proton in ^{178}Pb is bound by only 362(53) keV, indicating that ^{177}Pb probably lies beyond the proton drip line. In fact, ^{179}Pb may already be unbound since systematics [11] predict its proton binding energy to be 130(400) keV. A comparison between the experimental values of Q_α of ^{178}Pb, ^{180}Pb and ^{182}Pb deduced from this investigation and Ref. [2], and the predictions of various mass models and relations is shown in Table I.

By using the maximum likelihood method, the half-life of ^{178}Pb was calculated to be 122^{+290}_{-50} μs. With this half-life and the 7615-keV decay energy, an α-reduced width of 46^{+112}_{-20} keV was determined for ^{178}Pb using the formalism developed by

147

TABLE 1. Experimental Q_α (MeV) values compared with mass-model predictions

Mass model	^{178}Pb	^{180}Pb	^{182}Pb
M□ller - Nix (1988) [8]	7.59	7.17	6.78
M□ller, et al (1988) [8]	7.30	6.89	6.51
Comay-Kelson-Zidon (1988) [8]	8.52	8.25	7.60
Tachibana, et al. (1988) [8]	7.46	7.07	6.79
Janecke-Masson (1988) [8]	8.43	7.70	7.08
Masson-Janecke (1988) [8]	9.42	8.75	7.93
M□ller,et al. (1995) [9]	7.65	7.16	6.65
Experimental	**7.79**	**7.41 [2]**	**7.05 [2]**

Rasmussen [12]. The 46-keV width is close in value to those of neighboring even-even Pb nuclei [2]: ^{180}Pb (52±14), ^{182}Pb (61±7), and ^{184}Pb (60±6) keV. However, the large error bars do not allow us to extend the discussion in Ref. [2] concerning the magnitude of the Z = 82 gap for very neutron-deficient nuclei in the Pb mass region as indicated by α reduced width systematics.

In summary, we have evidence for the first identification of ^{178}Pb. Two decay chains were observed with a previously unknown alpha decay events correlated with the well-known [7,8] decay chain of ^{174}Hg. The two events had an E_α of 7615(30) keV and a half-life of $122 \, ^{+290}_{-50} \, \mu s$. More data on this nucleus is needed to reduce the error-bars on the half-life in order to make any conclusions regarding the magnitude of the Z=82 shell gap.

ACKNOWLEDGMENTS

We gratefully acknowledge the operations staff of the 88-Inch Cyclotron for providing the ^{78}Kr beam. Oak Ridge National Laboratory is managed by UT-Battelle, LLC under contract DE-AC05-00OR22725 with the U.S. Department of Energy. Nuclear physics research at the Lawrence Berkeley National Laboratory and at UNIRIB is supported by the U.S. Department of Energy through contracts Nos. DE-AC03-76SF00098 and FR-SV05-76OR00033, respectively.

REFERENCES

1. K. S. Toth, et. al., Z. Phys. A **355**, 225 (1996).
2. K. S. Toth, et. al., Phys. Rev. C **60**, 011302-1 (1999).
3. V. Ninov, K. E. Gregorich, and C. A. McGrath, in Proceedings of ENAM 98 Exotic Nuclei and Atomic Masses, Bellaire, Michigan, June 1998, edited by B. M. Sherrill, D. J. Morrissey, and C. N. Davids, AIP Conf. Proc. No. 455 (Woodbury, New York 1998), p. 704.
4. M. W. Rowe, et. al., in Proceedings of the International Symposium on Proton Emitting Nuclei, Oak Ridge, TN., Oct. 7-9, 1999, edited by J. C. Batchelder, AIP Conf. Proc. No. 518, (Melville, N.Y., 2000), p. 95.
5. K. S. Toth, et. al., Phys. Rev. C **58**, 1310 (1998).
6. G. L. Poli, et. al. , Phys. Rev. C **59**, R2979 (1999).
7. D. Seweryniak, et. al., Phys. Rev. C **60**, 031304-1 (1999).
8. J. Uusitalo, et al., Z. Phys. A **58**, 375 (1997).
9. P. E. Haustein, At. Data Nucl. Data Tables **39**, 289 (1988).
10. P. M□ller, J. R. Nix, W. D. Myers, and W. J. Swiatecki, At. Data Nucl. Data Tables **59**, 185 (1995).
11. G. Audi and A. H. Wapstra, Nucl. Phys. A **565**, 66 (1993).
12. J. O. Rasmussen, Phys. Rev. **113**, 1593 (1959).

Anisotropic α - Decay in Deformed Nuclei

D. S. Delion*, A. Insolia† and R. J. Liotta **

*Institute of Physics and Nuclear Engineering, Bucharest, Romania
†Dept. of Physics and Astronomy, Univ. of Catania and INFN, Catania, Italy
**KTH - Institute of Physics, Stockholm, Sweden

Abstract. A microscopic description of the alpha decay of the odd mass nuclei will be presented for axially deformed nuclei. Realistic mean field + pairing residual interaction in a very large single particle basis has been used. Systematics for At and Rn isotopes, as well as for ^{221}Fr, ^{241}Am and Pa - isotopes. It is found that the approach gives predictions in good agreement with experimental data for well deformed nuclei. New recent calculations for $^{241-243}Am$, $^{253-255}Es$ and $^{255-257}Fm$ isotopes will be discussed.

INTRODUCTION

It has been shown long time ago that in odd-mass actinides at very low temperature the alpha particles are emitted preferentially with respect to the direction of the total nuclear spin [1-5]. Recently, new experiments [6] have renewed the interest in this problem by reporting anisotropic emission in some At, near spherical isotopes, in connection with several theoretical descriptions of this effect.

Preferential emission of the alpha particles from deformed nuclei was first explained by Hill and Wheeler [7] and then by Bohr, Fröman and Mottelson [8] in terms of the penetration of the alpha cluster through the deformed Coulomb barrier. It was thus found that since for a prolate nucleus the barrier at the poles is thinner than at the equator, the probability to penetrate the barrier is larger along the nuclear symmetry axis. More recently, in order to explain observed anisotropies for almost spherical At isotopes, Berggren [9] proposed an alpha + core model. A quadrupole-quadrupole interaction acting between the already existing structureless alpha cluster and an odd-mass core was diagonalized in a weak coupling scheme. The strength of the interaction was adjusted to obtain the energy of the emitted alpha particle. Using this model several solutions with pronounced anisotropy were obtained [10]. No good comparison with the available data was obtained. Buck et al. [11] describe alpha decay from odd mass nuclei in a similar model, in which the depth of the alpha-core potential (taken as a square well), the alpha formation probability and the number of nodes in the radial wave function are fitted to the experimental data. Rowley et al. [12] followed the same philosophy, diagonalizing the quadrupole-quadrupole interaction in an extreme cluster model basis. Stewart et al. [13], using either semiclassical or coupled - channels transmission matrices without any formation mechanism, calculated also anisotropic α emission.

In a series of papers [14,15] we followed the traditional Hill and Wheeler line, but using a realistic deformed mean field with a large configuration space + pairing residual

CP681, *Proton-Emitting Nuclei: Second International Symposium; PROCON 2003*,
edited by E. Maglione and F. Soramel
© 2003 American Institute of Physics 0-7354-0150-0/03/$20.00

interaction in computing the preformation amplitude of the alpha cluster inside the nucleus. We estimated the penetration through the deformed Coulomb barrier within the framework of the WKB approximation. The anisotropy was explained mainly by the effect of the deformed barrier (see ref. [16] for an overview on the microscopic approach to the alpha decay problem).

The aim of the present talk is to give a short account about our work on the anisotropic alpha particle emission from odd-mass nuclei at low temperature. We will discuss some predictions of refs. [14,15] as well as some more recent calculations, yet unpublished, in connection with the experimental results obtained by Schuurmans et al. on anisotropy in At, Fr and Pa isotopes [17,18]. We will show that our predictions have been very well confirmed by the recent experimental findings in well deformed nuclei.

Finally, new experimental data on α-decay anisotropy should be soon available in the framework of a recently planned systematic investigation of the decay and fission in deformed nuclei [19]. In particular, new measurements have been planned for $^{241-243}$Am, $^{253-255}$Es and $^{255-257}$Fm isotopes. Calculations for such cases will be reported [20].

MICROSCOPIC DESCRIPTION OF THE ANISOTROPY

The mechanism describing the emission of the alpha particle used in refs. [14,15] consists in the classical two step process [21] : first the four nucleons cluster on a point at the nuclear surface with a given formation amplitude and afterwards this object penetrates the Coulomb barrier.

Let us consider the decay

$$B(I_i, K_i, M_i) \rightarrow A(I_f, K_f, M_f) + \alpha \tag{1}$$

where K_i, K_f and M_i, M_f are the projections of the initial and final total angular momenta in the intrinsic and laboratory frame, respectively [21].

We describe the mother and daughter nuclei within the BCS approximation. That is, the wave function of the nucleus X (A or B) is

$$|\phi^X\rangle = a_{k\Omega}^\dagger |(BCS)\rangle_\pi^X \otimes |(BCS)\rangle_\nu^X \tag{2}$$

where $\pi(\nu)$ label proton (neutron) degrees of freedom. The operator $a_{k\Omega}^\dagger$ is the creation operator of the unpaired nucleon with projection Ω on the intrinsic symmetry axis and k denotes the other quantum numbers. In the case of favoured transitions of odd mass nucleus the quantum numbers $k\Omega$ of the odd nucleon are unchanged during the decay.

The formation amplitude can be written as [14,15]

$$F(R, \vartheta, \varphi) = \sum_L f_{L\Omega}(R) Y_{L\Omega}(\vartheta, \varphi)$$

$$= \sum_L \int d\xi_\alpha d\xi_A [\phi_\alpha(\xi_\alpha)\phi^A(\xi_A)Y_L(R, \vartheta, \varphi)]_{J_B M_B}^* \phi^B(\xi_B) \tag{3}$$

where R is the distance between cluster and daughter nucleus and ξ are the internal coordinates. The intrinsic wave function of the alpha particle has a standard Gaussian form. Additional details on the evaluation of the multidimensional integral in eq. (3) can be found in ref.s [14,15].

Experimentally, nuclei are typically first produced, then separated, implanted on a foil of ferromagnetic material (cooled down to few $10^{-2}K$) and, eventually, oriented by applying a strong magnetic field. Anisotropy is thus measured with respect to the direction of the applied magnetic field [17,18].

If full alignment is not achieved in the orientation process of the implanted isotopes, the conditions are such that one has to average on the initial distribution of the angular momentum projections M_i. The total width is given by

$$\Gamma(\vartheta, \varphi) = \hbar v \left(\frac{R}{G_0(E,R)} \right)^2 \sum_l (F_l^2) W(\vartheta) \tag{4}$$

where F_l is the partial formation amplitude of the emitted alpha particle, i. e.

$$F_l = exp\left\{ -\frac{2l(l+1)}{\chi} \sqrt{\left(\frac{\chi}{kR} - 1 \right)} \right\}$$

$$\times \sum_{\Omega'} (-1)^{\Omega'} < I_i K_i l - \Omega' | I_f K_f > \sum_{l'} K^{\Omega'}_{ll'} f_{l'\Omega'}(R) \tag{5}$$

The matrix element $K_{ll'}$, as well as the quantities χ and $G_0(E,R)$ are defined as in ref. [14] (see also ref. [8] for additional detail). The microscopic formation amplitude enters into the calculation through the amplitude $f_{l'\Omega'}(R)$ in eq.(4).

The function W in eq. (4) determines the angular distribution of the emitted particle. After recoupling l and l' to the angular momentum L of the emitted alpha particle and assuming an axially symmetric nucleus, one gets

$$W(\vartheta) = \sum_L A_L P_L(cos\vartheta) \tag{6}$$

where the amplitudes A_L are given in terms of the F_l amplitudes of eq. (5) [15].

NUMERICAL CALCULATIONS AND DISCUSSION

We will apply the previous formalism to a few selected cases of anisotropic α decay from odd - even nuclei. In particular, we will discuss the Am, At, Fr, Pt, Es and Fm isotopes.

Application to the alpha-decay of ^{241}Am

We present in this section an application of the formalism developed above to the case of the favoured transition [15]

$$^{241}Am \rightarrow ^{237}Np + \alpha \tag{7}$$

for which $K_i = K_f = I_i$ and the Nilsson quantum numbers of both the mother and daughter nuclei are $\frac{5}{2}^-$ [523]. The diagonalization of the deformed Woods-Saxon potential is done using 18 major shells [22]. The deformation parameters were chosen as $\beta_2 = 0.22$, $\beta_3 = 0$ and $\beta_4 = 0.08$. This decay was investigated from the experimental point of view at the beginning of seventies [3] and it will serve in our future analysis as a reference case. The total width was computed according to presented formalism. We obtained $\Gamma_{th} = 2.09 \times 10^{-34}$ MeV which is quite close to the experimental value $\Gamma_{exp} = 3.34 \times 10^{-34}$ MeV. As for the case of even - even nuclei [15,16], the absolute alpha decay widths, for odd nuclei, are given within the right order of magnitude. For instance, in the case of $^{243}Am \rightarrow ^{239}Np + \alpha$, we obtained $\Gamma_{th} = 1.17 \times 10^{-33}$ MeV, to be compared with the experimental value $\Gamma_{exp} = 1.96 \times 10^{-33}$ MeV. However, the main goal of our analysis is to determine the influence of deformation on the angular distributions of the emitted alpha particles.

The model predicts a large enhancement of the anisotropy versus quadrupole deformation. The role of deformations with multipolarities higher than the quadrupole one is much less important [15]. The influence of the intrinsic structure of the mother and daughter wave functions can be estimated by studying the dependence of the angular distribution as a function of the angular momentum transfer L. We found that including only the $L = 0$ component, the total width Γ versus θ shows a variation which is 10% smaller with respect to the case in which all L components are included. This result puts in evidence the important role of the barrier deformation. As a matter of fact, the $L = 0$ part of the formation amplitude is isotropic and in such a case the calculated anisotropy has to be ascribed entirely to the barrier. A similar feature was found in even - even axially deformed nuclei [14]. Although higher L contributions seem to give rise to a small effect, one should not forget that without the inclusion of deformations and without a large basis included in the calculation of the single particle states, one would fail to reproduce the absolute value of the width by many orders of magnitude [16].

The function $W(\vartheta)$ is the relevant quantity regarding the anisotropy in alpha decay processes. Once the microscopic formation amplitude is calculated one can easily expand the function W in terms of even order Legendre polynomials, as shown in eq. (6)

We found that in the case in which all nuclei are assumed to be aligned with the maximum projection of the total angular momentum in the laboratory frame, M_i, the

TABLE 1. Function $W(\vartheta)$ at $\vartheta = 0$ and $\vartheta = \pi/2$ compared with the measured anisotropy for the favored transition $^{241}Am \rightarrow ^{237}Np + \alpha(\Omega^\pi = 5/2^-)$

ϑ	W_{th}	W_{exp}
0	1.500	1.610
$\pi/2$	0.736	0.714

coefficients A_L have all the same phase [15].

Applying the reduction procedure of ref. [3], the calculated W values can be compared with the experimental data [3] at $1/T = 90.5^0 K^{-1}$. We did this and obtained the results presented in table 1.

The agreement can be considered remarkable, specially if one considers that the absolute normalization is given by the formation amplitude entering in the evaluation of the A_L coefficients.

At - isotopes

With the basis and the residual interaction previously discussed we proceed to evaluate the absolute decay widths and the W-coefficients as a function of the deformation parameters for At isotopes. An important motivation for this study is the comparison with recent experimental data taken at ISOLDE (CERN) [17].

For many At - isotopes (starting from odd-proton nucleus ^{207}At with $I_i = I_f = \frac{9}{2}^-$ we have calculated the dependence on the deformation parameter β_2 of the A_l amplitudes as well as of the anisotropy of the decay. We have found within our model [15] that the amplitudes A_l (as well as the $W\vartheta$ coefficients) are almost similar for the different isotopes. The only relevant parameter is the nuclear deformation.

We have already reported [15] that both the A_l amplitudes and the anisotropy are strongly dependent on the deformation. The results seem to suggest that the anisotropy increases as the prolate deformation increases.

This is a possible key to read the experimental data [17]. We agree that this interpretation brings into the problem the possibility that the deformation is increasing approaching ^{211}At. Some criticism has been raised against this interpretation [17]. Anyway the data of ref.[17] suggest (in term of the model) a sharp change of nuclear properties (deformation from prolate to oblate case, for instance) to justify the dramatic decrease of the measured anisotropic ratio [17] down to values smaller than 1.

It is worthwhile to mention that the model has no free parameter and that the anisotropic ratio is a strongly dependent on the deformation. See, for instance the results in ref.s [15], for the almost spherical Rn - isotopes, with appreciable differences between cases like ^{207}Rn and the nucleus ^{207}At, in which the mass number is the same and any difference should be attributed only to different properties of the odd neutron and proton orbital entering in the problem. As general comment, we can say that the coefficient A_2 has positive values (in phase with $A_0 = 1$) for the prolate deformations and negative val-

TABLE 2. Deformations, experimental and computed coefficients A_2, for the $^{221}_{87}Fr$ and the indicated Pa isotopes (adapted from ref. [18]).

	β_2	β_3	A_2^{th}	A_2^{exp}
$^{221}_{87}Fr$	-0.069	0.0	-0.215	-0.375
	0.120	0.15	-0.373	
$^{227}_{91}Pa$	0.168	0.0	0.649	0.696
	0.168	0.1	0.748	
$^{229}_{91}Pa$	0.185	0.0	0.733	1.13
	0.185	0.08	0.808	

ues (opposite phase) for oblate ones. The other coefficients A_L with $L \neq 2$ are virtually negligible. The values of A_4 are one order of magnitude smaller than A_2. In spite of this, it is interesting to note that A_4 is positive and symmetric with respect to the deformation parameter β_2. A similar qualitative and even quantitative behaviour is found for the other At isotopes. Actually even for the odd-neutron case of ^{207}Rn ($I_i = I_f = \frac{5}{2}$) and the other Rn isotopes all the features discussed above are essentially the same.

The case of ^{221}Fr and Pa - isotopes

It is of great interest to refer the most recent results by Schuurmans et al. [17] for well deformed nuclei. For the ^{221}Fr, a ground state $K = 1/2$ was assumed with a prolate deformation [17], while a previous calculation [14,15] used an oblate ground state with $K = I = 5/2$ (referred in the first line of Table 2 as a theoretical predictions for the A_2^{th}).

For the Pa isotopes a prolate ground state was taken for the favoured $5/2^- \rightarrow 5/2^-$ ($^{227}_{91}Pa \rightarrow \alpha +^{223}_{89}Ac$) and $5/2^+ \rightarrow 5/2^+$ ($^{229}_{91}Pa \rightarrow \alpha +^{225}_{89}Ac$) transitions.

The results are reported in the Table 2. The agreement between the microscopic model and the experimental data is excellent. The calculation for the total widths gives $\Gamma = 0.64 \times 10^{-25}$ MeV and $\Gamma = 0.29 \times 10^{-28}$ MeV the $^{227}_{91}Pa$ and $^{229}_{91}Pa$, respectively. We found that the $L = 0$ part in the formation amplitude is the largest contribution. The neglect of the higher multipoles in the formation amplitude produces anisotropy only slightly smaller ($10 - 20\%$ in comparison with the case in which all multipolarities are included. This shows that the main role is played by the penetration through the deformed barrier. The effect is therefore expected to be larger for very well deformed nuclei.

The case of $^{241-243}Am$, $^{253-255}Es$ and $^{255-257}Fm$ isotopes.

Finally, we have applied this formalism to a few additional selected cases of anisotropic α-decay from axially deformed odd - even nuclei. In particular, we will discuss the $^{241-243}Am$, $^{253-255}Es$, $^{255-257}Fm$ isotopes [19,20]. Within the improved version of the model, we have considered here the case of ^{241}Am as a reference point.

TABLE 3. Theoretical predictions for the α anisotropy for the indicated isotopes. In the last column we report the ratio between the calculated width and the experimental value for a given isotopes.

^{A}X	A_2	A_4	$W(0^{o})$	$W(90^{o})$	Γ_{th}/Γ_{exp}
^{241}Am	1.083	0.059	2.143	0.481	0.26
^{243}Am	1.092	0.060	2.153	0.476	0.28
^{253}Es	1.467	0.180	2.647	0.334	0.81
^{255}Es	1.437	0.172	2.609	0.346	0.51
^{255}Fm	1.425	0.169	2.595	0.351	0.85
^{257}Fm	1.680	0.296	2.976	0.271	0.74

We think this will be quite useful from the theoretical and the experimental point of view as soon as the new data will become available in order to provide a sort of cross check on the model as well as on the experiment.

The new single particle basis with two harmonic oscillator (ho) parameters, initially applied to favoured α decay of even-even spherical and deformed nuclei in ref. [23], has been used for this calculation. The first part of the basis describes spectroscopic properties, while the second one is connected with the α-clustering properties, as it was recently shown in ref. [23]. The new basis was tested in a systematic calculation of the total α-decay widths for many nuclei, ranging from spherical to well deformed systems. A very good agreement with the experimental data was obtained [23].

For the nuclei we analyzed here, we used the same parametrisation as in ref. [23], i.e. the universal mean field diagonalized in a basis with two ho parameters, $\lambda = m\omega/\hbar$ and 0.7λ. The number of single particle states needed to reproduce the order of magnitude of the decay width turns out to be equal to 50. It is important to remark that the method of ref. [23] uses a spherical mean field to calculate the formation amplitude, i.e. only the f_{00} term is left in eq. (5).

Therefore, within the present model calculation, any anisotropy has to be ascribed entirely to the barrier penetration. Indeed, as it was shown in ref. [15], the amplitudes $f_{l'\Omega'}$ with $l' \neq 0$ contribute less than 10% to the total decay width. On the other end a large configuration mixing together with a realistic mean field are essential ingredients to get not only the absolute value of the total α width but also the correct values for the A_L amplitudes of eq. (6). This is the main issue in our approach. The f_{00} formation amplitude provides the correct absolute normalization within the present model without any *ad hoc* adjustable parameter.

The deformation parameters used in the present calculations were taken from the proposal [19] as well as the K^{π} quantum numbers of the considered odd nuclei [20]. Again, in our calculations we considered favoured decays with $I_i = I_f = K_i = K_f$ and, therefore, in eq. (5) only a single term with $\Omega' = 0$ is left.

Finally, it turns out, actually, that A_L amplitudes in eq. (6) practically do not depend upon the radius R [22]. We remind that $A_0 = 1$.

The A_L amplitudes and the resulting anisotropy $W(\vartheta)$ [see eq. (6)] are reported in Table 3, for the six nuclei considered in this paragraph [22]. When comparing the results of Table 3 with the future experimental data one should keep in mind that the calculations do assume angles equal to 0° and 90° as detection angles and that all nuclei are assumed fully aligned with the maximum projection of the total angular momentum in the laboratory frame. Additional experimental correction factors should be also in order for a proper comparison with the measurements.

The results show anyway a large anisotropy with a preferred polar emission, as expected for such large quadrupole deformation in axially symmetric prolate nuclei.

The expected anisotropy shows also a measurable dependence on the overall deformation parameters relevant for the axially deformed ground state description.

This is the case of the Am isotopes where an important additional contribution could arise from hexadecapole as well as from higher multipolarities [25]. Taking into account β_4 and β_6 deformations, according to ref. [25], we got $W(0^o) = 2.415, W(90^o) = 0.379$ for ^{241}Am and $W(0^o) = 2.361, W(90^o) = 0.399$ for ^{243}Am.

Thus, for the extreme case of the unfiltered $W(0^o)/W(90^o)$ ratio, if one compares with the results in Table 3, one gets anisotropies about 25% larger. The effect of hexadecapole deformation is therefore not negligible and it could result in a measured anisotropy comparable with that one expected for nuclei with larger intrinsic spin projection K.

In the last column one can see that the total decay width (integrated on angles) reproduces the experimental value within a factor of three (or four) in the worst cases ($^{241-243}Am$) and, actually, much better for all the other considered cases.

CONCLUDING REMARKS

In conclusion we have presented a realistic microscopic approach for the calculation of the formation amplitude for the alpha decay problem in axially symmetric deformed nuclei, within the well known approach by Mang and Rasmussen[17]. The main ingredients the use of a realistic deformed mean field, the large shell model space and the exact diagonalization of the deformed mean field. The predicted absolute values of the total alpha widths are reproduced within $10 - 30\%$. [16].

In the model we performed a systematic microscopic calculation of quantities related to alpha particle emission from oriented odd-mass nuclei and, in particular, for $_{85}At$, $_{86}Rn$, $_{87}Fr$, $_{91}Pa$ as well as Am and Es isotopes. The probability of emitting an alpha particle in the polar direction (or in the equatorial direction) is strongly dependent on the emission angle. For prolate (oblate) deformations polar (equatorial) emission is preferred. We also found that deformations higher than quadrupole can play an important role in some cases. Even in the region of near spherical nuclei the anisotropy was found to be measurable. In addition, we have found that the main role in the observed anisotropy is due to the deformed barrier penetration. This has been recently confirmed experimentally in the case of well deformed nuclei [18].

We emphasized in this study the importance of anisotropies in alpha decay processes as a tool to extract intrinsic deformation parameters in nuclei.

New experimental data should be soon available on the anisotropic decay as a systematic investigation of the decay and fission in deformed nuclei has been recently planned [19]. Calculations have been performed to provide theoretical predictions within the model for those cases [20].

REFERENCES

1. S.H.Hanauer, J.W.T.Dabbs, L.D.Roberts and G. W. Parker, Phys. Rev. **124** (1961) 1512;
    Q.O.Navarro, J. O Rasmussen ans D.A.Shirley, Phys. Lett. **2** (1962)353
2. A.J.Soinski, R.B.Frankel, Q.O.Navarro and D.D.Shirley, Phys. Rev. **C2** (1970) 2379
3. A.J.Soinski and D.D.Shirley, Phys. Rev. **C10** (1974) 1488
4. D. Vandeplassche, E. van Walle, C. Nuytten and L.Vanneste, Phys. Rev. Lett. **49** (1982) 1390
5. F. A. Dilmanian et al., Phys. Rev. Lett. **49** (1982) 1909
6. J. Wouters et al., Phys. Rev. Lett. **56** (1986) 1901;
    J. Wouters et al., Nucl. Instr. and Meth. **B26** (1987) 463;
    N. G. Nicolis et al., Phys. Rev. **C41** (1990) 2118
7. D.L. Hill and J. D. Wheeler, Phys. Rev. **89** (1953) 1102
8. P. O. Fröman, Mat. Fys. Skr. Dan. Vid. Selsk. **1** (1957) no. 3;
    A. Bohr, P. O. Fröman abd B. Mottelson, Dan. Mat. Fys. Medd. **29** (1955) no. 10
9. T. Berggren, Phys. Lett. **197B** (1987) 1;
    T. Berggren, Hyperfine Interactions **43** (1988) 407
10. T. Berggren, Phys. Rev. **C50** (1994) 2494
11. B. Buck, A.C. Merchant and S.M. Perez, J. Phys. **G18** (1992) 143
12. N. Rowley, G.D. Jones and M.W. Kermode, J. Phys. **G18** (1992) 165
13. T.L. Stewart et al., Phys. Rev. Lett. **77** (1996) 36; T.L. Stewart et al., Nucl. Phys. **A611** (1996) 332
14. A. Insolia, P. Curutchet, R. J. Liotta and D. S. Delion, Phys. Rev. **C44** (1991) 545;
    D.S.Delion, A.Insolia and R.J.Liotta, Phys. Rev **C46** (1992) 1346
15. D.S.Delion, A.Insolia and R.J.Liotta, Phys. Rev **C46** (1992) 884;
    D.S.Delion, A.Insolia and R.J.Liotta, Phys. Rev **C49** (1994) 3024
16. R. G. Lovas, R. J. Liotta, A. Insolia, K. Varga and D. S. Delion, Phys. Rep. **294** (1998) 265
17. P. Schuurmans et. al., (Nicole and Isolde Collaboration), Phys. Rev. Lett. **77** (1996) 4720
18. P. Schuurmans et. al., (Nicole and Isolde Collaboration), Phys. Rev. Lett. **82** (1999) 4787
19. *A study of spin depending phenomena in alpha decay and spontaneous fission of heavy actinide isotopes*; INTAS project "OPEN CALL 2000-0195"; and N. Severijns, private communication
20. D.S.Delion, A.Insolia and R.J.Liotta, submitted to Physical Review C
21. R.G. Thomas, Progr.Theor.Phys. **12** (1954) 253
    H. J. Mang, Ann. Rev. Nucl. Sci. **14** (1964) 1
22. R. Bengtsson et al., Phys. Scrip. **39** (1989) 196;
    S. Cwiok et al., Comp. Phys. Comm. **46** (1987) 379
23. D.S.Delion, A.Insolia and R.J.Liotta, Phys. Rev. **C 54** (1996) 292
24. D.S.Delion and A. Sandulescu, J. Phys. **G 28** (2002) 617
25. P. Moeller et al., At. Data and Nucl. Data Tab. **59** (1995) 185

BETA DELAYED PROTON EMISSION AND GAMMA SPECTROSCOPY

Rotational Bands in Rare-Earth Proton Emitters and Neighboring Nuclei

D. Seweryniak

Argonne National Laboratory, Argonne, IL 60439

Abstract. The anomalous proton-decay rates in ^{131}Eu and ^{141}Ho have been explained by the presence of large quadrupole deformation. Consistent with this hypothesis was the discovery of the decay branch in ^{131}Eu to the 2^+ state in the daughter nucleus. Direct evidence that ^{141}Ho and ^{131}Eu are deformed came from the observation of rotational bands. From the dynamic moment of inertia the deformation of β_2=0.25(4) was deduced for the ^{141}Ho ground state. The large signature splitting in the ground-state band in ^{141}Ho indicates that Coriolis mixing plays an important role in this nucleus. The comparison between the Particle-Rotor Model and the data indicates that ^{141}Ho has significant hexadecapole deformation and/or might be triaxial. The comparison between theory and experiment constrains other parameters used in the proton decay rate calculations from deformed nuclei such as the Coriolis interaction and pairing strength.

INTRODUCTION

Proton emission from deformed nuclei has been one of the focal points of proton-decay studies since the discovery of the highly deformed proton emitters ^{131}Eu and ^{141}Ho [1]. Properties of excited states in deformed rare-earth nuclei ^{147}Tm, the first proton emitter studied in-beam [2], ^{141}Ho [3], and ^{131}Eu allowed to extract the deformation and provided support for the single-particle configuration assignments for the proton emitting states. In the following section the rotational landscape of proton-rich rare-earth nuclei will be discussed. Subsequently, the in-beam results obtained for the deformed rare-earth proton emitters will be reviewed, followed by a discussion of Particle-Rotor model calculations.

PROPERTIES OF ROTATIONAL BANDS IN PROTON-RICH RARE-EARTH NUCLEI

All elements between Z=51 and Z=83 but two have known proton emitters. Just above the Z=50 shell proton emitters are spherical or weakly deformed. For heavier nuclei the proton drip line enters the island of large prolate deformation in the middle of the Z=50-82 and N=50-82 shells. When approaching the N=82 shell calculated shapes change rapidly to oblate and then to spherical close to the N=82 line. The following discussion will be limited to the transitional nuclei with 63≤Z≤69.

CP681, *Proton-Emitting Nuclei: Second International Symposium; PROCON 2003*,
edited by E. Maglione and F. Soramel
© 2003 American Institute of Physics 0-7354-0150-0/03/$20.00

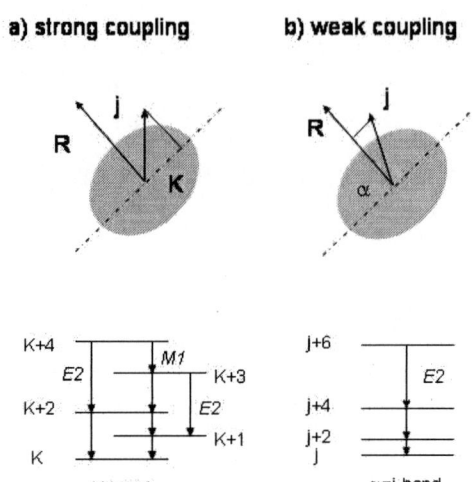

FIGURE 1. Two extreme cases of rotational bands based on (j,K) orbital in an odd-A nucleus: a) strongly coupled band and b) decoupled band.

The single-particle orbitals $3/2^+[411]$, $5/2^+[413]$, $7/2^-[523]$, $1/2^+[411]$ are predicted to constitute the ground state in Eu, Tb, Ho and Tm istopes, respectively, at deformations between $\beta_2=0.25$ and $\beta_2=0.30$. Below $\beta_2=0.25$ in the case of the $h_{11/2}$ proton orbital Coriolis effects become important and the strong-coupling picture cannot be applied. It is worth noting, that according to the microscopic-macroscopic calculations [4] the deformation changes from $\beta_2=0.25$ in ^{145}Tm to $\beta_2=-0.19$ in ^{147}Tm.

The Hamiltonian of an odd nucleon coupled to a rotor is composed of 3 major components: the kinetic energy of the rotor, the interaction of the odd particle with the deformed core, as described by the Nilsson model, and the Coriolis term which accounts for the coupling of the motions of the odd particle and the rotating core. In the limit when the Coriolis interaction is negligible compared to the interaction with the core, the projection of the single-particle angular momentum j on the symmetry axis, K, is a good quantum number. In this case a so-called strongly coupled band is built on the single-particle band head. The band consists of two E2 bands, which are referred to as signature partners, connected via M1 transitions (see Fig. 1a). The Coriolis interaction mixes states with K values which differ by one. All diagonal Coriolis matrix elements vanish except for K=1/2 bands. As a result the two signature partners in K=1/2

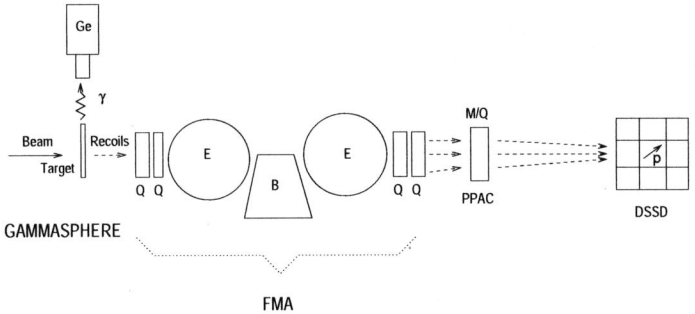

FIGURE 2. The setup for Recoil-Decay Tagging experiments at Argonne National Laboratory.

bands are shifted in energy leading to the signature splitting. The signature splitting is promoted to K>1/2 bands through mixing with K=1/2 bands. In the extreme case when Coriolis mixing dominates the Hamiltonian, a new coupling scheme emerges, the so-called weak coupling or rotational scheme. In this case, the projection of the particle angular momentum on the rotational axis, α, is conserved instead of K. The state with angular momentum I=j is pushed down in energy and becomes the band head. The decoupled E2 band is built on top of this state. The energies of states in this band are similar to the energies of the ground-state band of the even-even core (see Fig. 1b).

The strong coupling scheme is favorable when the deformation is large, j is small, and K is large. Weak coupling is realized in nuclei with small deformation for bands built on large j, small K single-particle states. In nuclei one very often encounters an intermediate situation. Many moderately deformed proton rich odd-Z nuclei with Z>50 and N<82 exhibit a strongly populated decoupled band based on the $h_{11/2}$ proton orbital.

IN-BEAM SPECTROSCOPY OF THE DEFORMED RARE-EARTH PROTON EMITTERS AND NEIGHBORING NUCLEI

Studies of excited states in deformed proton emitters require an efficient and very selective detection method. The most exotic proton emitters are produced with a cross section of the order of 100 nb. The total reaction cross section is about 500 mb. Assuming 20 γ rays per reaction, one γ ray out of 10^8 has to be selected to study excited states in the most exotic proton emitters.

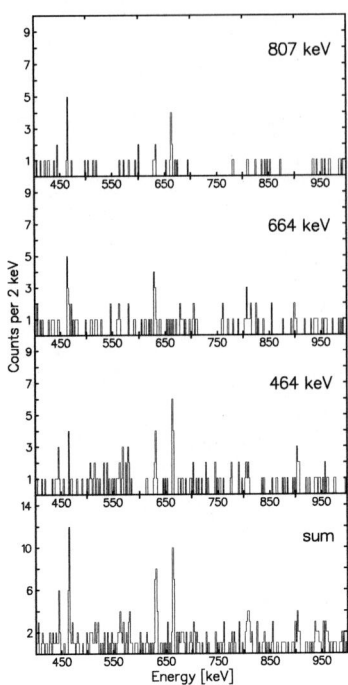

FIGURE 3. Proton tagged ^{147}Tm γ-ray spectra in coincidence with 464-keV, 664-keV and 807-keV transitions. The bottom panel contains the sum of the three gates.

Recoil-Decay Tagging

Recoil-Decay Tagging is the method of choice for studies of excited states in proton emitters. The method was used for the first time in connection with an array of Ge detectors in Ref. [5]. The experimental method and the setup used at Argonne National Laboratory is shown in Fig. 2. Proton emitters are produced using heavy-ion fusion-evaporation reactions. Prompt γ rays are detected in the GAMMASPHERE array of Compton suppressed Ge detectors, which consists of about 100 70% HPGe detectors, with a total efficiency of about 10% at 1.3 MeV. Reaction products are separated from beam particles and dispersed according to their mass-to-charge-state ratio in the Argonne Fragment Mass Analyzer (FMA). After passing through a focal plane detector, where A/Q is measured, the recoiling nuclei are implanted into a Double-Sided Si Strip Detector (DSSD) where they subsequently decay. The currently used DSSD is divided into 80 front strips and 80 back strips equivalent to 6400 pixels. Both the implantation and the decay take place in the same pixel. By using spatial and time information, decay particles and implants are correlated with each other. As a result, prompt γ rays can be tagged by the observed decays.

164

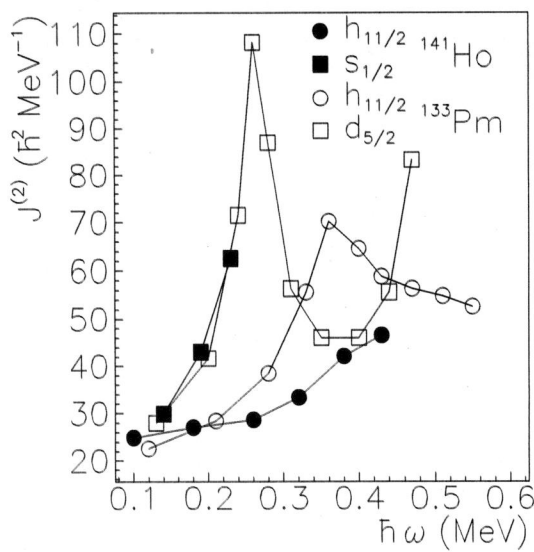

FIGURE 4. The dynamic moment of inertia deduced for the ground-state band and the isomer band in ^{141}Ho are compared with the $d_{5/2}$ and $h_{11/2}$ bands in ^{133}Pm.

^{147}Tm proton emitter

Excited states in ^{147}Tm were observed in one of the first RDT experiments with the modest array of Ge detectors AYEBALL coupled with the FMA [2]. In a subsequent 4-hour long test using GAMMASPHERE to detect prompt γ rays, γ-γ coincidences in ^{147}Tm were also measured. The γ-ray coincidence spectra obtained in this experiment are shown in Fig. 3. Originally, the 3 strongest transitions in the spectrum were assumed to form a band. The γ-γ coincidences measured in the second experiment confirmed this hypothesis as can be seen in Fig. 3. The energies of the transitions indicate their quadrupole character. The band was interpreted as the $h_{11/2}$ decoupled band. Several other transitions are present in Fig. 3. These transitions could constitute the extension of the $h_{11/2}$ band or form a side band. Another 1-2 long day experiment would help to firmly establish the high-spin portion of the band. Based on the energy of the $15/2^- \rightarrow 11/2^-$ transition, which should be very close to the energy of the 2^+ state in the even-even core, the deformation of $|\beta_2|=0.13$ was deduced from the Grodzins formula. The microscopic-macroscopic calculations predict $\beta_2=-0.19$ for ^{147}Tm [4].

^{141}Ho proton emitter

The observation of anomalous proton-decay rates in ^{141}Ho and ^{131}Eu [1] motivated a search for their excited states. Compared to ^{147}Tm the cross sections are about a factor

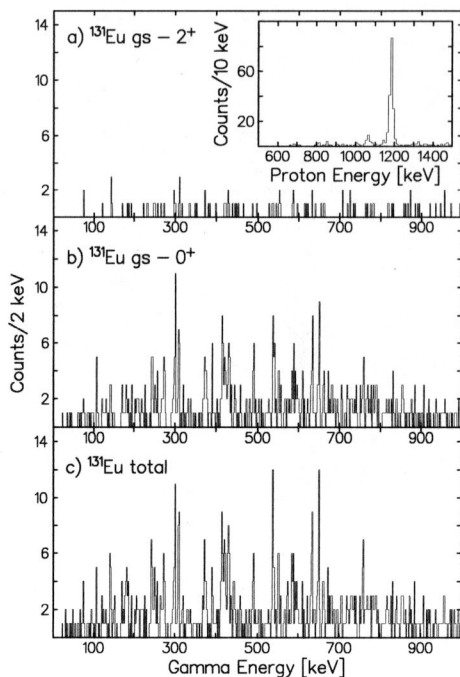

FIGURE 5. The γ-ray spectra correlated with the ground-state ^{131}Eu proton decay a) to the 2^+ excited state in the daughter nucleus, b) to the ground state and c) the sum. The inset shows the proton spectrum.

of 100 smaller. The results of the ^{141}Ho study and a detailed discussion can be found in Ref. [3]. Figure 4 shows the deduced moments of inertia for the ground-state band and the band based on the isomeric state in ^{141}Ho. According to the Nilsson model the $7/2^-$[523] configuration is the ground state in ^{141}Ho at the calculated deformation of β_2=0.29 [4]. Adiabatic proton-decay calculations [6] confirm this assignment and suggest the $1/2^+$[411] orbital for the isomer. The analysis of the moment of inertia as a function of rotational frequency supports these assignments. The first band crossing due to the alignment of the proton $h_{11/2}$ pair is blocked in the ground-state band. Using the Harris formula for the dynamic moment of inertia a deformation of β_2=0.25(4) was deduced for the ground-state band. However, the signature splitting between the two signature partners is surprisingly large. The detailed comparison of the ground-state band level energies with Particle-Rotor model calculations (see discussion below) indicates the need to include hexadecapole deformation and triaxiality. When a calculated value of β_4=-0.06 is used, the best fit is obtained for γ=-10°. This is supported by Total Routhian Surface calculations which indicate that the ^{141}Ho ground state is γ soft and develops triaxiality at higher angular momentum. It is worth noting that an upper limit of 1% for the proton decay to the 2^+ state in the daughter nucleus was obtained in this experiment.

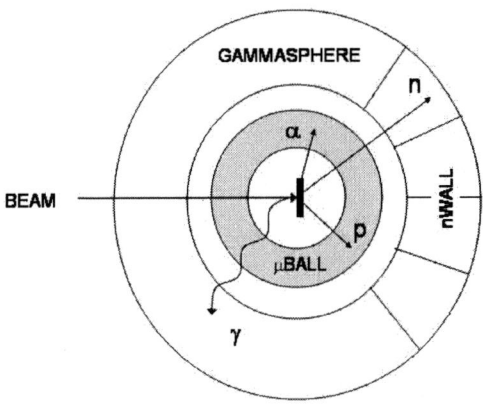

FIGURE 6. The experimental setup used to study proton rich nuclei around ^{141}Ho.

^{131}Eu proton emitter

The γ-ray spectra tagged with the ^{131}Eu ground-state proton decay and the decay to the 2^+ excited state are shown in Figs. 5a and b, respectively. The large number of γ rays in the spectrum indicates that at least three bands were populated with comparable intensity. Due to the fragmentation of the decay path the γ-γ coincidence spectra did not have enough statistics to construct an unambiguous level scheme. According to the Nilsson model the $5/2^+[413]$ and $3/2^+[411]$ orbitals are situated close to the Fermi surface in ^{131}Eu. The proton-decay rate calculations agree with the data for both configurations [1]. However, the measured branching ratio to the 2^+ state in the daughter nucleus agrees only with the $3/2^+[411]$ assignment [7]. Based on a comparison with systematics of rotational bands built on top of the $5/2^+[413]$ and $3/2^+[411]$ levels on the other side of the N=82 line, the lowest transitions in the bands should be around 80 keV and 110 keV, respectively. The 72-keV γ-ray line present in the ^{131}Eu spectrum favors the $3/2^+[411]$ assignment. Similarities between the spectra correlated with the ground-state to ground-state proton line and the fine structure line confirm that the two lines are indeed emitted from the same state.

Detection of light evaporated particles

The Recoil-Decay Tagging method can only be used for nuclei which proton or α decay. In most of the cases, excited states in nuclei in the immediate vicinity of proton emitters are not known. To fill this gap other methods have to be used. The best

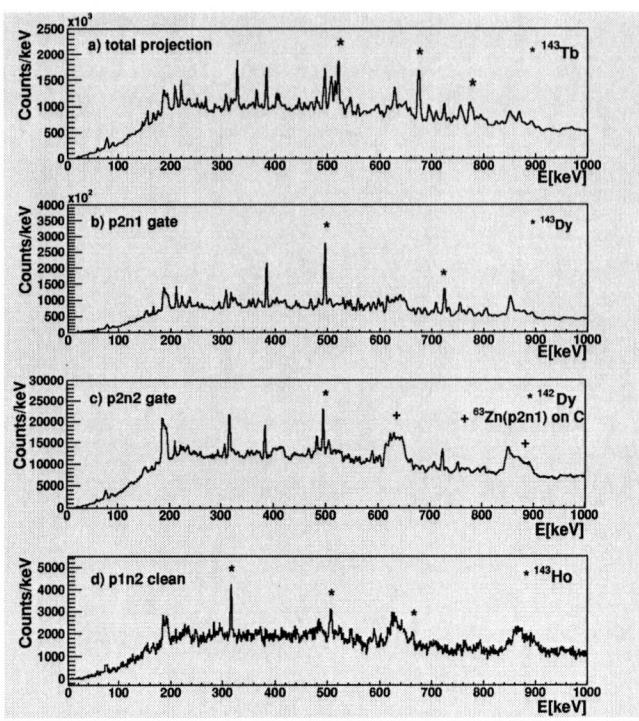

FIGURE 7. Gamma-ray spectra measured for the ^{92}Mo+^{54}Fe reaction. Individual panels correspond to a) total projection, b) two protons and 1 neutron detected, c) p2n2, d) p1n2 (contributions from channels with higher proton multiplicities and one neutron channels were subtracted out).

option at the moment to select weak reaction channels is the detection of light particles evaporated after heavy-ion fusion-evaporation reactions. GAMMASPHERE coupled with the Microball array of charge-particle detectors to detect evaporated protons and α particles and the Neutron Wall array of liquid scintillators for neutron detection was used to study nuclei around ^{141}Ho [8]. A schematic drawing of the detection system is shown in Fig. 6. The observed γ rays were assigned based on the intensity distribution as a function of the number of detected protons, α particles, and neutrons. In the preliminary analysis a set of γ-ray lines was associated with the p1n2 evaporation channel corresponding to the ^{143}Ho nucleus.

^{143}Ho nucleus

The calculated cross section for producing ^{143}Ho is about 50μb. Fig. 7 shows the γ-ray spectra measured in coincidence with different numbers of detected protons and neutrons. The spectrum in Fig. 7d is dominated by three γ-ray lines. All 3 transitions were emitted in coincidence and form a band. Based on the transition energies the band

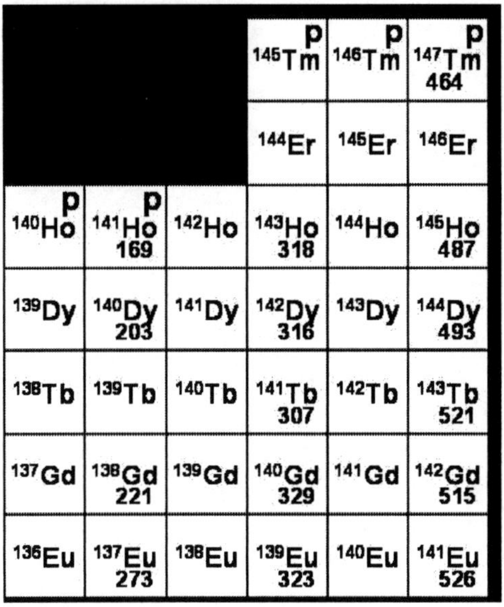

			p 145Tm	p 146Tm	p 147Tm 464
			144Er	145Er	146Er
p 140Ho	p 141Ho 169	142Ho	143Ho 318	144Ho	145Ho 487
139Dy	140Dy 203	141Dy	142Dy 316	143Dy	144Dy 493
138Tb	139Tb	140Tb	141Tb 307	142Tb	143Tb 521
137Gd	138Gd 221	139Gd	140Gd 329	141Gd	142Gd 515
136Eu	137Eu 273	138Eu	139Eu 323	140Eu	141Eu 526

FIGURE 8. The systematics of the $E(15/2^-)-E(11/2^-)$ and $E(2^+)$ energies in the vicinity of ^{141}Ho.

can be interpreted as the decoupled $h_{11/2}$ proton band. Figure 8 shows the energies of $15/2^- \rightarrow 11/2^-$ transitions in odd-Z nuclei and the energies of the 2^+ states in the even-even nuclei around ^{141}Ho. The energy of the first transition in ^{143}Ho agrees very well with the 2^+ energy in the daughter nucleus ^{142}Dy. The energies observed in the lighter N=76 isotones are similar which indicates that the deformation changes very little when more protons are added to the core. If this trend persists for even heavier isotones the structure of the proton emitter ^{145}Tm could be similar to ^{143}Ho.

THE PARTICLE-ROTOR MODEL

Rotational bands in odd-A nuclei, including odd-Z proton emitters, can be calculated using the Particle Rotor Model. The results of such calculations depend on the choice of input parameters. Due to a large number of parameters, which are not known very well and are very often strongly correlated, the parameter space should be constrained to enable meaningful comparison with experiment. In the calculations performed for the ^{141}Ho ground-state band a Woods-Saxon potential with the universal set of parameters was used to obtain single-particle energies. To take into account the effects of pairing the single-particle states underwent the BCS treatment. A proton pairing strength of 0.136 MeV was used in the calculations. As in other regions of the chart of nuclei the Coriolis interaction strength was attenuated by 15%. To illustrate the dependence of the calculations on different input parameters the calculated energies for the $7/2^-[523]$ band in ^{141}Ho are shown in Fig. 9. It can be seen from Fig. 9 that decreasing quadrupole

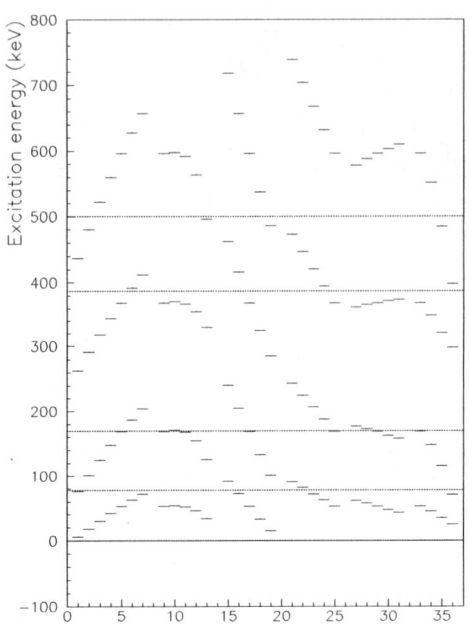

FIGURE 9. The calculated energies of the $9/2^-$, $11/2^-$, $13/2^-$, $15/2^-$ levels of the $7/2^-$[523] band in ^{141}Ho, relative to the $7/2^-$ band head, as a function of (from left to right): β_2 = (0.25, 0.26, 0.27, 0.28, 0.29, 0.30, 0.31), γ=(0°, -5°, ..., -20°), pairing strength = (110%, 105%, 100%, 95%, 90%), Coriolis attenuation = (100%, 97.5%, 95%, 92.5%, 90%), E(2^+) = (180, 190, 200, 210, 220 keV), β_4 = (0.00, -0.02, -0.04, -0.06). Only one parameter was varied at a time. Remaining parameters were fixed at the underlined values. The horizontal lines correspond to the measured level energies.

deformation, increasing negative hexadecapole deformation, increasing negative γ deformation facilitates transition to the decoupling regime marked by arapid decrease in the energy of the $11/2^-$ and $15/2^-$ states with respect to the other states. For small β_2 the splitting between different K orbitals decreases, which enhances K-mixing. For larger β_4 values single-particle states are compressed and the mixing increases as well. The non-axial degree of freedom brings in additional $\Delta K=2$ mixing. Obviously, increasing the Coriolis strength has the same effect. It is worth noting that changing the pairing strength has a very strong effect on level energies. It is primarily because the Coriolis matrix elements are attenuated by a pairing factor which depends sensitively on the pairing strength. The disagreement between non-adiabatic calculations in Ref. [6, 9] and the decay rates in ^{141}Ho could be attributed to the lack of the pairing attenuation factor in the calculations.

OUTLOOK

The return of GAMMASPHERE to Argonne National Laboratory promises another exciting round of RDT studies of deformed proton emitters. Studies of excited states in the odd-odd proton emitter ^{146}Tm would allow a better understanding of its complex proton-decay fine structure. A study of ^{145}Tm, albeit a very difficult case due to a small cross section and short half life, would help to determine the deformation in this nucleus and to elucidate the proposed particle-vibration picture of this nucleus [10]. The properties of the ground-state band in ^{117}La would help to determine the single-particle configuration of the proton decaying band head. Because of a very small cross section the case of ^{135}Tb seems to be beyond the reach of the present experimental facilities.

ACKNOWLEDGMENTS

This work was supported by the U.S. DOE, under Contract No. W-31-109-ENG-38.

REFERENCES

1. Davids, C. N., Woods, P. J., Seweryniak, D., Sonzogni, A. A., Batchelder, J. C., Bingham, C. R., Davinson, T., Henderson, D. J., Irvine, R. J., Poli, G. L., Uusitalo, J., and Walters, W. B., *Phys. Rev. Lett.*, **80**, 1849 (1998).
2. Seweryniak, D., Davids, C. N., Walters, W. B., Woods, P. J., Ahmad, I., Amro, H., Blumenthal, D. J., Brown, L. T., Carpenter, M. P., Davinson, T., Fischer, S. M., Henderson, D. J., Janssens, R. V. F., Khoo, T. L., Hibbert, I., Irvine, R. J., Lister, C. J., Mackenzie, J. A., Nisius, D., Parry, C., and Wadsworth, R., *Phys.Rev.*, **C 55**, R2137 (1997).
3. Seweryniak, D., Woods, P., Ressler, J. J., Davids, C. N., Heinz, A., Sonzogni, A. A., Uusitalo, J., Walters, W. B., Caggiano, J. A., Carpenter, M. P., Cizewski, J. A., Davinson, T., Ding, K. Y., Fotiades, N., Garg, U., Janssens, R. V. F., Khoo, T. L., Kondev, F. G., Lauritsen, T., Lister, C. J., Reiter, P., Shergur, J., and Wiedenhöver, I., *Phys. Rev. Lett.*, **86**, 1458 (2000).
4. Möller, P., Nix, J. R., Myers, W., and Światecki, W., *At. Data Nucl. Data Tables*, **59**, 185 (1995).
5. Paul, E. S., Woods, P., Davinson, T., Page, R. D., Sellin, P. J., Beausang, C. W., Clark, R. M., Cunningham, R. A., Forbes, S. A., Fossan, D. B., Gizon, A., Gizon, J., Hauschild, K., Hibbert, I. M., James, A. N., LaFosse, D. R., Lazarus, I., Schnare, H., Simpson, J., Wadsworth, R., and Waring, M. P., *Phys. Rev.*, **C 51**, 78 (1995).
6. Esbensen, H., and Davids, C., *Phys. Rev.*, **C 63**, 014315R (2001).
7. Sonzogni, A. A., Davids, C. N., Woods, P. J., Seweryniak, D., Carpenter, J. J., M. P.and Ressler, Schwartz, J., Uusitalo, J., and Walters, W. B., *Phys. Rev. Lett.*, **83**, 1116 (1999).
8. Seweryniak, D., Carpenter, M. P., Cullen, D., Davids, C. N., Freeman, S. J., Mukherjee, G., Woods, P. J., Chiara, C. J., Reviol, W., Sarantites, D. G., Hammond, N., Janssens, R. V. F., Khoo, T. L., Kondev, F., Lauritsen, T., Lister, C. J., Fallon, P., Gorgen, A., Machiavelli, A. O., and Ward, D., *private communication* (2003).
9. Kruppa, A., Barmore, B., Nazarewicz, W., and Vertse, T., *Phys. Rev. Lett.*, **84**, 4549 (2000).
10. Davids, C., and Esbensen, H., *Phys. Rev.*, **C 64**, 034317R (2001).

Gamma-Ray Spectroscopy
Beyond the Proton-Drip Line

C.-H. Yu*, J.C. Batchelder†, C.R. Bingham**, C.J. Gross‡, R. Grzywacz‡
and K. Rykaczewski‡

*Physics Division, Oak Ridge National Laboratory, Oak Ridge, TN 37831, U.S.A.
†Oak Ridge Associated Universities, Oak Ridge, TN 37831, U.S.A.
**Department of Physics and Astronomy, University of Tennessee, Knoxville, Tn 37996
‡Oak Ridge National Laboratory, Oak Ridge, TN 37831, U.S.A.

Abstract. A series of experiments were performed to establish excited states in proton emitters ^{151}Lu, ^{109}I and ^{113}Cs. These experiments used (HI, p2n) reactions to populate the proton emitters and employed the recoil decay tagging technique to identify gamma rays in the corresponding proton-unbound nucleus. Gamma-decay sequences were established in these nuclei and their configurations were tentatively assigned based on systematic trends. This paper reviews the three experimental studies and discusses the data in comparison with those of their neighboring nuclei.

1. Introduction

As a result of fast-improving experimental equipment and techniques, the study of nuclear ground-state proton radioactivity has made tremendous progress in the past decade. Such progress has led to the exciting possibility of in-beam studies of nuclei beyond the proton-drip line. Excited states in these proton-unbound nuclei may provide important information on nuclear structure at extreme conditions, and thus help understand fundamental nuclear interactions.

The development of modern electromagnetic devices (*e.g.* recoil mass separator) coupled with a new generation of highly efficient γ-ray detector arrays (*e.g.* Gammasphere) makes it possible to utilize the "Recoil-Decay Tagging" (RDT) technique, which correlates the prompt-γ rays detected at the target position with the charged-particle radioactivities detected at the focal plane of the electromagnetic device. These very selective correlations result in gamma spectra with little background and enable the study of excited states in proton-unbound nuclei.

This paper reviews the RDT studies of proton emitters ^{151}Lu, ^{109}I and ^{113}Cs. More detailed descriptions of these experiments have been published previously and can be found in Refs. [1, 2, 3].

2. Excited States in ^{151}Lu

Excited states of ^{151}Lu were populated using the ^{96}Ru(^{58}Ni, p2n) reaction. The ^{58}Ni beam was provided by the tandem accelerator of the Holifield Radioactive Ion Beam

CP681, Proton-Emitting Nuclei: Second International Symposium; PROCON 2003,
edited by E. Maglione and F. Soramel

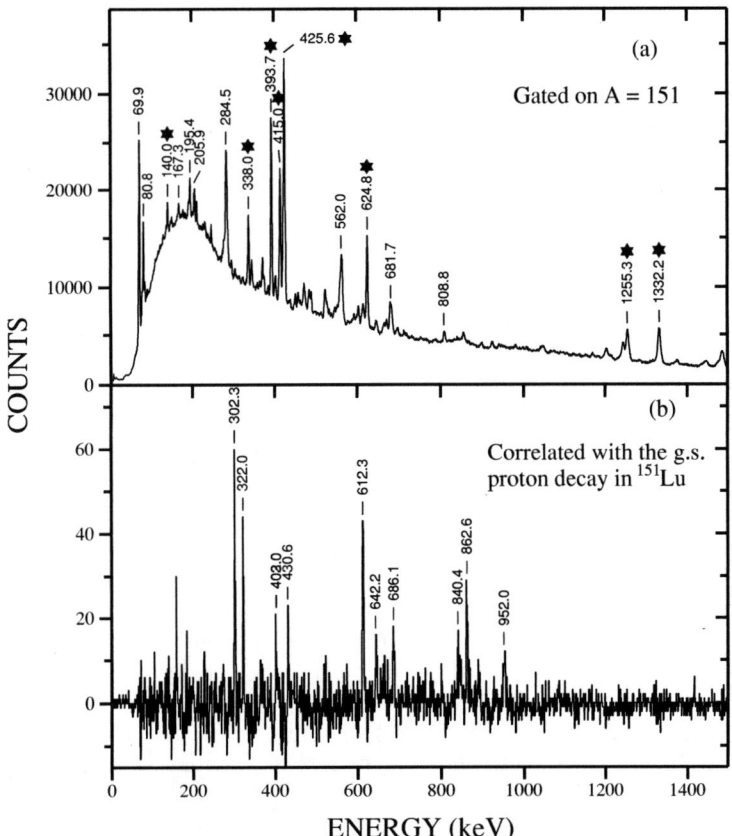

FIGURE 1. (a) Gamma-ray spectrum corresponding to $A = 151$ nuclei produced in the ^{96}Ru + ^{58}Ni reaction. Peaks marked with stars are known[6] transitions in ^{151}Tm. (b) Gamma-ray spectrum correlated with the 1.23-MeV ground state proton decay of ^{151}Lu.

Facility (HRIBF) at the Oak Ridge National Laboratory. The target was a layer of 540-$\mu g/cm^2$ isotopically enriched ^{96}Ru metal deposited on a 2-mg/cm^2 Au supporting foil which faced the beam. The average beam current used in the experiment was about 4.5 pnA and the total beam-on-target time amounted to about 95 hours. Six clover Ge detectors of the ORNL CLARION array were used to detect prompt gamma-rays, and the Recoil Mass Separator (RMS) at HRIBF was used to select the desired mass of $A = 151$. A gas-filled position sensitive avalanche counter (PSAC) was placed at the focal plane to detect the recoils and identify different groups of masses. More detailed descriptions of the RMS and PSAC can be found in Ref. [4]. Behind the PSAC, a 65-μm thick, $40mm \times 40mm$ Double-sided Silicon Strip Detector (DSSD) was used to detect the recoiling ions and their charged-particle decays. More detailed descriptions on the DSSD can be found in Ref. [5].

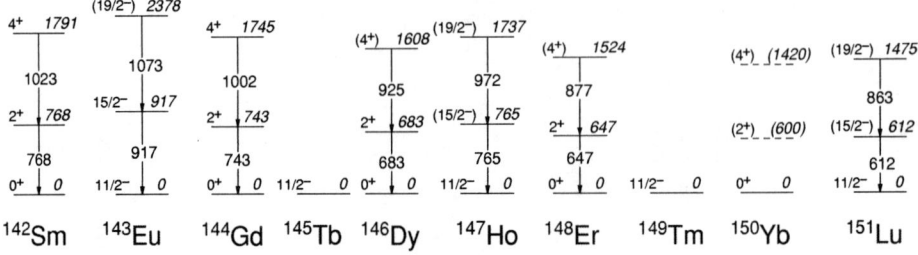

FIGURE 2. Proposed level scheme of ^{151}Lu in case of strong feeding bypassing the possible isomeric state (see text), together with those of lighter N=80 isotones. No excited states are known in ^{150}Yb, and dashed levels shown in the figure for this nucleus are extrapolations from the systematic trend of the lighter even-even isotones. Data for ^{142}Sm, ^{143}Eu, ^{144}Gd, ^{146}Dy, ^{147}Ho and ^{148}Er are taken from Refs.[9, 10, 11, 12, 13, 14].

Previous measurements[7, 8] of the $h_{11/2}$ ground-state proton decay in ^{151}Lu established the proton energy to be $E_p = 1.233(3)$ MeV and the half-life to be $T_{1/2} = 85(10)$ ms. Our measurement resulted in a $T_{1/2}$ of 80(2) ms, which agrees with the previous value, but is considerably more precise. In order to identify gamma rays belonging to ^{151}Lu, we first examined the gamma spectrum gated on mass A=151. This mass-gated γ spectrum is shown in Fig. 1(a), and is dominated by the known[6] transitions in ^{151}Tm ($3p$ channel). Next, the spectrum of γ rays that are correlated with the 1.23-MeV proton decay in ^{151}Lu was obtained and shown in Fig. 1(b). All strong peaks in Fig. 1(a) are absent in Fig. 1(b). Since the spectrum shown in Fig. 1(b) is correlated with the ground-state proton decay of ^{151}Lu, the γ rays shown in this spectrum can be firmly assigned to ^{151}Lu. The fact that the transitions shown in Fig. 1(b) are almost all invisible in Fig. 1(a) also eliminates the possibility that these transitions are residuals of the strong peaks in Fig. 1(a).

The transitions shown in Fig. 1(b) are very weak (only about 0.2% compared to γ rays in ^{151}Tm, the strongest $A = 151$ nucleus populated in the reaction). Although all ten transitions indicated in Fig. 1(b) can be confidently assigned as depopulating excited states of ^{151}Lu, the ordering of these transitions is less transparent. Due to low statistics, gating on these transitions in the mass-gated $\gamma\gamma$ matrix produced no useful result. Therefore, spin and parity assignments for levels associated with these γ rays have to be made based on comparisons with states in the neighboring nuclei, and on considerations of our measured relative γ-ray intensities.

The placement of these gamma rays is also complicated by a possible high-spin isomer, which was inferred[1] from the systematic trend of the neighboring nuclei of ^{151}Lu. If the isomer exists, the observed gamma rays can be assigned as directly feeding the ground state only if there are strong prompt transitions bypassing the isomeric level. In this case, the strong γ rays shown in Fig. 1(b) could be the lowest cascade of transitions feeding the $\pi h_{11/2}$ ground state. In particular, we could assign the 612-keV and 862-keV transitions as the $15/2^- \rightarrow 11/2^-$ and $19/2^- \rightarrow 15/2^-$ transitions. Such an assignment would fit the systematic trend of the $N = 80$ isotones, see Fig. 2. When comparing the low-spin transitions in ^{147}Ho to its even-even core, ^{146}Dy, we see that the $15/2^- \rightarrow 11/2^-$ and $19/2^- \rightarrow 15/2^-$ transitions in ^{147}Ho closely resemble the $2^+ \rightarrow 0^+$ and the $4^+ \rightarrow 2^+$ transitions in ^{146}Dy. This suggests that the low-spin levels in ^{147}Ho are formed simply by coupling the odd $h_{11/2}$ proton to the 2^+ and 4^+ states of the even-even core. Meanwhile, the 4^+ and 2^+ energies decrease with increasing proton number. If we extrapolate these decreasing energies to Z=70, we would put the 2^+ level of ^{150}Yb at about 600 keV, and the 4^+ level at about 1400 keV. Our tentative assignment of the $15/2^- \rightarrow 11/2^-$ and $19/2^- \rightarrow 15/2^-$ transitions in ^{151}Lu would then resemble the lowest two transitions in its even-even core, ^{150}Yb, similar to the way ^{147}Ho resembles ^{146}Dy.

3. Excited States in ^{109}I

In a previous RDT experiment using the EUROGAM-I array and the Daresbury recoil separator, five γ-ray transitions (505-, 596-, 729-, 845-, and 908 keV) were reported [15] to form the yrast decay sequence in the proton emitter ^{109}I. This result was based on a very weak proton-correlated γ-ray spectrum (see Fig. 7b of Ref. [15]), as well as on $\gamma\gamma$ coincidence spectra that were not correlated with protons. The lack of statistics in this previous study was partly due to a beam energy that was not optimized to produce ^{109}I.

To verify, and possibly extend the information on excited states in ^{109}I, we carried out an RDT experiment using the same ^{54}Fe(^{58}Ni,p2n) reaction as the previous study, but at the optimized beam energy of 220 MeV. The ^{58}Ni beam was provided by the ATLAS superconducting linear accelerator at Argonne National Laboratory and the target was an isotopically enriched, 500-$\mu g/cm^2$, self-supporting ^{54}Fe foil. Prompt γ rays were detected with Gammasphere Ge detector array. Recoiling ions passed through the Fragment Mass Analyzer (FMA, tuned to $A = 109$), and their masses were determined at the focal plane by a position-sensitive parallel-grid avalanche counter. These ions were then implanted into a $40mm \times 40mm$, 60-μm thick DSSD for the identification of charged-particle decays.

The ^{109}I ground-state proton decay has been previously established[16, 17, 18] to have a proton energy and half-life of $E_p = 829 \pm 4$ keV and $T_{1/2} = 100 \pm 5 \mu$s, respectively. This proton decay was easily identified in the present experiment. The total number of the ^{109}I protons observed from the experiment was about 6000, of which about 3500 were correlated with prompt γ rays, and these events were used in the final analysis. In particular, the proton-correlated $\gamma\gamma$ events were very useful for the construction of ^{109}I decay scheme.

The analysis of the proton-correlated $\gamma\gamma$ matrix showed that the 593-, 717-, 881-, and 1057-keV transitions form the yrast decay sequence in ^{109}I. Other weaker transitions either feed into or decay out of this sequence. Examples of gated coincidence spectra are presented in Fig. 3. Based on coincidence relationships as well as intensity arguments, a decay scheme for ^{109}I was established and is shown in Fig. 4. Since the statistics achieved in the present experiment were too low for any angular correlation analysis, the spin and parity assignments for the shown levels were based on systematic trends in nuclei neighboring ^{109}I. In ^{111,113,115,117}I [19, 20, 21, 22], the level schemes are dominated by a yrast-decay sequence which is built on the $\pi h_{11/2}$ single particle state.

175

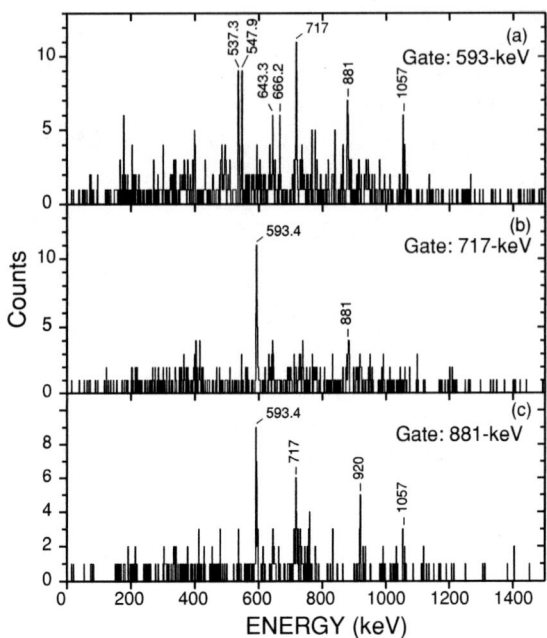

FIGURE 3. Gamma-ray spectra obtained by gating on the (a) 593-keV, (b) 717-keV, and (c) 881-keV transitions in the $\gamma\gamma$ coincidence matrix that is correlated with the 829-keV ground-state proton decay in ^{109}I.

This state then decays through several $9/2^+$ and $7/2^+$ states to the $5/2^+$ ground state. For transitions observed in ^{109}I, their energy and intensity characteristics strongly suggest that the 593-, 717-, 881-, and 1057-keV sequence is the band built on the $\pi h_{11/2}$ state. The intensity of this sequence dominates that of the total decay, very similar to the dominant intensities of the $\pi h_{11/2}$ bands in ^{111,113,115,117}I. The spin and parity assignments of transitions shown in Fig. 4 and the ensuing discussions are, therefore, based on the assumption that this sequence is the $\pi h_{11/2}$ band. It should be noted that this band is completely different from that proposed in the previous study [15].

Previous systematic analyses[23, 24] of proton emitters have shown that with a small spectroscopic factor, ^{109}I could be deformed. In fact, using a multi-particle theory for proton transitions, Bugrov and Kadmensky[25] could reproduce the half-life of ^{109}I with a quadrupole deformation of $\beta \approx 0.10 - 0.15$. Although no direct measurement of deformation could be obtained from the present experiment, a comparison of excited states in ^{109}I with those of its neighboring nuclei can be used to gain insight of the evolution of deformation with changing mass. In Fig. 5(a), the $15/2^- \rightarrow 11/2^-$ transition energies (closed circles) of the $\pi h_{11/2}$ bands for a series of odd-A iodine isotopes are presented as a function of neutron number. For comparison, the $2^+ \rightarrow 0^+$ transition energies (open circles) of the even-even Te core isotopes, and the $15/2^- \rightarrow 11/2^-$ transition energies (diamonds) of the $\nu h_{11/2}$ bands for the odd-A Te isotopes are also shown in the same figure. The light iodines lie in a transitional region. The low-spin part of the $h_{11/2}$ bands

176

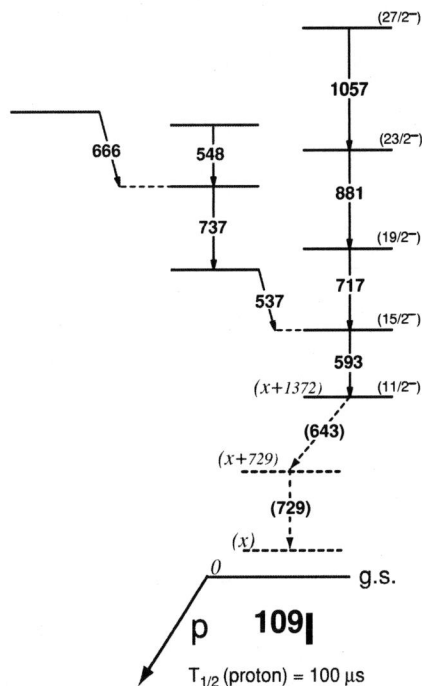

FIGURE 4. Proposed level scheme of ^{109}I based on the present work. See text and ref [2] for detailed arguments for the placement of transitions and their spin and parity assignment.

in these nuclei can be understood as formed by coupling the $h_{11/2}$ protons to their respective even-even Te cores, which are weakly or moderately deformed vibrators. The ground state quadrupole deformation of Te and I isotopes has been predicted[26] to decrease with decreasing neutron number, see Fig. 5(b). For iodine isotopes, the trend of the $15/2^- \rightarrow 11/2^-$ transition energies agrees with the predicted trend of the quadrupole deformation, *i.e.* the rising $E_\gamma(15/2^- \rightarrow 11/2^-)$ corresponds to a decreasing β_2 with decreasing N. Such a trend shows that although ^{109}I may be deformed, the deformation is small and follows the deformation pattern in the iodine isotopes.

Fig. 5 also shows two interesting features: (1) The increasing $E_\gamma(2^+ \rightarrow 0^+)$ as a function of N for the telluriums is not expected according to the prediction of increasing β_2 as a function of N. (2) When an $h_{11/2}$ proton is added to the tellurium core, the N dependence of the $E_\gamma(2^+ \rightarrow 0^+)$ is reversed: The $E_\gamma(15/2^- \rightarrow 11/2^-)$ for iodines decreases with N, a trend that is opposite to that shown by the telluriums [see open and closed circles in Fig. 5(a)]. An earlier study[32] of heavier iodine isotopes already pointed out this phenomenon, and suggested that a microscopic collective model with a higher-order perturbation expansion of the particle-core interaction may offer an explanation. It is interesting to note that when adding an extra $h_{11/2}$ neutron to the core, the odd-A telluriums show an $E_\gamma(15/2^- \rightarrow 11/2^-)$ that closely follows the trend of the even-even

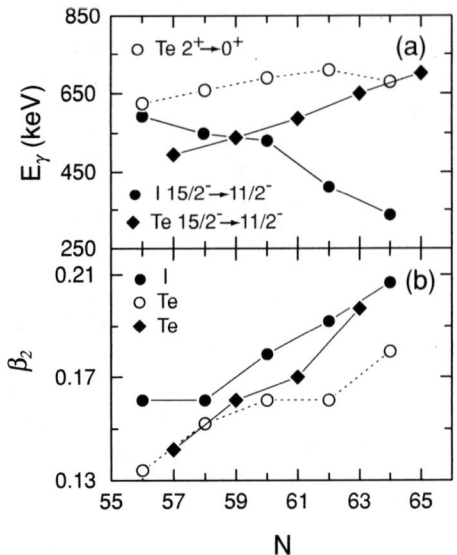

FIGURE 5. (a) Experimental $15/2^- \rightarrow 11/2^-$ transition energies for the iodine isotopes (closed circles), $2^+ \rightarrow 0^+$ transition energies for the even-even tellurium isotopes (open circles) and the $15/2^- \rightarrow 11/2^-$ transition energies for the odd-A tellurium isotopes (diamonds) as a function of N. Data for [109]I are from the present work, for [111-119]I are from Refs. [19, 20, 21, 22, 27], and those for [108-117]Te are from Refs. [2, 28, 29, 30, 31]. (b) Calculated[26] quadrupole deformation, β_2, for odd-A iodine isotopes at ground states.

telluriums [see open circles and diamonds in Fig. 5(b)]. The addition of an $h_{11/2}$ proton seems to have a much larger polarization to the Te core than the addition of an $h_{11/2}$ neutron. This could be due to the fact that the proton Fermi level is much closer to the N=50 closed shell, and thus the nuclear properties are more sensitive to the change of valence particles.

4. Decay Sequences in [113]Cs

Two experiments were performed to establish excited states in [113]Cs, both using the [58]Ni([58]Ni, p2n) reaction at a beam energy of 230 MeV. The first experiment was performed at the HRIBF of Oak Ridge National Laboratory, and used the same experimental setup described in Section 2. Ten gamma rays in [113]Cs were identified from this experiment. The lack of proton-correlated $\gamma\gamma$ coincidence data, however, made it difficult to determine the positions of these gamma rays in the decay scheme. The HRIBF experiment also resulted in a ground-state proton-decay half-life of $18.3\pm0.3\mu$s, which is in agreement with previous measurements [33, 34]. A follow-up experiment was performed at the ATLAS of Argonne National Laboratory and used the same experimental setup described in section 3. In addition to confirming the 10 gamma rays identified in

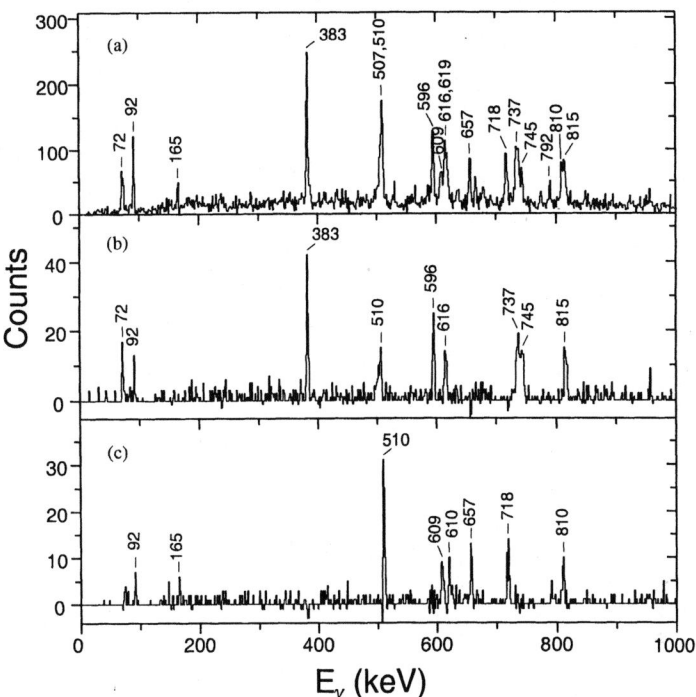

FIGURE 6. (a) Total projection of the $\gamma\gamma$ matrix correlated with the ground-state proton decay in ^{113}Cs produced in the ATLAS experiment using the ^{58}Ni(^{58}Ni, p2n) reaction. (b) Sum of proton-correlated spectra gated on γ transitions in the yrast band (band 1) established in ^{113}Cs. (c) Sum of proton-correlated spectra gated on γ transitions in band 2 established in ^{113}Cs.

the HRIBF experiment, the ATLAS experiment identified a few new low-energy transitions in ^{113}Cs. The higher gamma-detection efficiency of Gammasphere also allowed proton-correlated $\gamma\gamma$ coincidence analysis, which was important for establishing the decay scheme of these gamma rays.

Fig. 6(a) shows the total projection of the $\gamma\gamma$ matrix correlated with the 0.96-MeV ground-state proton decay in ^{113}Cs. The analysis of this matrix resulted in two decay sequences in ^{113}Cs, and a tentative level scheme is shown in Fig. 7. Of the two decay sequences, band 1 is stronger than band 2. Fig. 6(b) shows the sum of proton-correlated spectra gated on clean transitions in this band. In heavier Cs isotopes in this mass region, proton $h_{11/2}$ bands have been observed to be the yrast decay sequences. As a result, the observed band 1 in ^{113}Cs is most likely associated with the $h_{11/2}$ proton excitations. Based on comparisons to heavier Cs isotopes, the bandhead of band 1 is tentative assigned as $I=11/2^-$. The ordering of transitions in this band is based on coincidence relationships as well as intensity arguments. The 72-keV transition is strongly in coincidence with band 1, and is most likely one of the transitions that connect band 1 to the proposed positive-parity, $I=3/2$ ground states. The actual connections between band 1 and the ground state, however, could not be established from the present

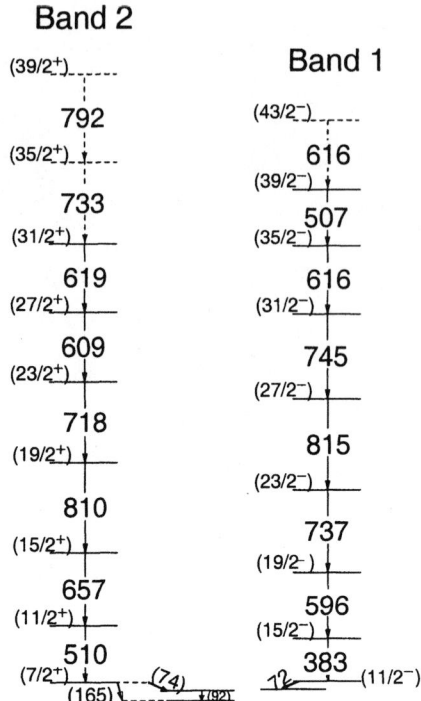

FIGURE 7. Decay sequences established from the RDT study of ^{113}Cs using a ^{58}Ni(^{58}Ni, p2n) reaction. The ordering of the transitions in both bands are based on coincidence relationships and intensity arguments, and are not firm. The tentative spin and parity assignment are based on systematic trend in Cs isotopes.

study.

Fig. 6(c) shows the sum of proton-correlated spectra gated on clean γ transitions in band 2. This band is weaker than band 1 and the ordering of transitions in the band is based on coincidence relationships as well as intensity arguments. A possible configuration assignment for this band is the positive-parity, $\pi f_{7/2}d_{5/2}$ band. Such a band has been observed[35] in ^{117}Cs (no excited states in ^{115}Cs are known). Another possibility is that band 2 is the signature partner of band 1, and the $M1/E2$ intraband transitions were too weak to be observed in our experiments. The tentative spin and parity assignment shown in Fig. 7, however, are based the assumption that band 2 is the positive-parity, $\pi f_{7/2}d_{5/2}$ band. Similar to band 1, a few lower-energy transitions, *e.g.* the 165-, 74- and 92-keV transitions, were observed to be in coincidence with this band, and they are most likely the transitions that link band 2 to the proposed $3/2^+$ ground state. However, the low statistics did not allow a firm establishment of these connections.

^{113}Cs is one of the proton emitters that have a relatively small experimental spectroscopic factor, and were therefore suggested to be deformed at ground state. Indeed, theoretical calculations by Bugrov and Kadmenski [25] indicate that a quadrupole deformation of $\beta_2 \approx 0.2$ is necessary in order to explain the observed short half-life. Our measurement of excited states in ^{113}Cs is consistent with this assessment. For example, the proposed $15/2^- \rightarrow 11/2^-$ transition energy of 383 keV in ^{113}Cs is smaller than its lighter N=58 isotones [E_γ ($15/2^- \rightarrow 11/2^-$)=555- and 1080 keV for ^{111}I [19] and ^{109}Sb [36], respectively]. A quantitative measurement of deformation in ^{113}Cs, however, is not possible at this point.

5. Summary

Excited states were identified in proton emitters ^{151}Lu, ^{109}I and ^{113}Cs using (HI, p2n) reactions. In each case, the recoil decay tagging technique was employed and gamma rays were unambiguously assigned to the corresponding proton emitter. In the case of ^{109}I and ^{113}Cs, proton correlated $\gamma\gamma$ coincidence analyses were possible, and rotational decay sequences were established. The yrast decay sequences of both ^{109}I and ^{113}Cs were understood to be associated with the proton $h_{11/2}$ configuration, and based on comparisons to their neighboring isotone and isotopes, they confirm the previous suggestion that these two proton emitters are weakly to moderately deformed.

ACKNOWLEDGMENTS

Oak Ridge National Laboratory is managed by UT-Battelle, LLC, for the U.S. D.O.E under contract DE-AC05-00OR22725. The contributions of the HRIBF, ATLAS and Gammasphere staff to these experiments are gratefully acknowleged.

REFERENCES

1. Yu, C. H., et al., *Phys. Rev. C*, **58**, R3042 (1999).
2. Yu, C. H., et al., *Phys. Rev. C*, **59**, R1834 (1999).
3. Gross, C., et al., "In-Beam Gamma-Ray Spectroscopy in the Ground-State Proton Emitter ^{113}Cs," in *International Conference on Exotic Nuclei and Atomic Masses*, edited by B. M. Sherrill et al., AIP Conference Proceedings 455, American Institute of Physics, New York, 1998, p. 444.
4. Gross, C. J., et al., *Nucl. Instrum. and Methods A*, **450**, 12 (2000).
5. Sellin, P. J., et al., *Nucl. Instrum. and Methods A*, **311**, 217 (1992).
6. Singh, B., *Nucl. Data Sheets*, **80**, 263 (1997).
7. Hofmann, S., et al., *Z. Phys. A*, **305**, 111 (1982).
8. Sellin, P. J., et al., *Phys. Rev. C*, **47**, 1933 (1993).
9. Lach, M., et al., *Z. Phys. A*, **319**, 235 (1984).
10. Gui, S. Z., et al., *Z. Phys. A*, **305**, 297 (1982).
11. de Angelis, G., et al., *Z. Phys. A*, **341**, 371 (1992).
12. Nolte, E., *Z. Phys. A*, **306**, 211 (1982).
13. Piiparinen, M., et al., *Z. Phys. A*, **343**, 367 (1992).
14. Müller-Veggian, M., et al., *Z. Phys. A*, **330**, 343 (1988).
15. Paul, E. S., et al., *Phys. Rev. C*, **51**, 78 (1995).
16. Faestermann, T., et al., *Phys. Lett. B*, **137**, 23 (1984).
17. Heine, F., et al., *Z. Phys. A*, **340**, 225 (1991).
18. Sellin, P. J., et al., *Phys. Rev. C*, **47**, 1933 (1993).

19. Paul, E. S., et al., *Phys. Rev. C*, **61**, 064320 (2000).
20. Waring, M. P., et al., *Phys. Rev. C*, **51**, 2427 (1995).
21. Paul, E. S., et al., *J. Phys. G*, **18**, 837 (1992).
22. Waring, M. P., et al., *Phys. Rev. C*, **48**, 2629 (1993).
23. Woods, P. J., and Davids, C. N., *Annu. Rev. Nucl. Part. Sci.*, **47**, 541 (1997).
24. Åberg, S., et al., *Phys. Rev. C*, **56**, 1762 (1997).
25. Bugrov, V. P., and Kadmensky, S. G., *Sov. J. Nucl. Phys.*, **49**, 967 (1989).
26. Möller, P., et al., *At. Data Nucl. Data Tables*, **59**, 185 (1995).
27. Liang, Y., et al., *Phys. Rev. C*, **45**, 1041 (1992).
28. Lane, G. J., et al., *Phys. Rev. C*, **55**, 1559 (1997).
29. Moon, C. B., et al., *Z. Phys. A*, **357**, 127 (1997).
30. Moon, C. B., et al., *Z. Phys. A*, **349**, 1 (1994).
31. Duyar, C., et al., *Z. Phys. A*, **348**, 63 (1994).
32. Piel, W. F., et al., *Phys. Rev. C*, **31**, 456 (1985).
33. Batchelder, J. C., et al., *Phys. Rev. C*, **57**, R1042 (1998).
34. Page, R. D., et al., *Phys. Rev. Lett.*, **72**, 1798 (1994).
35. Smith, J. F., et al., *Phys. Rev. C*, **63**, 024319 (2001).
36. Janzen, V. P., et al., *Phys. Rev. Lett.*, **72**, 1160 (1994).

First Observation of Excited States in ^{140}Dy

W. Królas[1,2,3], R. Grzywacz[4,5], K. P. Rykaczewski[4,5], J. C. Batchelder[6],
C. R. Bingham[4,7], C. J. Gross[4,8], D. Fong[2], J. H. Hamilton[2], D. J. Hartley[7],
J. K. Hwang[2], Y. Larochelle[7], T. A. Lewis[4], K. H. Maier[1,4],
J. W. McConnell[1], A. Piechaczek[9], A. V. Ramayya[2], K. Rykaczewski[10],
D. Shapira[4], M. N. Tantawy[7], J. A. Winger[11], C. -H. Yu[4], E. F. Zganjar[9],
A. T. Kruppa[1,12], W. Nazarewicz[4,7,13], and T. Vertse[1,12]

[1]Joint Institute for Heavy Ion Research, Oak Ridge, TN 37831, USA
[2]Dept. of Physics and Astronomy, Vanderbilt University, Nashville, TN 37235, USA
[3]H. Niewodniczański Institute of Nuclear Physics, PL 31-342 Kraków, Poland
[4]Physics Division, Oak Ridge National Laboratory, Oak Ridge, TN 37831, USA
[5]Institute of Experimental Physics, Warsaw University, PL 00-681 Warsaw, Poland
[6]UNIRIB Oak Ridge Associated Universities, Oak Ridge, TN 37831, USA
[7]Dept. of Physics and Astronomy, University of Tennessee, Knoxville, TN 37996, USA
[8]Oak Ridge Institute for Science and Education, Oak Ridge, TN 37831, USA
[9]Dept. of Physics and Astronomy, Louisiana State University, Baton Rouge, LA 70803, USA
[10]Oak Ridge High School, Oak Ridge, TN 37830, USA
[11]Dept. of Physics and Astronomy, Mississippi State University, Mississippi State, MS 39762, USA
[12]Institute of Nuclear Research, Hungarian Academy of Sciences, H 4001 Debrecen, Hungary
[13]Institute of Theoretical Physics, Warsaw University, PL 00-681 Warszawa, Poland

Abstract. The mass $A = 140$ products of the ^{54}Fe (315 MeV) + ^{92}Mo reaction were selected by a recoil mass spectrometer and studied in a recoil – delayed γ-γ coincidence experiment. A new 7 μs isomer was identified in the drip line nucleus ^{140}Dy. Five cascading γ transitions were assigned to the decay of the $I^\pi = 8^-$ {$v9/2^-[514] \otimes v7/2^+[404]$} K isomer via the ground state band transitions. The established structure of ^{140}Dy is discussed with reference to the fine structure in proton decay of ^{141}Ho.

The ^{140}Dy [1, 2] is located at the borderline of bound nuclei. It is a daughter nucleus for the proton radioactive ^{141gs,m}Ho [3, 4, 5]. The knowledge of the structure and shape of ^{140}Dy is essential for the understanding of proton emission rates from the ground and isomeric state of ^{141}Ho since it determines the potential tunneled by the proton. Comparison of the properties of excited levels in ^{140}Dy to the structure of less exotic N=74 isotones can reveal the effects caused by the proximity of the proton drip line.

In an experiment performed at the Holifield Radioactive Ion Beam Facility at Oak Ridge National Laboratory the ^{140}Dy ions were populated in the $2p4n$ channel of the ^{54}Fe (315 MeV) + ^{92}Mo reaction, and selected by the Recoil Mass Spectrometer [6]. The RMS acceptance allowed us to collect the recoils produced within a half of the nominal 1 mg/cm^2 target thickness. The recoils, separated according to their mass $A = 140$ and charge $Q = +27$, with the energy of 92 MeV ($\pm$10%), were implanted in a passive catcher placed in the center of the Clover Germanium Detector Array for Recoil Decay Spectroscopy [7] which consisted of 4 segmented Clover Germanium detectors.

CP681, *Proton-Emitting Nuclei: Second International Symposium; PROCON 2003*,
edited by E. Maglione and F. Soramel
© 2003 American Institute of Physics 0-7354-0150-0/03/$20.00

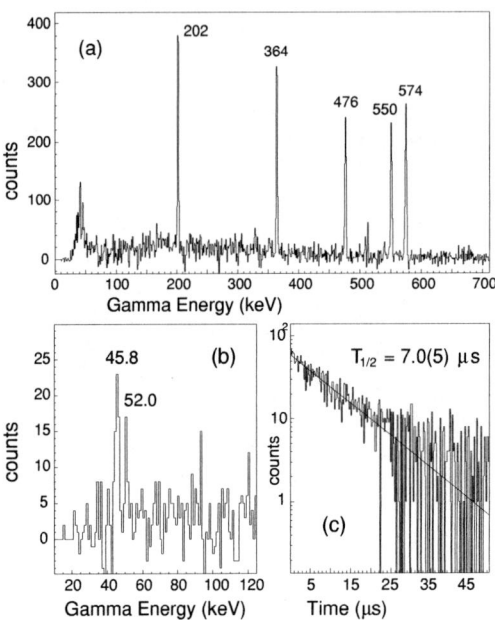

FIGURE 1. ^{140}Dy γ lines from the decay of the $I^\pi = (8^-)$ isomer obtained from double γ coincidence data by adding five spectra gated on the labeled transitions - panel (a). Dysprosium K_α and K_β X rays in coincidence with the sequence of five new γ lines from triple γ coincidence data - panel (b). Decay pattern produced by double-gating on five transitions - panel (c).

A Multichannel Plate Detector [8] provided a reference time for the recoil implantation and enabled a recoil – delayed γ-γ coincidence study.

We have identified a new cascade of γ-rays at 202, 364, 476, 550 and 574 keV correlated with the implantation of the selected $A = 140$ recoils, see Fig. 1(a). The five γ lines are in coincidence with each other. Analysis of triple γ coincidence data revealed that the group of transitions is also in coincidence with Dy K X-rays, see Fig. 1(b), which places them in one cascade in ^{140}Dy. The spectrum of time differences between the MCP signals and five γ transitions shows a characteristic decay pattern with a half-life of $7.0 \pm 0.5\mu$s, see Fig. 1(c).

Basing on the intensities and energies of the transitions a level scheme shown in Fig. 2 (right part) was constructed. It resembles a rotational band in a deformed nucleus fed by an isomeric level. A comparison to the systematics of level schemes of less exotic $N = 74$ even-even isotones of ^{134}Nd, ^{136}Sm and ^{138}Gd displayed also in Fig. 2 (left part) shows striking similarity. This leads us to the interpretation of the isomeric level at 2166 keV as an $I^\pi = (8^-)$ $\{\nu 9/2^-[514] \otimes \nu 7/2^+[404]\}$ K isomer decaying via the $8^+ \to 6^+ \to 4^+ \to 2^+ \to 0^+$ cascade belonging to the ground-state band in ^{140}Dy. A similar decay scheme of the ^{140}Dy $K^\pi = 8^-$ isomer was also proposed by the authors of another independant experiment [2, 9].

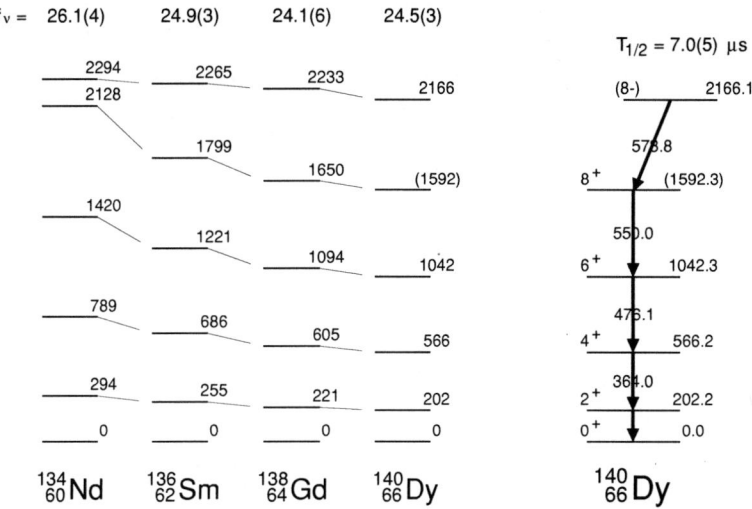

FIGURE 2. The established decay scheme of the $K^\pi = 8^-$ isomer in ^{140}Dy (right panel). The systematics of the 2^+, 4^+, 6^+ and 8^+ ground-state band states and the $K^\pi = 8^-$-isomer in even-even $Z \geq 60$, $N = 74$ isotones (left panel).

The isomeric level is placed at an excitation energy close to the 2150 keV predicted in [10] for this two-quasineutron configuration. The hindrance per degree of K-forbidness f_ν for the new ^{140}Dy isomeric transition was determined to be 24.5(3), a value very close to the ones found for less exotic $N = 74$ isotones as displayed in the upper part of Fig. 2. This indicates that the configuration and properties of this K isomer in ^{140}Dy are not affected by the proximity of the proton drip line. Interestingly, the observed $E(2^+)$ energy of 202 keV in ^{140}Dy is significantly higher than the 138 keV known in the "$N = 82$ mirror" nucleus ^{156}Dy.

Accounting for about 0.5 mg/cm^2 active target thickness we measured a 40 μb cross-section for population of the 8^- isomer. This value is comparable with the 30 μb total ^{140}Dy production cross-section predicted by the HIVAP code [11] and indicates that major part of the high-spin states populated in the evaporation residue deexcite via the 8^- isomer. Such a sizable population of the isomer is surprising since one would expect more deexcitation strength to go through the ground state band thus bypassing the non-yrast isomer. In case of the neighboring $N = 74$ isotone ^{138}Gd it has been found that only a 1.4% fraction of the evaporation residue cross section populates the analog $K^\pi - 8^-$ isomer [12].

Following the global systematics of the ground to the first-excited 2^+ state transitions in even-even nuclei of Raman et al. [13] the observed excitation energy of the first 2^+ state gives a deformation parameter $\beta_2 \approx 0.24$ for ^{140}Dy. This is a somewhat smaller quadrupole deformation than the previously anticipated values (see e.g. [4, 10]) but it is close to the $\beta_2 = 0.25$ derived from the observed level schemes of ^{141gs,m}Ho [5].

The experimental level scheme of ^{140}Dy provided reliable input for the predictions of proton emission rates from ^{141gs,m}Ho. Moreover, the precisely known energy of 2^+ state in ^{140}Dy was an essential aid in the search for a fine structure in ^{141}Ho proton decay [14, 15].

Further experimental efforts are needed to better understand the structure of drip line nucleus ^{140}Dy. In-beam gamma spectroscopy studies should be aimed at extending the ground state band above the 8^+ level and confirming the ordering of the g.s. band transitions which is now based entirely on systematics. In particular the ordering of two highest energy transitions can be reversed. Also, it would be interesting to identify a $K^\pi = 8^-$ band built on top of the 8^- isomer. In principle this can be accomplished with the isomer decay tagging technique, however, the low cross-section for ^{140}Dy production is a limiting factor.

This work was supported by the U.S. D.O.E. through Contracts No. DE-FG02-96ER40983, DE-FG05-88ER40407, DE-FG02-96ER40978, DE-FG02-96ER41006, DE-AC05-76OR00033, DE-FG02-96ER40963, by the Polish KBN Grants No. 2P03B 08617 and 2P03B 04516, by the Hungarian OTKA Grants No. T026244 and T029003, and by NATO Grant No. PST.GLG.977613. The Joint Institute for Heavy Ion Research is supported by its members, University of Tennessee, Vanderbilt University and Oak Ridge National Laboratory. ORNL is managed by UT-Battelle, LLC, for the U.S. Department of Energy under Contract No. DE-AC05-00OR22725.

REFERENCES

1.   Królas, W. et al., *Phys. Rev.* **C65**, 031303 (2002).
2.   Cullen, D.M. et al., *Phys. Lett.* **B529**, 42-49 (2002).
3.   Davids, C.N. et al, *Phys. Rev. Lett.* **80**, 1849-1852 (1998).
4.   Rykaczewski, K. et al., *Phys. Rev.* **C60**, 011301 (1999).
5.   Seweryniak, D. et al., *Phys. Rev. Lett.* **86**, 1458-1461 (2001).
6.   Gross, C.J. et al., *Nucl. Instrum. Methods Phys. Res.* **A450**, 12-29 (2000).
7.   Batchelder, J.C. et al., *Nucl. Instrum. Methods Phys. Res.* **A**, in press.
8.   Shapira, D., Lewis, T.A., and Hulett, L.D, *Nucl. Instrum. Methods Phys. Res.* **A454**, 409-420 (2000).
9.   Cullen, D.M. et al., "Identification of Excited States in ^{140}Dy", proceedings to this conference.
10. Xu, F. R., Walker, P.M., and Wyss, R., *Phys. Rev.* **C59**, 731-734 (1999).
11. Reisdorf, W., *Zeit. Phys.* **A300**, 227-240 (1981).
12. Cullen, D.M. et al., *Phys. Rev.* **C58**, 846-850 (1998).
13. Raman, S., Nestor, C.W. JR., and Tikkanen, P., *At. Data and Nucl. Data Tables* **78**, 1-128 (2001).
14. Rykaczewski, K. et al., *AIP Conf. Proc.* **638**, 149-154 (2002).
15. Rykaczewski, K. et al., "Fine Structure in One-Proton Emission", proceedings to this conference.

Identification of excited states in ^{140}Dy

D.M. Cullen*, M.P. Carpenter, C.N. Davids[†], A.M. Fletcher, S.J. Freeman*,
R.V.F. Janssens[†], F. Kondev**, C.J. Lister[†], L.K. Pattison*, D. Seweryniak[†],
J.F. Smith*, A. Bruce[‡], K. Abu Saleem[†§], I. Ahmad, A. Heinz, T.L. Khoo,
E.F. Moore[†], G. Mukherjee[†¶], C. Wheldon[∥] and A. Woehr[††]

*Department of Physics and Astronomy, University of Manchester, Manchester M13 9PL, UK.
†Physics Division, Argonne National Laboratory, Argonne, Illinois 60439, USA.
**Technology Development Division, Argonne National Laboratory, Argonne, Illinois 60439, USA.
‡School of Engineering, University of Brighton, Brighton, BN2 4GJ, UK.
§Department of Physics, Illinois Institute of Technology, Chicago, Illinois 60616, USA.
¶Department of Physics, University of Massachusetts, Lowell, MA 01854, USA.
∥Department of Physics, University of Surrey, Guildford GU2 7XH, UK.
††University of Maryland, College Park, Maryland 20742, USA.

Abstract. Excited structures in the proton-rich nucleus ^{140}Dy were established following the decay of a 7.3 ± 1.5 μs, $K^{\pi} = 8^{-}$ isomer. The isomer decays into the yrast line at the 8^{+} state, revealing a rotational band with a deduced deformation of $\beta_2 = 0.24(3)$. The isotope ^{140}Dy is the daughter of the deformed proton emitter ^{141}Ho. The new information obtained here supports the role of deformation in proton emission and the previous assignments of single-particle configurations to the two proton emitting states in ^{141}Ho. In addition, the reduced hindrance factor measured for the isomer is consistent with the trend observed in the $N = 74$ isotones and shows no deviation which could be attributed to the proximity of ^{140}Dy to the proton drip line.

Introduction

The proton drip line is readily delineated, above $Z = 50$, by nuclei which decay by the emission of a proton. So-called proton emitters have been identified in almost all odd-Z systems from Sb (Z=51) to Bi (Z=83) [1]. In most cases, proton emission is understood in terms of simple quantum tunnelling through a one-dimensional barrier in a spherical nucleus. Hence, proton decay has become a potent spectroscopic tool to characterise states located near the Fermi surface in nuclei at the very limits of stability.

Proton radioactivity was recently observed in ^{141}Ho and ^{131}Eu [2]. Based on the measured half-lives, the proton emitting states were interpreted as requiring the presence of a sizeable quadrupole deformation [3]. In ^{131}Eu, additional information was obtained by not only observing proton decay to the ground state of the daughter nucleus, ^{130}Sm, but also to the first excited 2^{+} level [4]. By utilising both the excitation energy of the 2^{+} state and the branching ratio, the spin and intrinsic configuration of the proton emitting state was unambiguously determined [4]. In ^{141}Ho, at the time of the experiment, no such decay to the 2^{+} level had been observed, and the assignment of the $7/2^{-}[523]$ and $1/2^{+}[411]$ Nilsson configurations to the ^{141}Ho ground and isomeric states, respectively, was deduced solely from the measured decay rates [2, 5]. Confirmation of sizeable deformation as well as complementary information on the structure of these states was obtained from in-beam γ-ray studies utilising the Recoil Decay Tagging (RDT)

CP681, *Proton-Emitting Nuclei: Second International Symposium; PROCON 2003*,
edited by E. Maglione and F. Soramel
© 2003 American Institute of Physics 0-7354-0150-0/03/$20.00

technique where rotational bands were established on top of both proton-emitting states [6]. The deduced deformations, $\beta_2 \approx 0.25(4)$, are consistent with those inferred from the proton decay rates.

While both the configuration and deformation of the ^{141}Ho proton-emitting states are consistent with the results of the in-beam γ-ray study, critical information about the ^{140}Dy daughter nucleus was still missing before this study. For example, proton emission to the 2^+ level in ^{140}Dy was no established, however, an upper limit for the branching ratio between the 2^+ and 0^+ feedings into ^{140}Dy had been placed at 1% [6] for decays from the assigned $7/2^-$ ^{141}Ho, ground state. Based on this limit, calculations using the adiabatic formalism of Ref. [7] place a lower limit of ≈ 190 keV for the excitation energy of the 2^+ state in ^{140}Dy. To confirm the implicit theoretical assumption that both the parent and daughter have the same deformation, it would be useful to measure the yrast band of ^{140}Dy, in order to establish the excitation of the 2^+ level as well as to deduce the associated deformation. The isotope, ^{140}Dy is, however, difficult to study with conventional in-beam γ-ray spectroscopy, because its production cross-section using heavy-ion fusion evaporation reactions is small and of order, $\sigma \leq 10\mu$b. The problem of small cross sections was overcome for ^{141}Ho by utilising the RDT technique, a powerful method which correlates prompt-γ radiation with the characteristic charged-particle decay of the nucleus of interest [6]. Unfortunately, ^{140}Dy decays only via β^+ emission and, thus, the RDT technique cannot be applied for in-beam studies of this isotope.

Recently, Cullen *et al.* [8] suggested that the yrast band of ^{140}Dy could be identified, at least up to the 8^+ level, by measuring γ rays emitted following the decay of a predicted $K^\pi = 8^-$ isomer. Indeed, $K^\pi = 8^-$ isomeric states exist in all of the even-even $N = 74$ isotones, $^{128}_{54}$Xe [9], $^{130}_{56}$Ba [10], $^{132}_{58}$Ce [11], $^{134}_{60}$Nd [12], $^{136}_{62}$Sm [13] and $^{138}_{64}$Gd [14]. The associated half-lives range from nanoseconds (Xe) to milliseconds (Ba,Ce). Recent calculations [15] have predicted the presence of a similar $K^\pi = 8^-$ isomeric state in ^{140}Dy. The measurement of the isomer's half-life and excitation energy is also of interest as it provides one of the few tests of the applicability of the concept of K-forbiddenness at the proton drip line and gives additional information about the shape of ^{140}Dy.

Experiment at the University of Jyväskylä, Finland.

An experiment to establish excited states in ^{140}Dy was first attempted at the University of Jyväskylä. The experiment was part of our successful campaign based on the correlation of prompt and delayed γ-ray transitions across isomeric states [8, 16, 17, 18]. A ^{54}Fe + ^{92}Mo reaction was used at beam energies of 226- and 236-MeV and the full details of the experiment are given in Ref. [16]. The experiment was successful in establishing the prompt feeding and delayed decay of two new isomeric states in the $N = 77$ isotones ^{142}Tb and ^{144}Ho, but failed to establish the $K^\pi = 8^-$ isomeric state in ^{140}Dy.

Experiment at Argonne National Lab., USA.

A second experiment, with the same beam (at a higher energy of 245 MeV) and target combination, was performed at Argonne National Laboratory. The experimental setup used the Fragment Mass Analyser (FMA) [19] and the results are discussed more fully in Ref. [20]. The recoiling nuclei were implanted into a large area silicon detector at the focal plane of the FMA. Seven 7 HPGe detectors of various sizes and efficiencies surrounded this Si detector, see Fig. 1. Mass 140 isobars only represented $\approx 5\%$ of the fusion-evaporation cross section; the two dominant masses were A=142 and 143. In order to minimise contributions from these unwanted reaction channels, slits were used to allow only 2 charge states of mass 140 residues to be detected at the focal plane. As a result, a beam intensity as high as 20 pnA was accommodated.

FIGURE 1. The experimental setup at the focal plane of the FMA.

The analysis of the γ-ray coincidence matrix produced with the requirement that the γ rays were emitted within 20 μs of the detection of an implanted ion yielded a set of five γ rays (202.1, 363.2, 475.8, 549.4, and 573.6 keV) in coincidence with one another. A spectrum produced from a sum of coincidence gates placed on all five transitions is shown in Fig. 2.

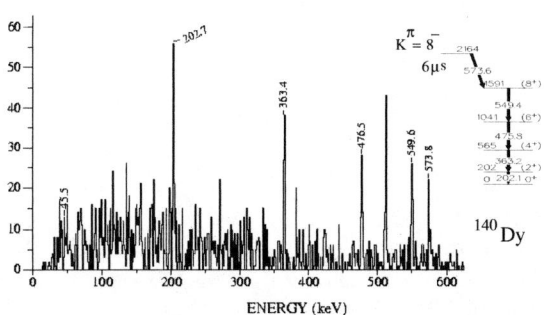

FIGURE 2. Spectrum and level scheme for the $K^{\pi} = 8^-$ isomer decay in ^{140}Dy.

The half-life of the isomer was established to be 7.3$\pm$1.5 μs which compares well with the systematic trend of the lifetimes of the $K^{\pi} = 8^-$ isomeric states across the $N = 74$ chain. Figure 2 shows the ordering of the γ rays from the decay of the $K^{\pi} = 8^-$ isomer in ^{140}Dy and Fig. 3 illustrates the systematic trend for the 0^+, 2^+, 4^+, 6^+, 8^+

and $K^\pi = 8^-$ states across the known $N = 74$ chain, from ^{130}Ba to ^{138}Gd, together with the new data on ^{140}Dy. It can be seen that the new points continue the smooth trend exhibited by earlier data.

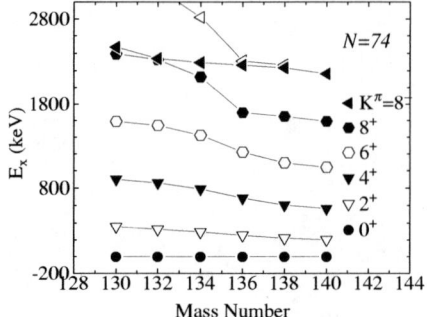

FIGURE 3. Systematics for $K^\pi = 8^-$ isomer decays in the $N = 74$ nuclei.

The $\beta_2 = 0.24$ deformation of the ground-state band in ^{140}Dy was deduced using the approach proposed by Grodzins [21]. In addition, the deformation of the daughter, calculated with the Grodzins prescription, is nearly the same as that extracted from the rotational band built on the ground state in ^{141}Ho ($\beta_2 = 0.25(4)$) [6]. Recent Woods-Saxon constrained shape polarisation calculations [15] predict the existence of a $K^\pi = 8^-$ isomeric state in ^{140}Dy at an excitation energy of 2150 keV with a deformation $\beta_2 = 0.26$. Thus, the experimental excitation energy of 2164 keV and the associated deformation of $\beta_2 = 0.24(3)$ are in good agreement with these theoretical expectations.

The branching ratio between the proton decay from the $7/2^-$ state in ^{141}Ho to the ground state and first 2^+ state in ^{140}Dy has been calculated with the adiabatic formalism of Ref. [7]. In the calculations, values of $\beta_2 = 0.244$, obtained from the Grodzins formula, and $\beta_4 = -0.046$, scaled from the Möller prediction, were used for ^{140}Dy. (Möller *et al.* [22] predict deformations $\beta_2 = 0.267$ and $\beta_4 = -0.05$ for ^{140}Dy.) The resulting branching ratio is 0.7%, consistent with the experimental upper limit of 1% obtained in Ref. [6]. As mentioned above, in the study of ^{131}Eu this branching ratio enabled a determination of the intrinsic configuration of the proton emitting state. Such information was required, because the decay rates for the two candidate configurations, were by chance, nearly the same, while the calculated branching ratios differed significantly. For ^{141}Ho, the measured decay rates led to an unambiguous assignment of the configurations to the two proton emitting states [2, 5].

In order to compare the isomeric decay of the $K^\pi = 8^-$ state in ^{140}Dy with the decays of other $K^\pi = 8^-$ isomers in the neighbouring $N = 74$ isotones, the reduced hindrance factor, f_ν, has been calculated. This factor is defined as

$$f_\nu = \{T_{1/2}^\gamma / T_{1/2}^W\}^{1/\nu}, \tag{1}$$

where $T_{1/2}^\gamma$ is the partial γ-ray half life, $T_{1/2}^W$ is the Weisskopf single-particle estimate, ν is the degree of K forbiddenness, $\nu = \Delta K - \lambda$, and λ is the multipolarity of the transition depopulating the isomeric state. The reduced hindrance per degree of K-forbiddenness for the direct, 573.6-keV, $E1$ transition from the $K^\pi = 8^-$ isomeric state to the ground-state band in ^{140}Dy is deduced to be 24.6. This value is essentially the same as those obtained for all other N=74 even-even isotones ($^{134}_{60}$Nd 26.1, $^{136}_{62}$Sm 24.9 and $^{138}_{64}$Gd 24.1), with the exception of ^{130}Ba (43.5), see below for a further discussion of this point. The fact that, in addition to the smooth energy systematics of Fig. 3, the f_ν factor is also very close to that of the other $N = 74$ isotones can then be viewed as a clear indication that the interpretation of the new long-lived state as a K isomer is justified, and that the proximity of the proton drip line has little effect upon the stability and properties of this K-isomer. Furthermore, the fact that K is approximately a good quantum number in ^{140}Dy can in turn be regarded as strong experimental evidence for an axially symmetric nuclear shape. This validates the approach used here to extract the prolate deformation parameter and adds credibility to the discussions presented above.

The near equality of the hindrance factors can also be used to propose a configuration for the ^{140}Dy isomer. In the mass 180 region, it has been shown that there is a correlation between the reduced hindrance factors and the configuration changes involved in the decay of similar K-isomers [23, 24, 25]. The equality of the hindrance factors, therefore, supports the $7/2^+[404] \otimes 9/2^-[514]$ two-quasineutron configuration for the $K^\pi = 8^-$ isomer in ^{140}Dy, matching the 8^-, ν^2 configuration established in the $N = 74$ isotones, ^{134}Nd [12], ^{136}Sm [13] and ^{138}Gd [14]. As already pointed out by Bruce *et al.* [14], the larger f_ν value for ^{130}Ba could reflect a change in the configuration and/or possibly in the shape associated with the isomer. Additional confirmation of the configuration of the $K^\pi = 8^-$ isomer in ^{140}Dy could be obtained from $M1/E2$ branching ratios of the collective rotational band built on the isomer as in Refs. [17, 26]. This information is presently not available.

In summary, excited states in the proton-rich nucleus ^{140}Dy have been established following the decay of a $K^\pi = 8^-$ isomeric state. The excitation energy of the isomer is established as 2.16 MeV with a half-life of 7.3$\pm$1.5 μs. The isomer decays into the yrast line at the 8^+ state, revealing a rotational band based on an axially symmetric shape with a deduced deformation of $\beta_2 = 0.24(3)$. ^{140}Dy is the daughter nucleus of the deformed proton emitter ^{141}Ho, and the new information obtained in this work supports previous single-particle assignments to the two proton emitting states in ^{141}Ho. Based on the measured 2^+ energy and the deduced deformation, the adiabatic calculation of Ref. [7] predicts a branching ratio for proton decays to the first excited 2^+ state in ^{140}Dy of 0.7%. This value is consistent with the experimental upper limit of 1%. In addition, the reduced hindrance factor deduced for the isomer is consistent with the trend observed in the $N = 74$ isotones and shows no deviations which could be attributed to the proximity of ^{140}Dy to the proton drip line.

ACKNOWLEDGMENTS

D.M.C. acknowledges receipt of an EPSRC Advanced Fellowship AF/100225 and two of us (A.F. and L.K.P.) acknowledge receipt of EPSRC studentships. This work is supported by the U.S. Department of Energy, Nuclear Physics Division, under Contract No. W-31-109-ENG-38. Similar results were also observed by W. Krolas *et al.* Ref. [27] and also in these proceedings.

REFERENCES

1. P. J. Woods and C. N. Davids, Annu. Rev. Nucl. Part. Sci. **47** (1997) 541.
2. C.N. Davids, *et al.*, Phys. Rev. Lett. **80** (1998) 1849.
3. V.P. Bugrov and S.G. Kadmensky, Sov. J. Nucl. Phys. **49** (1989) 967.
4. A.A. Sonzogni, *et al.*, Phys. Rev. Lett. **83** (1999) 116.
5. K. Rykaczewski, *et al.*, Phys. Rev. C **60** (1999) 011301.
6. D. Seweryniak, *et al.*, Phys. Rev. Lett. **86** (2001) 1458.
7. H. Esbensen and C.N. Davids, Phys. Rev. C **63** (2001) 014315.
8. D.M. Cullen, *et al.*, Nucl. Phys. **A682** (2001) 264c.
9. L. Goettig, *et al.*, Nucl. Phys. **A357** (1981) 109.
10. H.F. Brinkman, C. Heiser, K.F. Alexander, W, Neubert, and H. Rotter, Nucl. Phys. **81**, 233 (1966).
11. D. Ward, R.M. Diamond, and F.S. Stephens, Nucl. Phys. **A117**, 309 (1968); Yu. V. Sergeenkov, Nuclear Data Sheets **65**,277 (1992).
12. D.G. Parkinson, I. A. Fraser, J.C. Lisle, and J.C. Willmott, Nucl. Phys. **A194**, 443 (1972).
13. A.M. Bruce, P.M. Walker, P.H. Regan, G.D. Dracoulis, A.P. Byrne, T. Kibédi, G.J. Lane, and K.C. Yeung, Phys. Rev. **C50**, 480 (1994).
14. A.M. Bruce, *et al.*, Phys. Rev. **C55**, 620 (1997).
15. F.R. Xu, P.M. Walker, and R. Wyss, Phys. Rev. **C59**, 731 (1999).
16. C. Scholey, *et al.*, Phys. Rev. C **63** (2001) 034321.
17. D.M. Cullen, *et al.*, Phys. Rev. C **58** (1998) 846.
18. D.M. Cullen, *et al.*, Phys. Rev. C **66**, 034308 (2002).
19. C.N. Davids, *et al.*, Nucl. Instru. Meth. Phys. Res. B **70** (1992) 358.
20. D.M. Cullen, *et al.*, Phys. Lett. **B**529 (2002) 42.
21. L. Grodzins, Phys. Lett. **2**, 88 (1962).
22. P. Möller, J.R. Nix, W.D. Myers and W.J. Swiatecki, At. Data Nucl. Data Tables **59** (1995) 185.
23. F.G. Kondev, G.D. Dracoulis, A.P. Byrne, T. Kibédi, S. Bayer and G.J. Lane, Phys. Rev. C **54** (1996) R459.
24. G.D. Dracoulis, F.G. Kondev, A.P. Byrne, T. Kibédi, S. Bayer, P.M. Davidson, P.M. Walker, C. Purry and C.J. Pearson, Phys. Rev. C **53** (1996) 1205.
25. F.G. Kondev, G.D. Dracoulis, A.P. Byrne and T. Kibédi, Nucl. Phys. A **632** (1998) 473.
26. P.M. Walker and G.D. Dracoulis, Nature **399**, 35 (1999).
27. W. Krolas *et al.*, Phys. Rev. C **65** (2002) 031303(R).

Beta-Delayed Multi-Particle Emission Studies at ISOLDE

M.J.G. Borge

Instituto de Estructura de la Materia, CSIC, Serrano 113bis, E-28006-Madrid, Spain

Abstract. We report here on the recent β-decay studies made at ISOLDE/CERN to determine the multiparticle breakup mechanism of excited states in light nuclei. The β-2p emission in ^{31}Ar is resolved. Mirror beta transitions in the A=9 chain are compared and a large asymmetry factor is deduced for the transitions to high excitation energy in ^{9}Be (11.8 MeV) and ^{9}B (12.2 MeV) fed in the β-decay of ^{9}Li and ^{9}C respectively. It is shown that the asymmetry is not due to experimental problems or differences in the mechanisms of breakup or in the spin of the states. only differences in the partial decay branches of the breakup channels has been found.

INTRODUCTION

The main characteristic of beta-decay far from stability is the amount of decay channels open. This is due to the quadratic increase of the isobaric mass differences and the reduction in the separation energies for emitting nucleons or clusters of nucleons when moving away from the valley of stability.

The process of β-delayed particle emission has been the subject of much study during the last few decades, as it allows to uniquely determine the decay-pattern from the energy of the emitted particle, if the final state is known. As the decay pattern depends on the structure of the states involved, the decay properties of great interest in themselves are also an important tool to investigate nuclear structure.

Beta-delayed multi-particle emission occurs when the nucleus breaks up into more than two particles. The main interest in β-delayed multi-particle emission is the fact that the mechanism of the break-up is not fully determined by energy and momentum conservation. In three body break-up there are three binary subsystems and each sub-system have resonances controlling the break-up. Either the break-up proceeds via each of these resonances sequentially or the beta-daughter breaks up directly into the three body continuum. Sequential and direct break-up represent limiting multi-particle mechanisms as direct processes and compound nucleus are for nuclear-reaction mechanisms. The break-up process is therefore determined by the width of the resonance in the binary subsystem, the height of the barrier and the structure within the decaying state that could single out a specific channel. *Learning about this structure is one of the main goals in studying β-delayed multi-particle emission.*

In a binary decay, to which reduces a sequential 3-body break-up, the process is described by the R-matrix theory, which depends on the barrier penetrability and on the reduced width. This method has been applied to the decays of $\beta p\alpha\alpha$ from ^{9}C [1] and $\beta n\alpha\alpha$ of ^{9}Li [2] assuming sequential decay. But the obtained spectroscopic factors

CP681, *Proton-Emitting Nuclei: Second International Symposium; PROCON 2003,*
edited by E. Maglione and F. Soramel
© 2003 American Institute of Physics 0-7354-0150-0/03/$20.00

are strongly dependent of the interactions used.

The case of resonances of width comparable with the energy available in any of the two-body subsystems has been discussed in [3] and references therein. The author argues that in this case the break-up of the binary system proceeds so fast that the presence of the third particle is still felt. They have developed a formalism for expanding the decay amplitude in a set of functions (hyperspherical harmonics) that fully describes the final state of three particles. This formalism has been applied to the break-up resonances in ^{6}Be, ^{6}He, ^{6}Li, ^{9}Be and ^{9}B [4]. All these breakups except for the ^{6}Be(0^+) state has also been successfully described applying the R-matrix theory previously mentioned. As the mechanism involved are so different the fact that both theoretical approaches are able to describe most of the data poses a problem. It was often mentioned in the literature that a kinematically complete study where all particles were detected and their energy and relative angles measured should disentangle the mechanism [3].

Therefore we have decided to build a set-up with the latest strip detectors achieving large solid angle and high granularity to efficiently detect events of high multiplicity. This allow us to measure with high accuracy the energy and direction of the emitted particle in the decay. One can thus determine the correlations and the detailed kinematics of the multi-particle emission process. To successfully use the system it is needed to have point-like sources of high purity as the ones delivered by ISOL-type installations.

This technique has been applied to study the multi-particle decay modes of two proton drip line nuclei ^{31}Ar ($T_Z=-5/2$) and ^{9}C ($T_Z=-3/2$) and the mirror nucleus of the latter ^{9}Li ($T_Z=3/2$).

The elusive character of the two proton radioactivity (predicted by Goldanskii in 1960 [5] and observed in ^{45}Fe [6, 7] in 2002) has caused quite some interest in the $\beta 2p$ decay channel. As the escape of the delayed proton pair require tunnelling through the potential barrier, the observation of energy and angular correlation between the protons could give information on the nucleon-nucleon interaction. However, this decay mode will compete with sequential emission where the first proton feeds a proton emitting state. The question of the mechanism of $\beta 2p$ concerns the relative importance of sequential and direct emission of the two protons.

Eight nuclei have been observed to beta decay by 2p emission from the IAS [8]. As they are in the region, A = 22-43, where most of the intermediate levels open to the Q_β-window are narrow, it is easy to experimentally determine the decay mechanism. Experimentally the different mechanisms can be distinguished by looking at the individual proton energies of the 2p events and their angular distributions. In sequential emission the individual proton energies depend on the intermediate state. Thus, the spectrum will peak at a certain energy while in direct emission the proton spectrum is continuous. The angular distribution in direct emission is far from being isotropic while the angular dependence introduced by the momentum coupling in the sequential case is negligible. All $\beta 2p$ emitters have Z > N, so the IAS collects a large part of the beta-strength. This explains why $\beta 2p$ has only been observed prior to our studies from the IAS and the mechanism of 2p-emission from the IAS has been resolved for the ^{22}Al and ^{26}P [9] decays.

The nucleus $^{31}_{18}$Ar$_{13}$ ($T_Z = -5/2$) was the most promising case to improve our fundamental knowledge of $\beta 2p$ processes. It has been studied in some detail since it was one of the candidates for direct 2p radioactivity. Several experiments [10, 11, 12] at GANIL had identified the main decay branches (βp, β2p) and determined a half-life value of $15.1^{+1.3}_{-1.1}$ ms. However, these experiments suffered from a contamination of ^{29}S, also a βp emitter, and from an extended source size. The first experimental determination of individual proton energies in β2p decay [13] indicated that the decay mechanism was complex.

The mechanism of two proton emission was studied via the individual proton energy distribution and by looking to the angular dependence of the two-proton events. The one-proton projections show individual peaks one narrow and with an energy determined by the excitation energy in the intermediate nucleus in this case ^{30}S and the second one broaden as it is emitted by a recoiling nucleus (for details in the individual energies and angular projections see ref. [14]). The angular distributions are compatible with isotropic emission. Many β2p transitions were identified and the 2p-emission was independent of isospin and mainly sequential. A scatter plot of the Q_{2p} versus individual proton energies showed the presence of diagonal lines indicating the presence of a common intermediate level in the proton daughter. More details about this study can be found in the proceedings of the previous PROCON conference [8] and references therein.

A summary of the main results obtained at ISOLDE (CERN) to study the multiparticle breakup following the decays of ^{9}C and ^{9}Li in order to compare mirror beta transitions is described in the following.

STUDY OF MIRROR TRANSITIONS IN THE A = 9 CHAIN

Due to the stability of the α-particle and the fact that ^{8}Be is unbound, the multi-particle break-up dominates the decay of very light proton rich nuclei. As an example ^{9}C β-decays to states in ^{9}B that are all well above the p$\alpha\alpha$ threshold. The mirror partner, ^{9}Li has also large decay branches ($\approx$50%) to n$\alpha\alpha$ final states. The latter has been fully explained in terms of sequential decays through both the ^{5}He and ^{8}Be channels [2], although components directly to n$\alpha\alpha$ states have also been suggested [15]. The beta decay of ^{9}C has been studied previously [1], comparison of mirror transitions in the A=9 chain shows unusual large asymmetries. An asymmetry value δ ($\delta = (ft)^+/(ft)^- - 1$) of 1.2 ± 0.5 is determined [1, 2] in the decays to the 5/2$^-$ states at excitation energy $\sim$2.4 MeV. The large uncertainty in the δ value mainly arises from incomplete knowledge of the ^{9}C decay.

A potentially larger asymmetry — and a much more interesting one — seems to exist in the decays to highly excited states at $\sim$12 MeV excitation energy fed with branching ratios of 2–3 %. The decay of ^{9}Li was investigated in ref. [2] giving for the mirror level around 11.8 MeV a value of $B_{GT} = 5.6 \pm 1.2$. In the study of Ref. [15] the beta feeding at high excitation energy was assigned to the 11.28 MeV state in ^{9}Be (see Fig. 1). Only a lower limit on B_{GT} value is given in [1] for the corresponding transition from ^{9}C. The possible asymmetry for the decay to the states around 12 MeV is interesting not only due to the fact that the individual B_{GT} values are large (with large overlap in wave-functions, an unambiguous interpretation is much easier made), but also due to the special role played by this transition for the ^{9}Li decay. It seems to belong to a class of high-B_{GT} transitions observed [16] at the neutron drip line and has been suggested to be due either

195

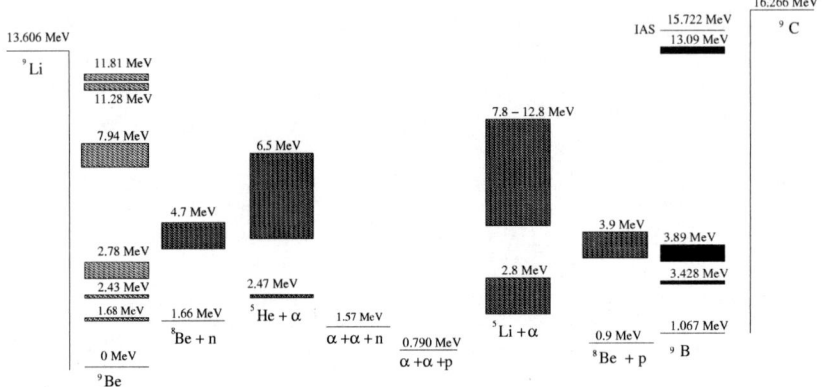

FIGURE 1. Partial level scheme of the nuclei in the A=9 chain taken from [19]. It is also display the ground state and first excited state of the nuclei that could participate in the breakup process. All energies are referred to the g.s. of ^{6}Be and the values are taken from [19].

to a lowering of the giant Gamow-Teller resonance [17] or to the occurrence of "two-neutron $\rightarrow$ deuteron" transitions [18]. Knowing whether the mirror transition on the proton rich side has a similar strength would help greatly in identifying what causes the large transition strengths. Among the four cases known with high B_{GT}-values ^{6,8}He and ^{9,11}Li the study in the mirror system can only be done in the A = 9 isobars as in the other systems the mirror nucleus is unbound.

Fig. 1 shows the decay schemes of the mirror nuclei ^{9}Li and ^{9}C as they were known [19] prior to the experiments we report upon in this contribution.

Using a CaO target, the ^{9}C nuclei were extracted via the $A = 25$ CO$^+$ sideband at ISOLDE/CERN with a yield of about 10^3 ions/s. The molecular beam passed through a hole in an annular detector and was implanted in a thin carbon foil (for details of the set-up see [20, 21]). The C-foil was in the diagonal of a set of two telescopes. Two different sets of telescopes were used to study the decay. In the first one (schematically shown on the left part of Fig. 2), the telescopes consisted both of a DSSSD in the front and a thick Si surface barrier detector in the back. A photograph of the system is shown on the right part of Fig. 2. When faulty strips and shadowing effects from the frame of the C-foil were taken into account effective solid angles for these telescopes were Ω_1=10.0 %, Ω_2=7.6 % of 4π with an energy range of 250 - 13.500 keV for protons and beyond 450 keV for alphas. In order to detect the feeding to the ^{9}B ground state (gs) that breaks up in a proton of 164 keV and 2α one of the DSSSD was substituted for a 25μm-100mm^2 Si detector. For first time these protons were directly measured. A more precise ground state branch of b(gs) = 54.1(15)% was obtained in agreement with previous values [1, 22]. The corresponding B_{GT}=0.0184(6) agrees well with the value of the mirror transition in ^{9}Li (B_{GT}=0.019(2)) [2], thus no asymmetry is deduced for the ground state - ground state transitions in the A=9 chain.

The ^{9}C nucleus decays to levels in the daughter nucleus ^{9}B that break into three charged particles: a proton and two α-particles. These three particles were measured in coincidences. No proton alpha discrimination was possible with this system except for

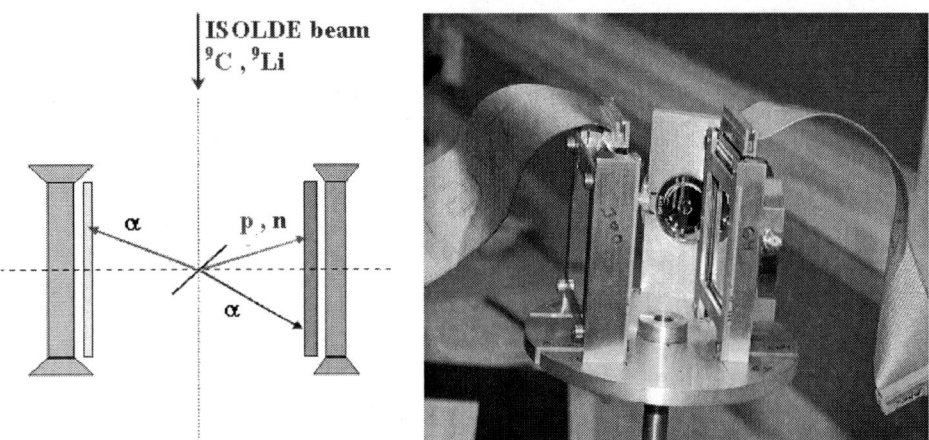

FIGURE 2. Experimental Setup. On the left a schematic view of the setup optimized for triple coincidences. It consisted of two telescope with a DSSSD in front and a thick Si surface barrier detector in the back for the detection of betas and other energetic particles traversing the DSSSD. The ISOLDE low energy high purity beam goes through the 4 mm hole of an annular detector and is stopped in a thin C-foil forming a point-like radioactive source. A photograph of the setup is shown on the right hand side.

high energy protons that reach the back detector of the telescope (see Fig.2). However, in the decays where all three particles were detected, particle identification could be deduced from the requirement of momentum conservation since the break-up of ^{9}B take place almost at rest. In the analysis events with multiplicity three or higher were selected.

Fig. 3 shows an overview of the triple coincidences obtained for the ^{9}C decay with the second setup. To be compared with equivalent picture shown in ref [21] for the data obtained with the first setup.

On the left part the possible levels in ^{9}B available for allowed beta transitions within the Q_{EC}-window are schematically drawn. On the right part the sum energy of the three particle is represented versus the individual energies of each of them ($p\alpha\alpha$). Thus each event is represented by three points laying on the same horizontal line. The projection onto the y-axis shows the beta feeding to the different states if efficiency corrections for the different channels are taken into account. If one concentrate in the part of high energy the presence of diagonals is clearly observed. Due to the reduced angular resolution of this setup compare with the one shown in Fig. 2 the decay through the '8Be(gs) channel is enhanced. The lines of slopes 9/5 and 9/8 correspond to the energy of the first emitted particle of energy E_1 when the decay proceeds through the ^{5}Li and ^{8}Be channels respectively. $E_{sum} = (M_1 + M_2)E_1/M_2 + x$, where x = 92 keV for ^{8}Be (gs) , slope 9/8, and x = 1.97 MeV for ^{5}Li(gs). M_1 stand for the mass of the first emitted particle and M_2 for the mass of the recoiling system. As stressed above the presence of diagonal lines indicates that the decay mainly occurs sequentially, so R-matrix formalism can be used to described the breakup of the level at high excitation energy around 12 MeV in ^{9}B.

A satisfactory description of the three-body decay from a narrow energy region ($\pm$ 50 keV) around the 12.2 MeV resonance is obtained within the sequential model involving the ground and the first-excited states of ^{5}Li and ^{8}Be. To get a detail explanation of the

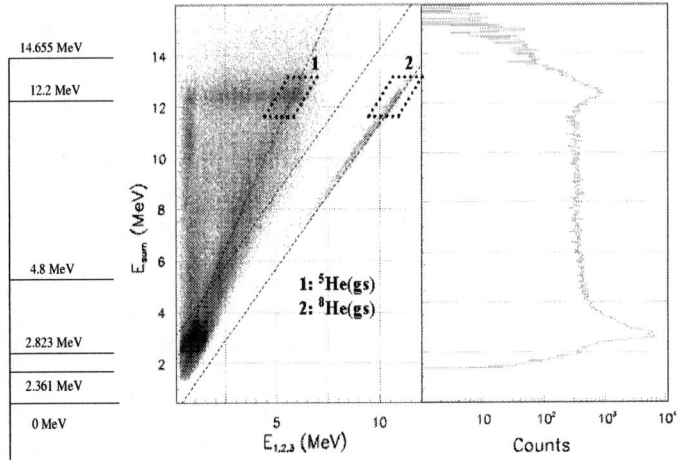

FIGURE 3. Scatter plots of the E_{sum} of the 3-particle breakups versus individual particle energies from the ^{9}C β-decay obtained with the setup of Fig. 2 where one of the DSSSD is substituted by a $25\mu m$-$100mm^2$ Si detector. On the left the levels in the daughter nucleus ^{9}B available for the decay are schematically presented. On the right hand side the projection onto the E_{sum} corresponding to the beta feeding if corrections for angular acceptance are done.

analysis method see [21] The angular distribution of the events taken with 9/5 slope, dominated by the ^{5}Li($3/2^-$) branch allows to fix the spin of the 12.2 MeV level to 5/2 [21].

Table 1: Partial breakup channels of the 12.2 MeV state in ^{9}B

$^8Be(0^+) + p$	$^8Be(2^+) + p$	$^5Li(3/2^-) + \alpha$	$^5Li(1/2^-) + \alpha$	Ref.
0.090±0.010	0.25±0.07	0.60±0.06	0.06±0.04	This work
0.085±0.014	0.18±0.03	0.74±0.08		[22]

Table 1 summarized the partial decay branches obtained from the breakup of the 12.2 MeV state in ^{9}B. Our results compare well with the recent ones of Gete et al. [22], but their setup was less efficient for triple coincidences what manifest in the fact that they cannot distinguish the contribution from the $^5Li(1/2^-) + \alpha$ channel. The high energy protons from the break up of the 12.2 MeV state into $^8Be(0^+)$ are used to determine a B_{GT} value of 1.20(15) for this level.

The study of the ^{9}Li β-decay was done with the same setup as ^{9}C to avoid in the comparison of mirror transitions the systematic errors that could come from the setup. An intense ^{9}Li-ion beam was produced at ISOLDE/CERN by spallation and fragmentation of a Ta-foil target by a 1.4 GeV proton beam from PS-Booster. The beam was stop in a C-foil placed 45° with respect to the axis of the two set of telescopes, each with a DSSSD in the front one of 64 μm thick and the other of 300 μm thick respectively (see Fig. 2). In these standard DSSSDs detectors the deadlayers from strip contacts and

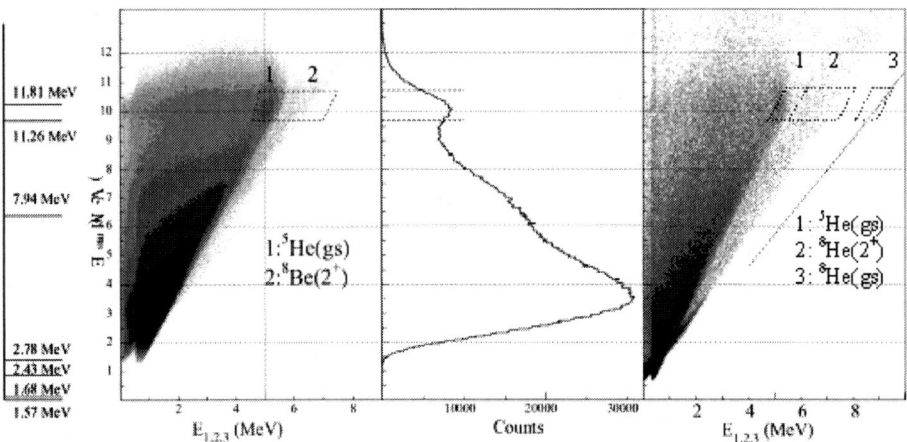

FIGURE 4. Scatter plots of the reconstructed E_{sum} of the $n\alpha\alpha$ particles versus individual particle energies from the ^{9}Li β-decay for the two experiments done. Notice the effect at low energies of the used of DSSD-detectors with ultrathin windows. On the left the levels in the daughter nucleus ^{9}Be available for the decay are schematically presented. In the central part the E_{sum} projection corresponding to the first experiment is displayed. The dashed lines show the interval used in the analysis.

doping layers were equivalent to 630 nm of Si what constitute a serious limitation for detecting low-energy α-particles present in the β-decay of ^{9}Li. Later a test experiment was done with two newly designed DSSSDs with a reduced deadlayer equivalent to 100 nm of Si (both 60 μm thick). The effect in the energy spectrum of the use of ultrathin window DSSSDs is shown in scatter plots of Fig.4 where up to 1 MeV of sensitivity is gained in the total energy of the breakup.

The β-decay of ^{9}Li feeds in 50 % of the cases the ground state of ^{9}Be and the rest the excited states that decay by the emission of a neutron (P_n = 50.8(2) % [23]) and two charged particles (alphas). The analysis of the data is based in the detection of the two alphas, by the use of momentum conservation the neutron energy and direction can be deduced and the complete kinematics reconstructed. It should be noticed that the high granularity of the setup is crucial to obtain a reasonable resolution of the deduced neutron energy. This is the first time that this kind of analysis is attempt. The only previous study of ^{9}Li β-decay where $\alpha\alpha$ coincidences were measured [15] did not have sufficient angular resolution and no detailed analysis of the breakup mechanism was performed. Fig. 4 displays the overview of the reconstructed triple coincidences in ^{9}Li versus the individual particle energies for both experiments. One can see on the right hand the scatter plot from the test run with the ultrathin window DSSSD Detectors. A gain of 1 MeV in the reconstructed E_{sum} is achieved by reducing the deadlayers in the detectors. The observation of the diagonal lines in the decay of ^{9}Li is worse resolved as one of the particles (the neutron) is not detected, but reconstructed from the conservation of momentum. In this occasion only the decay through the ^{5}He(gs) is clearly

seen (dashed square). The threshold for particle break-up is situated at 1.57 MeV,

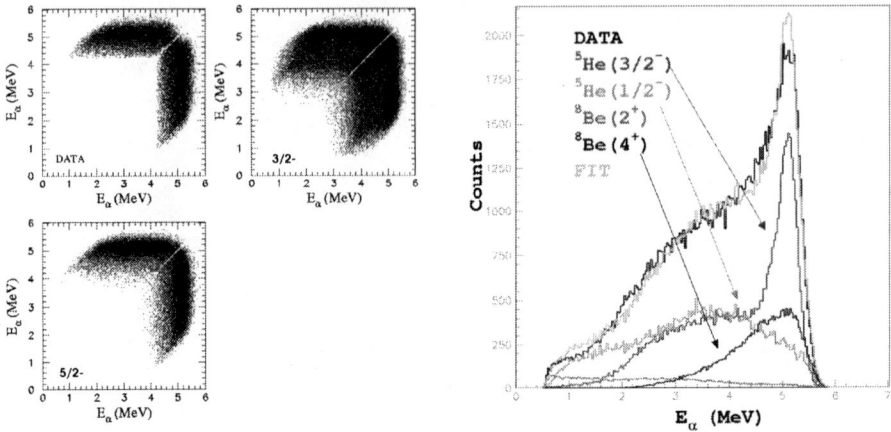

FIGURE 5. Analysis of the 11.8 MeV state in ^{9}Be through the region $9.7 \leq E_{sum} \leq 10.7$ MeV. On the left the two dimension plot showing the $E_{\alpha 1}$ versus $E_{\alpha 2}$ energy spectrum for the decay channel through the ^{5}He(gs) is compared with the simulations for two possible spin values for the 11.8 MeV state. Clearly the data favours spin $5/2^-$. On the right the α-spectrum is compared to the fit obtained from the four main partial decay branches of the 11.8 MeV state.

therefore the sum energy is the energy of the state minus this value. The projection onto the sum energy shows that in the region of interest (11-12 MeV) there is a peak centred around $E_{sum} = 10.1$ MeV corresponding to the feeding of the 11.81 MeV level (11.81 MeV-1.57 MeV). An R-matrix fit to the region in between the dashed lines using the position and width of the 11.28 and 11.81 MeV states in ^{9}Be given in Ref. [19] shows that the β-decay of ^{9}Li mainly feeds the 11.81 MeV state. The earlier controversy [2, 15] on the feeding pattern to this region of ^{9}Be is therefore resolved. This result is consistent with the assignment of spin $7/2^+$ given for the 11.28 MeV state observed in (p,p') [24] and (e,e') [25] inelastic scattering studies.

The main analysis is done therefore in the region $9.7 \leq E_{sum} \leq 10.7$ MeV. Once knowing which is the high excited level fed in the β-decay of ^{9}Li, one can determine the spin of the state by looking to the angular correlations between the emitted alpha particles. The events in the rectangular region label as '1' in Fig. 4 correspond to the decay mainly through the ^{5}He(gs) channel. A two dimension plot ($E_{\alpha 1}$ versus $E_{\alpha 2}$) of these data is shown on the left hand side of Fig. 5. Monte-Carlo simulations of the α-energies for the branch ^{9}Be$(J^\pi) \longrightarrow ^5He(3/2^-) \longrightarrow n + \alpha$ were done for $J^\pi = 5/2^-$ and $3/2^-$. The data clearly favours $J^\pi = 5/2^-$. The analysis of the angular distribution between the two alphas corroborate this result, for more details see Ref. [26].

To extract the partial branching ratios of the 11.81 MeV state through the different channels two different fits have been done. One to the energy of the alphas coming from the interval of E_{sum} chosen for the analysis and the other to the two dimension plot of $E_{\alpha 1}$ versus $E_{\alpha 2}$. The data was fitted to the distributions of alphas coming from Monte-Carlo simulations of the decay through five different channels which contribution are given in Table 2. A fit to the α-spectrum from the first run is shown on the right hand side of Fig. 5, as already explained the contribution of the channel through $^8Be(0^+)$ was not detected during this first run of high statistics due mainly to the deadlayers of the detectors. The statistics in the test run was less, but the energy thresholds were lower so the channel $^8Be(gs) + n$ could also be observed. The branching ratios obtained for the different channels in the two experiments for the different fits are average and given in Table 2 with conservative error bars that include both the statistical error as well as the difference in value among the results. These final values are in good agreement with the earlier determination of the branching ratio to the $^8Be(2^+)$ of 12(4) % and the upper limit on the branching ratio to ^8Be of 3 % [27]. It is interesting to notice that the break up through the $^8Be(4^+)$ channel is of considerable importance.

Table 2: Partial breakup channels of the 11.8 MeV state in ^9Be

$^8Be(0^+) + n$	$^8Be(2^+) + n$	$^8Be(4^+) + n$	$^5He(3/2^-) + \alpha$	$^5He(1/2^-) + \alpha$
0.02±0.01	0.10±0.06	0.12±0.08	0.28±0.06	0.48±0.07

As already mentioned a fit in a restricted E_{sum}-interval very close to the centroid position of the level is enough to obtain the B_{GT}-value. A test of the quality of the fit has been done and is discussed in detailed in Ref. [26]. So to extract the B_{GT}-value one has to find the absolute branching ratio for a well defined part of the spectrum. This was taken from the high part of the singles alpha spectrum in the energy interval of 4.5-5.5 MeV. that correspond to $30(5)\times 10^{-4}$ of the total. The uncertainty coming mainly from the separation of noise and low energy alphas. Crucial in this analysis the low thresholds of the ultrathin window DSSSD detectors as the α spectrum from the ^9Li decay peaks below 600 keV. Considering that different correction factors: mixing of alphas from other states in the interval considered, energy dependence of the phase space factor and the fact that not all the 11.8 keV peak falls in the energy intervals as well as the half-life of ^9Li and the P_n-value [23], the resulting B_{GT} value is 5.3(10) [26].

The comparison with the mirror transition in the decay of ^9C gives a asymmetry parameter $\delta = (B_{GT}^- / B_{GT}^+) - 1 = 3.4(10)$. This is the highest asymmetry ever observed.

CONCLUSIONS

In this contribution we have revised recent results from multiparticle emission studies in light nuclei. We have studied the multiparticle breakup of the high excited states of the A=9 chain populated in the beta decay of ^9Li and ^9C. We have identified the main levels at high energy fed in these mirror beta decays as 11.8 MeV (Γ=400 keV) in ^9Be and 12.2 MeV (Γ=450 keV) in ^9B respectively. The position and width of the former was taken from the literature [19] and the position and width of the latter was deduced in this work. The distribution of the emitted particles indicates that the breakup is mainly sequential in both cases and R-matrix model was used to described the breakup in different channels.

The study of the angular distributions between the first emitted particle and the direction of the other two in the central of mass-system has allowed to determine the spin of these states to be 5/2 in both cases.

For first time the partial decay branches for these mirror states have been studied with the same setup to avoid systematic effects. In the decay of both mirror levels the emission of an alpha first dominates as expected if one consider that these states fed with high B_{GT} values should have an structure close to the parents and far from the α-cluster structure of ^{8}Be. Nevertheless the big differences in the detail of the partial decay branches of these mirror states indicates that the radial part of the wave functions are very different and probably responsible of the large asymmetry factor δ deduced for this mirror transitions. This δ-factor of 3.4(10) is the largest ever obtained.

REFERENCES

1. D. Mikolas et al., Phys. Rev. **C37**, 766 (1988).
2. G. Nyman et al., Nucl. Phys. **A510**, 189 (1990).
3. A.A. Korsheninnikov, Sov. J. Nucl. Phys. **52**, 827 (1990).
4. O.V. Bochkarev et al., Nucl. Phys. **A505**, 215 (1989); Sov. J. Nucl. Phys. **52**, 964 (1990); Sov. J. Nucl. Phys. **55** 955 (1992).
5. V.I. Gol'danskii, Nucl. Phys. **19** 482 (1960).
6. J. Giovinazzo et al., Phys. Rev. Lett. **89**, 102501 (2002).
7. M. Pfützner et al., Eur. Phys. J. **A14**, 279 (2002).
8. M.J.G. Borge, "Beta delayed two-particle emission" in PROCON ' Physics, New York 2000, pp. 264-274.
9. M.D. Cable et al., Phys. Rev. **C30**, 1276 (1984).
10. V. Borrel et al., Nucl. Phys. **A473**, 331 (1987).
11. V. Borrel et al., Nucl. Phys. **A531**, 353 (1991).
12. D. Bazin et al., Phys. Rev. **C45**, 69 (1992).
13. M.J.G. Borge et al., Nucl. Phys. **A515**, 21 (1990).
14. H.O.U. Fynbo et al., Nucl. Phys.**A677**, 38 (2000).
15. M. Langevin et al., Nucl. Phys. **A366**, 449 (1980).
16. M.J.G. Borge et al., Z. Phys. **A340**, 255 (1991).
17. H. Sagawa, I. Hamamoto and M. Ishihara, Phys. Lett. **303B**, 215 (1993).
18. A. Poves et al., Z. Phys. **A347**, 227 (1994).
19. Ajzenberg-Selove, F., et al., Nucl. Phys. **A490**, 1-133 (1988).
20. O. Tengblad et al.,"Beta decay asymmetry in mirror nuclei: A=9" in ENPE 99, edited by B. Rubio, M. Lozano, and W. Gelletly, AIP Proceedings 495, American Institute of Physics, New York 1999, pp. 19-22.
21. U.C. Bergmann et al., Nucl. Phys. **A692**, 427 (2001).
22. E. Gete et al., Phys. Rev. **C61**, 064310 (2000).
23. G. Audi et al., Nucl. Phys. **A624**, 1 (1997).
24. S. Dixit et al., Phys. Rev. **C43**, 1758-1776 (1991).
25. J.P. Glickman et al., Phys. Rev. **C43**, 1740-1757 (1991).
26. Y. Prezado et al., Phys. Lett. B in preparation.
27. C.L. Cocke and P.R. Christensen, Nucl. Phys. **A111**, 623 (1968).

β-delayed p and γ decay of ^{58}Zn and comparison with ^{58}Ni(^{3}He,t)^{58}Cu reaction

A. Jokinen[*†], IS403 Collaboration[†] and the ISOLDE Collaboration[†]

[*]Department of Physics, PB 35 (YFL), FIN-40014 University of Jyväskylä
& Helsinki Institute of Physics, PB 64, FIN-00014 University of Helsinki
[†]CERN, CH-1211 Geneva 23, Switzerland

Abstract. Symmetry tests between (p,n) type charge-exchange reactions and beta decay are discussed. An example of such comparison in A=58 system is given with special emphasis to test reliability of charge- exchange data as a probe for Gamow-Teller strength. Data collected so far is reviewed and the recent beta decay experiment at ISOLDE-CERN is outlined.

INTRODUCTION

Within a framework of charge symmetric nuclear interaction, an isopin is a good quantum number and symmetric structures are expected between nuclei with same isospin projection $\pm T_z$. Corresponding states in different nuclei with same T_z are called isobaric analog states. More than symmetric states, also symmetric transitions are observed if an inital and/or final state is among analog states.

GAMOW-TELLER PROBES

Beta decay strength and experimental observable are coupled via an equation (1):

$$(1 + \delta_r) ft = \frac{C}{(1 - \delta_c)\langle F \rangle^2 + (\frac{g_A}{g_V})^2 \langle GT \rangle^2} \tag{1}$$

Thus the measurement of the branching ratios, half-lives and Q-values can provide direct information on the Gamow-Teller strength. Unfortunately large part of the decay strength is situated outside of the available decay energy window and the relevant information is partly non-accessible. Situation gets better when approaching more exotic nuclei as the mass diffence between the parent and the daughter nucleus increases.

In early 80's it was proposed that same information, i.e. Gamow-Teller strength can be obtained from (p,n)-type measurements performed at small angles and at intermediate energies (100-150 MeV/u). The cross section of the charge-exchange reaction is related to Gamow-Teller strength according to the following relation (2) proposed in [1]:

$$\sigma(0°) = K N_{\sigma\tau} |J_{\sigma\tau}(0°)|^2 B(GT) \tag{2}$$

CP681, *Proton-Emitting Nuclei: Second International Symposium; PROCON 2003*,
edited by E. Maglione and F. Soramel
© 2003 American Institute of Physics 0-7354-0150-0/03/$20.00

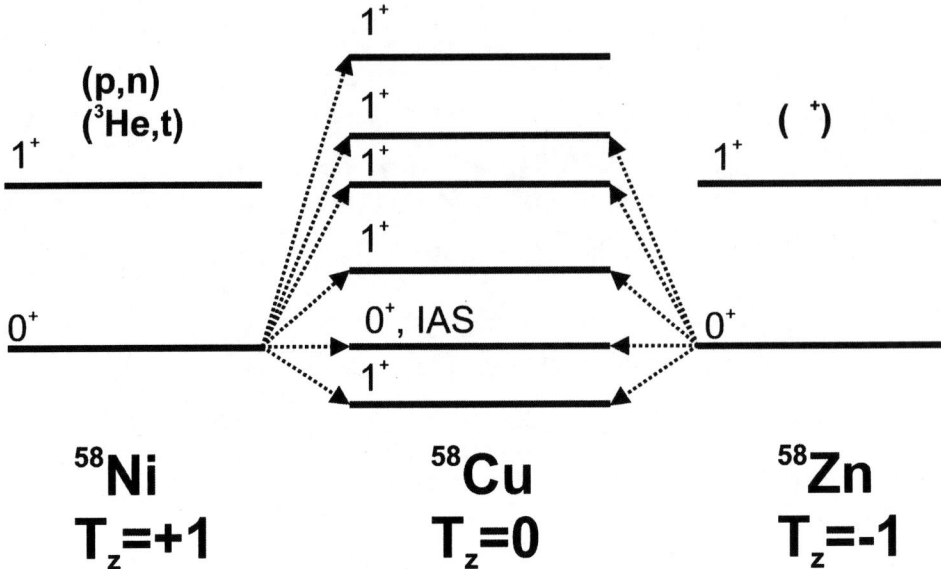

FIGURE 1. A=58 levels system in isospin space after correction for Coulomb energy difference

Equation above includes many assumptions, and one may ask if the information extracted is reliable. Although one-step process is expected to dominate, the question of more complicated processes remains. While selected energy range and small angles enhance $L = 0$ component, the role of higher multipoles can be questioned, especially for weak transitions. However, such a relation has made it possible to probe Gamow-Teller strength beyond decay energy window resulting in, for example, observation of Gamow-Teller giant resonance.

COMPARISON OF GT-PROBES

Comparison between isospin symmetric processes have been scarce so far. The simplest system where such a comparison is possible, is between $T = 1/2$ mirror nuclei. In sd-shell a relatively good agreement has been obtained recently, while comparing the ^{27}Al$(^{3}$He,t$)^{27}$S reaction and ^{27}Si beta-decay [2]. For $T = 1$ systems, the symmetry tests have been examined only for light sd-shell nuclei, see for example Ref. [3] for comparison of ^{38}Ar(p,n)^{38}K and ^{38}Ca beta decay. In what follows we describe the first results for comparison of $T = 1$ system beyond sd-shell.

Charge-exchange data on A=58 system

A=58 system provides an interesting possibility to compare beta decay and hadronic charge-exchange probes beyond *sd*-shell. A stable ^{58}Ni is a good target material for charge-exchange reaction studies. Such reaction on ^{58}Ni ($T_z = +1$) populates states in self-conjugate ($N = Z = 29$) ^{58}Cu ($T_z = 0$) nucleus. Same states are fed by the beta decay of ^{58}Zn ($T_z = -1$) providing a direct comparison to hadronic probes. Due to relatively high Q-value ($Q_{EC} \approx 9.4 MeV$) of ^{58}Zn, the comparison covers wide excitation energy range. Figure 1 illustrates the idea of such a comparison in A=58 system.

$^{58}Ni(p,n)^{58}Cu$

Charge-exchange reaction data on ^{58}Ni target at intermediate energies starts from pioneeringt work of [4, 5]. They applied (p,n) reaction at 120 MeV/u and 160 MeV/u. On that time they reached an energy resolution of the order of 300 and 600 keV with previously mentioned incident energies, respectively. With (p,n) reaction experimental difficulties are related to neutron detection, which was based on time-of-flight over 100 m flight path.

$^{58}Ni(^{3}He,t)^{58}Cu$

A great improvement in energy resolution and sensitivity to weak branches was obtained when ^{58}Ni(^{3}He,t)^{58}Cu -reaction insted of ^{58}Ni(p,n)^{58}Cu -reaction was applied. In the first experiment, an energy resolution of the order of 50 keV was obtained resulting in a fine structure of the Gamow-Teller strength [6, 7].

Beta-decay of ^{58}Zn

Beta decay of ^{58}Zn was first time studied by means of β-γ spectroscopy in late 90's at ISOLDE [8]. On that experiment the half-life of ^{58}Zn was the first time measured and the feeding to the ground state and the lowest excited states were determined. By combining these data with the relatively accurate Q_{EC}-value of 9370(50) keV, obtained in a double charge exchange reaction ^{58}Ni(π^{+},π^{-})^{58}Zn [9], Gamow-Teller strength to the lowest states could be determined. The comparison of beta decay, hadronic probes and shell model predictions are compiled in the Table 1.

Beta-decay of ^{58}Zn revisited at ISOLDE-CERN

Compared to the first beta decay study at ISOLDE, experimental conditions have improved remarkably. Laser ionization, which was applied in the previous experiment, too, has gone through consilidation leading to higher stability. Application of high resolu-

TABLE 1. Comparison of Gamow-Teller strength to a ground state and an excited state at 1051 keV and the beta decay half-life as extracted from different sets of experimental data.

	B(GT;g.s.)	B(GT;1051 keV)	$T_{1/2}$ [ms]	Ref.
(p,n)	0.264(6)	0.72(4)	82	[4]
(^{3}He,t)	0.264(6)	0.46(10)	86	[7]
^{58}Zn(β^+)	0.31	0.54(26)	86(18)	[8]

tion separator allows to reduce unwanted isobaric contaminants, which are created in the cavity of the laser ion source through surface ionization. Beta-gamma spectroscopy with new miniball cluster detectors increased the efficiency almost one order of magnitude. Finally, earlier unobserved beta-delayed protons were searched for with Si-ball setup consisting of 36 50×50 mm^2 Si-detectors, each of them further segmented to four individual detectors resulting in total of 144 charged-particle detector channels. Description of the Si-ball setup is given elsewhere in this Proceedings [10]. Figure 2 presents the preliminary beta gated gamma spectrum obtained with the new setup. Gamma-transitions at 203 and 848 keV are clearly visible and further information is expected, when proper projections are applied.

In addition to pure improvement of experimental conditions, the new study of beta decay was motivated by the new, relevant results. An improved study of ^{58}Ni(^{3}He,t)^{58}Cu reaction was already mentioned. In addition, beta decay of ^{58}Cu, which is used to normalize all charge-exchange studies, was remeasured in two separate experiments. A high-precisoin gamma-spectroscopy setup was applied at IGISOL-facility in the Department of Jyväskylä [11]. A total absorption technique was applied at GSI on-line isotope separator [12]. The ground state beta branchings obtained were 80.8(7)% and 81.2(5)% at IGISOL and at GSI, respectively. A high accuracy of both experiments and good agreement provides a confident normalization for charge-exchange reaction studies.

SUMMARY AND OUTLOOK

We have performed the first comparison between the beta decay and hadronic probes for Gamow-Teller strength beyound sd-shell. Due to the low statistics of the first beta decay study, we have repeated the beta decay study of ^{58}Zn at ISOLDE with much improved experimental conditions. We also aim for similar tests with higher T-values. One of the $T_z = 3/2$ candidates is beta decay of ^{47}Mn to ^{47}Cr, which can be compared to the ^{47}Ti(^{3}He,t)^{47}V reaction. While approaching larger values of T, the beta decaying partner tends to be far from stability. This effect is underlined with increarsing mass number of the studied case. Generally such nuclei are situated close to the proton dripline and and the importance of beta delayed proton channels gets higher. To obtain information on the beta-decay strength, an detailed information on the beta delayed protons is necessity.

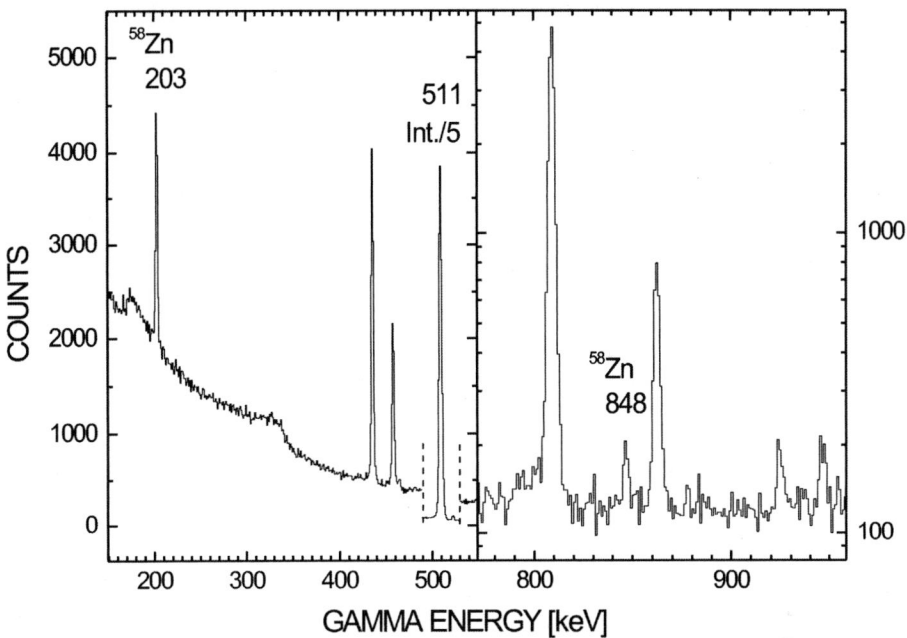

FIGURE 2. Partial gamma-spectrum at A=58 obtained in coincidence with beta-particles. Remark the scale difference in two layouts. In addition, 511 keV annihilation line is reduced by factor of 5 in intensity.

ACKNOWLEDGMENTS

A.J. would like to thank the Academy of Finland for the financial support.

REFERENCES

1. Goodman, C. D., Goulding, C. A., Greenfiled, M. B., Rapaport, J., Bainum, D. E., Foster, C. C., Love, W. G., and Petrovich, F., *Phys. Rev. Lett.*, **44**, 1755 (1980).
2. Fujita, Y., Akimune, H., Daito, I., Fujimura, H., Fujiwara, M., Harakeh, M. N., Inomata, T., Jänecke, J., Katori, K., Tamii, A., Tanaka, M., Ueno, H., and Yosoi, M., *Phys. Rev. C*, **59**, 90 (1999).
3. Anderson, B. D., Baldwin, A. R., Baumann, P., Brown, B. A., Didierjean, F., Foster, C. C., Garcia, L. A. C., Huck, A., Knipper, A., Madey, R., Manley, D. M., Marguier, G., Ramdhane, M., Ravn, H., Richard-Serre, C., Walter, G., and Watson, J. W., *Phys. Rev. C*, **54**, 602 (1996).
4. Rapaport, J., Taddeucci, T., Welch, T. P., Gaarde, C., Larsen, J., Horen, D. J., Sugarbaker, E., Koncz, P., Foster, C. C., Goodman, C. D., Goulding, C. A., and Masterson, T., *Nucl. Phys. A*, **310**, 371 (1983).
5. Rapaport, J., Taddeucci, T., Welch, T. P., Horen, D. J., McGrory, J. B., Gaarde, C., Larsen, J.,

Sugarbaker, E., Koncz, P., Foster, C. C., Goodman, C. D., Goulding, C. A., and Masterson, T., *Phys. Lett. B*, **119**, 61 (1982).

6. Fujita, Y., Fujita, H., Adachi, T., Berg, G. P. A., Caurier, E., Fujimura, H., Hara, K., Katanaka, H., Janas, Z., Kamiya, J., Kawabata, T., Langanke, K., Martinez-Pinedo, G., Noro, T., Roeckl, E., shimbara, Y., Shinada, T., van der Werf, S. Y., Yoshifuku, M., Yosoi, M., and Zegers, R. G. T., *Eur. Phys. J. A*, **13**, 411 (2002).

7. Fujita, Y., Akimune, H., Daito, I., Fujiwara, M., Harakerh, M. N., Inomata, T., Jänecke, J., Katori, H., Nakada, H., Nakayama, S., Tamii, A., Tanaka, M., Toyokawa, H., and Yosoi, M., *Phys. Lett. B*, **365**, 29 (1996).

8. Jokinen, A., Oinonen, M., Aysto, J., Baumann, P., Dendooven, P., Didierjean, F., Fedoseyev, V., Huck, A., Jading, Y., Knipper, A., Koizumi, M., Koster, U., Lettry, J., Lipas, P., Liu, W., Mishin, V., Ramdhane, M., Ravn, H., Roeckl, E., Sebastian, V., Walter, G., and the ISOLDE Collaboration, *Eur. Phys. J. A*, **3**, 271 (1998).

9. Seth, K. K., Iversen, S., Kaletka, M., Barlow, D., Saha, A., and Soundranayagam, R., *Phys. Lett. B*, **173**, 397 (1986).

10. Fraile, L., "The ISOLDE Silicon Ball," in *these Proceedings*, 2003.

11. Peräjärvi, K., Dendooven, P., Gorska, M., Huikari, J., Jokinen, A., Kolhinen, V. S., La Commara, M., Lhersonneau, G., Nieminen, A., Nummela, S., Penttilä, H., Roeckl, E., Wang, J. C., and Äystö, J., *Nucl. Phys. A*, **696**, 233 (2000).

12. Janas, Z., Karny, M., Fujita, Y., Batist, L., Cano-Ott, D., Collatz, R., Dendooven, P., Gadea, A., Gierlik, M., Hellström, M., Hu, Z., Jokinen, A., Kirchner, R., Klepper, O., Moroz, F., Oinonen, M., Penttilä, H., Plochocki, A., Roeckl, E., Rubio, B., Shibata, M., Tain, J. L., and Wittman, V., *Eur. Phys. J. A*, **12**, 143 (2001).

Beta-delayed proton decay of a high-spin isomer of ^{94}Ag

I. Mukha[*†], H. Grawe[*], J. Döring[*], C. Plettner[*], L. Batist[**], A. Blazhev[*‡],
C. Hoffman[§], Z. Janas[¶], R. Kirchner[*], M. La Commara[‖], C. Mazzocchi[*],
E. Roeckl[*], S. L. Tabor[§] and M. Wiedeking[§]

[*]Gesellschaft für Schwerionenforschung mbH, D-64291 Darmstadt, Germany
[†]on leave from Kurchatov Institute, RU-123182 Moscow, Russsia
[**]St. Petersburg Nuclear Physics Institute, RU-188350 Gatchina, Russia
[‡]University of Sofia, BG-1164, Sofia, Bulgaria
[§]Florida State University, Tallahassee, FL-32306, USA
[¶]Warsaw University, PL-00681 Warsaw, Poland
[‖]Universita di Napoli "Federico II", I-80126 Napoli, Italy

Abstract.
We have observed the high-spin states in ^{93}Rh populated by delayed-proton emission following the β decay of ^{94}Ag. The existence of a second isomer in ^{94}Ag with I≥17, $T_{1/2}$=0.42(5) s, whose β-decay energy is at least 15.5 MeV, has been established. This state features most likely the highest spin ever observed for β-decaying nuclei.

MOTIVATION

In the β^+/EC decay of the N=Z nucleus ^{94}Ag, preliminary evidence for a long-lived ($T_{1/2}$=0.3(2) s), high-spin (I≥17) isomer in ^{94}Ag, in addition to the known (7$^+$) isomer, was obtained in a previous GSI-ISOL experiment [1]. The decay of the two ^{94}Ag isomers was studied by measuring $\beta - \gamma$ coincidences, which yielded information about excited states in the daughter nucleus ^{94}Pd. In this contribution, we report that the existence of the (I≥17) isomer has been confirmed in a recent GSI-ISOL experiment, where both β-delayed γ rays [2] and protons [3] were detected.

EXPERIMENT AND RESULTS

The ^{94}Ag nuclei were produced in the fusion-evaporation reaction ^{58}Ni(^{40}Ca,p3n), utilizing a 4.8 MeV/u, 80 particle-nA ^{40}Ca beam from the heavy-ion accelerator UNILAC at GSI. The target consisted of a 2.6 mg/cm^2 thick ^{58}Ni foil enriched to 99.8%. The reaction products were stopped in the FEBIAD-B3C ion source [4] which suppressed the release of the isobaric ^{94}Pd contamination by a factor of 30 without reducing the ^{94}Ag yield. The mass-94 nuclei were separated at the GSI on-line mass separator [5]. The average yield of mass-separated ^{94}Ag was about 1.5 atom/s. The collection point was surrounded by high-granularity arrays of Ge and Si detectors [3]. The β^+/EC-delayed proton decay of ^{94}Ag was measured by recording β-p, p-γ, β-p-γ and p-γ-γ events.

CP681, Proton-Emitting Nuclei: Second International Symposium; PROCON 2003,
edited by E. Maglione and F. Soramel
© 2003 American Institute of Physics 0-7354-0150-0/03/$20.00

Seventeen Ge detectors were mounted in three composite detectors, i.e., one Cluster [7], two Super-Clover detectors, and two single Ge detectors. Their total photo-peak efficiency was 3.2% for 1.33 MeV γ rays. The Si detector, developed for decay studies of exotic nuclei at FRS and ISOL facilities [6], was used for detecting β-particles and protons. It consisted of three large Si detectors. Each detector had an area of 60x60 mm^2, was 1 mm thick and had 32 strips with two-channel read-outs. The energy resolution was about 25 keV. The Si detectors were positioned inside a cylindrical vacuum chamber with 1 mm thick Aluminium wall, and were arranged around the collection point covering 65% of 4π solid angle. A thin plastic tape served for the collection of activity from the GSI-ISOL facility. In order to suppress unwanted activity of daughter nuclei, the tape system was used to quickly remove the implanted source. The activity was implanted continuously (grow-in mode) into the periodically removed tape, whereas after the tape movement the radioactive source decay was measured as well (decay mode).

The spectrum of charged particles following β decay of ^{94}Ag is shown in Fig. 1. It has a pronounced peak around 0.5 MeV due to detection of positrons from β decay. The spectrum falls off exponentially with increasing energy till 2.5 MeV. In the higher energy region, the spectrum behavior changes dramatically. It has a broad bump around 3.5 MeV, which is shown as solid-line histogram in the inset of Fig. 1, in comparison with the spectrum of β-delayed protons from the ^{94}Ag decays as measured with a ΔE-E telescope of Si detectors [8] (the dashed histogram). As one can see, the high-energy part of the measured spectrum matches the spectrum of β-delayed protons from ^{94}Ag.

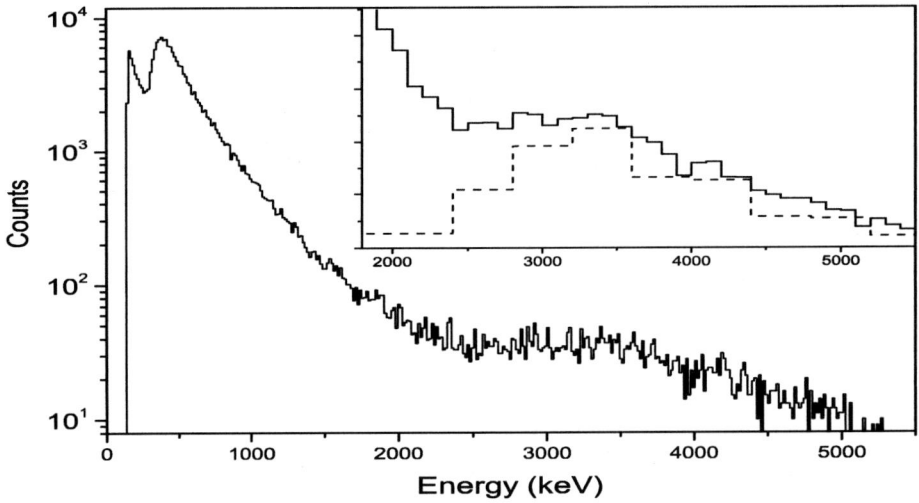

FIGURE 1. Spectrum of β particles and protons from the ^{94}Ag decay, measured by the Si detector.

The spectrum of γ rays measured in coincidence with low-energy (0.3<E<1 MeV) β particles is shown in the upper panel of Fig. 2. The spectrum agrees with the previous measurement [1], where almost all γ lines have been assigned to the decay of ^{94}Ag populating excited states of ^{94}Pd. However, when we select γ rays measured in coincidence with the high-energy region (E>2.5 MeV) where protons are expected to dominate (see Fig. 1), the spectrum changes dramatically, see the middle panel of Fig. 2. This γ-ray spectrum shows more than 20 peaks, and most of them match the γ de-excitation pattern of high-spin states in ^{93}Rh, which are known from in-beam γ-spectroscopy [10]. Thus we have observed the feeding of high-spin states in ^{93}Rh, populated by proton emission

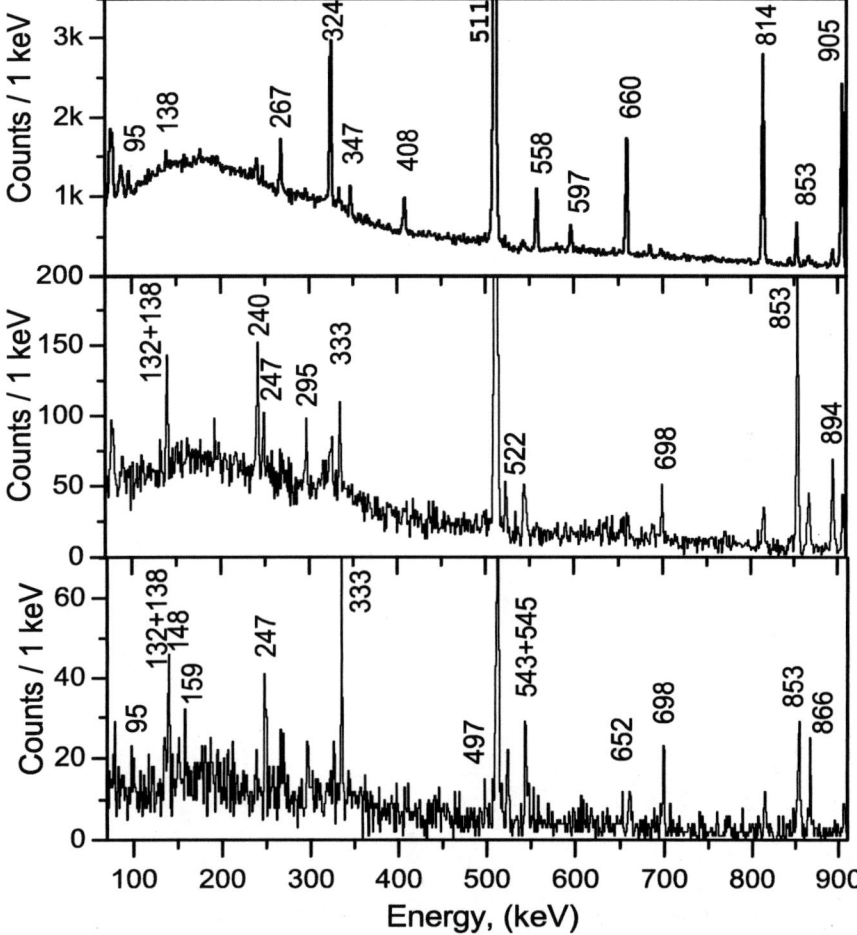

FIGURE 2. Energy spectrum of γ rays from ^{94}Ag measured: in coincidence with β-particles (upper panel); in coincidence with protons (middle panel); in triple coincidence with protons and 138, 247, 522 and 698 keV γ rays (lower panel). Energies of known transitions in ^{94}Pd and ^{93}Rh are indicated in keV.

from excited states of ^{94}Pd. In the lower panel of Fig. 2, the γ spectrum is displayed as being projected from the triple p-γ-γ events and using the sum of 138 keV, 247 keV, 522 keV, 698 keV γ-ray gates. An analysis of the p-γ-γ coincidences helped us to establish γ cascades and to make the ^{93}Rh level assignments.

The resulting scheme of ^{93}Rh levels populated in the βp decay of ^{94}Ag is shown in Fig. 3. The levels are displayed in two yrast bands with even and odd parities shown in the center and on the right-hand side, respectively. On the left-hand side, we show the non-yrast levels observed, see also [9]. The yrast-band assignments are adopted from those made in [10]. The 853 keV and 5445 keV levels are the only exceptions. On the basis of the larger intensity of the 853 keV γ-line relative to the 866 keV one, we conclude that these two transitions must be inverted in comparison with the earlier assignment [10]. The analogous conclusion is drawn for the 5445 keV state on the basis of the relative intensities of the 247 keV and 698 keV γ lines. In addition to [10], we have observed two weak γ lines at 132 keV and 148 keV belonging to the odd-parity band (see the lower panel of Fig. 2). As they have the lowest intensity, they have to be placed on the top of the 6387 keV level. Any further γ rays of higher energy would have escaped observation at this level of intensity.

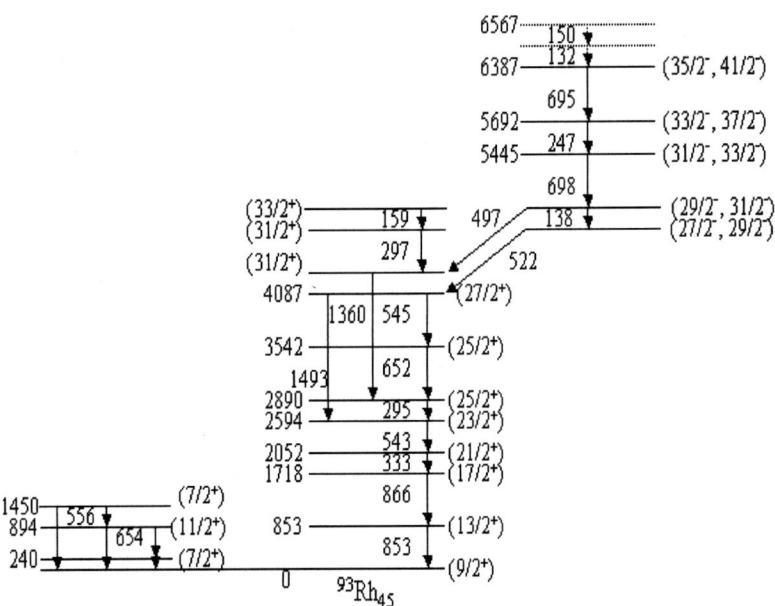

FIGURE 3. States in ^{93}Rh populated in β decay of ^{94}Ag via emission of delayed protons. Energies of states and of γ rays are given in keV. Tentative spin-parity assignments are shown in parentheses.

The high-spin levels of ^{93}Rh fed in β decay give further evidence for the second, high-spin isomer in ^{94}Ag with I$\geq$17, which has been tentatively identified in [1]. Another estimate of its spin can be deduced inspecting the highest-spin levels observed in ^{93}Rh. Assuming a (39/2$^-$, 45/2$^-$) spin-parity range for the 6567 keV level, we obtain I$\geq$18 for the daughter state in ^{94}Pd, and hence I$\geq$17 for the ^{94}Ag isomer. A remarkable feature of the observed decay pattern is a preferred feeding of odd-parity states at high excitation energies in ^{93}Rh, which points to odd-orbital momenta of the emitted protons when assuming even parity for the parent states [1].

The energy spectrum of β particles and protons measured with the Si detector in coincidence with the known 698 keV γ transition in ^{93}Rh is shown in Fig. 4. The spectrum has a maximum at an energy around 3 MeV, with the low-energy bump corresponding to β particles. The 698 keV γ transition (see the ^{93}Rh level scheme in Fig. 3) indicates a population of the 5445 keV, $(31/2^-, 33/2^-)$ state in ^{93}Rh (thus discriminating against p-γ events from the decay of the (7^+) isomer). As the proton separation energy in ^{94}Pd is S$_p$=4.47 MeV [11], the end-point of the proton spectrum of about 5.5 MeV gives an estimate for the highest excitation energy of ^{94}Pd states fed in β decay of the second high-spin isomer in ^{94}Ag, which is at least 15.5 MeV. As can be seen from the comparatively modest intensity of the β component in Fig. 4, the electron-capture (EC) is an important β-decay mechanism leading to proton emission. We have

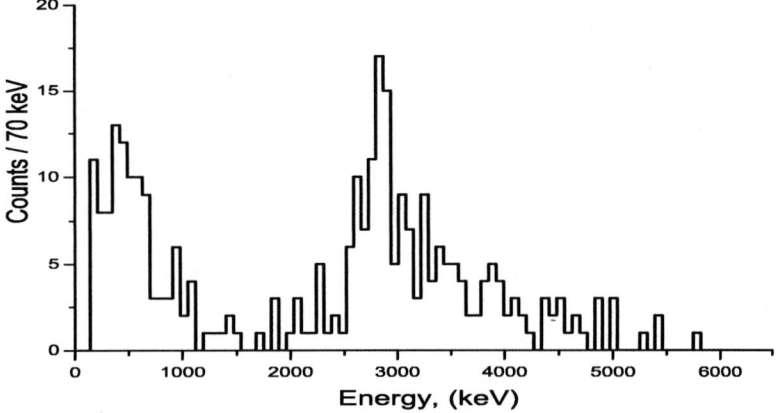

FIGURE 4. Protons and β-particles measured in coincidence with the 698 keV γ-transition in ^{93}Rh which stems from a 5.445 MeV level in this nucleus.

used the β^+/EC-delayed proton spectrum gated by γ rays for estimating the Q$_{EC}$ value of the (I$\geq$17) isomer in ^{94}Ag. For this purpose we assumed that (i) a Gamow-Teller transition of ^{94}Ag into a highly-excited state in ^{94}Pd proceeds either by β^+ emission or electron-capture, (ii) the ^{94}Pd excited state de-excites by a proton emission into excited states in ^{93}Rh, (iii) which in turn emit γ rays. Thus both (β-p-γ) and (p-γ) events from β^+ and EC processes can be measured, respectively. The theoretical dependence of the β^+/EC ratio from β-decay energy, E$_\beta$, was taken from [13]. Thus the Q$_{EC}$ value of the (I$\geq$17) isomer of ^{94}Ag can be determined as the sum Q$_{EC}$=E$_\beta$+S$_p$+E*+1.022 (MeV), where E* is the excitation energy of the ^{93}Rh level fed by proton emission. For the case of a βp transition with E$_p$=3 MeV and E$_\gamma$=247 keV (E*=5692 keV), the β^+/EC proton ratio is of 1.0(4), which yields E$_\beta$=1.6(2) MeV and consequently, Q$_{EC}$=15.7(2) MeV. Actually, it represents an estimate of the lower limit, as the 5692 keV state in ^{93}Rh, which de-excites by the 247 keV γ ray, can be populated also from a higher-lying state in ^{93}Rh. Assuming Q$_{EC}$=12.6 MeV for the ground state of ^{94}Ag [12], we conclude that the excitation energy of the (I$\geq$17) isomer is at least 3 MeV, and it should be open to several exotic decay modes, such as direct one-proton or α-particle emission.

The half-life values of both ^{94}Ag isomers were obtained by fitting the activity time-distributions measured in the grow-in and decay modes. The half-life value of the (7^+) isomer was fitted by using four β-coincident 814 keV, 905 keV, 267 keV and 324 keV γ-lines yielding the average value of $T_{1/2}$=0.62(3) s, which differs from the previous value of 0.36(3) s derived from the grow-in mode of measurements [1]. We expect that the present value, being derived from both the grow-in and decay modes, is more accurate though a detailed comparison will be done elsewhere [3]. On the basis of the assumed decay scheme, the proton-coincident 138 keV, 247 keV, 295 keV, 333 keV, 522 keV, 698 keV and 543 keV γ lines were chosen for the half-life evaluation of the $(I \geq 17)$ isomer in ^{94}Ag, yielding a half-life value of 0.42(5) s, which agrees with the previous value 0.3(2) s [1] within the experimental errors.

Shell-model calculations have predicted numerous spin-gap isomers in the proton and neutron $(p_{1/2}, g_{9/2})$ model space [14], but so far only the I^{π}=(21/2$^+$) state in ^{95}Pd was observed [15]. These calculations fail to predict a high-spin isomer in ^{94}Ag, with the closest candidate being a 21$^+$ level, which represents the highest spin in the model space [1]. Only inclusion of excitations of configuration space up to (3p-3h) in the ^{100}Sn core can reproduce this type of isomerism [16].

SUMMARY AND OUTLOOK

We have obtained experimental evidence for a $I \geq 17$, $T_{1/2}$=0.42(5) s isomer in ^{94}Ag, whose β-decay energy is at least 15.5 MeV. This relatively long-lived state has presumably the highest spin ever observed for β-decaying nuclei. It is expected to be open to several exotic decay modes like direct one-proton or α-particle emission. Though none of the latter radioactivity was observed, further experiments with such an exotic nucleus may bring new exciting results.

We thank W. Hüller and K. Burkard for their help operating the GSI–ISOL facility.

REFERENCES

1. M. La Commara et al., Nucl. Phys. A**708**, 167 (2002).
2. C. Plettner et al., in preparation.
3. I. Mukha et al., GSI Scientific Report 2003, Section "Nuclei far off Stability", in print; Nucl. Phys. A, to be submitted.
4. R. Kirchner et al., Nucl. Instrum. Meth. Phys. Res. B**70**, 186 (1992).
5. K. Burkard et al., Nucl. Instrum. Meth. Phys. Res. A**126**, 135 (1997).
6. I. Mukha et al., GSI Scientific Report 2003, Section "Instruments and Methods", in print.
7. J. Eberth et al., Nucl. Instrum. Meth. Phys. Res. A**369**, 135 (1997).
8. L. Batist et al., in preparation.
9. K. Schmidt et al., Eur. Phys. J. A**8**, 303 (2000).
10. H. A. Roth et al., J. Phys. B**21**, L1 (1995).
11. G. Audi and A.H. Wapstra, Nucl. Phys. A**595**, 409 (1995).
12. L. Batist et al., Nucl. Phys. A, in print.
13. N. B. Gove, M. J. Martin, Nucl. Data Tables **10**, 205 (1971).
14. K. Ogawa, Phys. Rev. C**28**, 958 (1983).
15. E. Nolte and H. Hick, Z. Phys. A**305**, 289 (1982).
16. F. Nowacki, private communication, and Nucl. Phys. A**704**, 223c (2002).

NUCLEAR ASTROPHYSICS

Spectroscopy of ^{19}Na by ^{18}Ne+p elastic scattering at Louvain-la-Neuve

Carmen Angulo

Centre de Recherches du Cyclotron, Université catholique de Louvain, 2 chemin du cyclotron,
B-1348 Louvain-la-Neuve, Belgium, e-mail: angulo@cyc.ucl.ac.be

Abstract. The second excited state ($J^{\pi} = 1/2^{+}$, $\ell = 0$) of the proton-rich nucleus ^{19}Na has been observed for the first time using the elastic scattering reaction ^{18}Ne+p in inverse kinematics. An intense (4×10^{6} pps) ^{18}Ne radioactive beam was produced at the Radioactive Beam Facility at Louvain-la-Neuve to bombard a polyethylene target. The recoil protons were detected at 20 different angles in the LEDA segmented silicon detector. The resulting elastic scattering cross sections have been analysed using the R-matrix method. We find that the second excited state in ^{19}Na lies at the center-of-mass energy $E_{\text{c.m.}} = 1.066 \pm 0.003$ MeV with respect to the ^{18}Ne+p threshold, and has a proton width $\Gamma_p = 101 \pm 3$ keV. The Coulomb shift between the $1/2^{+}$ mirror levels in ^{19}O and ^{19}Na is 0.73 MeV, among the largest values currently observed in light exotic nuclei.

INTRODUCTION

The structure and decay modes of light exotic nuclei are a major source of interests in nuclear physics research. Proton-rich light nuclei are a remarkable case since the level scheme is not known for many species. This paper presents the experimental study of the ^{19}Na nucleus [1], which is proton unbound by 321 ± 13 keV [2]. Even though it is not far from stability, almost no spectroscopic information is available: there are no spin assignments, and only the location of the ground state and of the first excited state have been determined by nuclear reactions involving stable beams [3, 4]. The ^{19}Na nucleus has also some interest for nuclear astrophysics. The ^{18}Ne(2p,γ)^{20}Mg reaction may play a role on long-lived waiting point isotopes near the proton drip lines for the rp-process reaction flow, although very high densities would be required [5]. But only the Q-value of the ^{18}Ne(p,p)^{18}Ne reaction is of relevance as the width of its resonances does not enter the reaction rate because production and decay of ^{19}Na are in an equilibrium [6].

The aim of the present experiment was to measure the location and the width of the second excited state of ^{19}Na, which according to the mirror nucleus ^{19}O ($E_x = 1.4717 \pm 0.0004$ MeV) should have a spin $J^{\pi} = 1/2^{+}$ [2]. We have used the elastic scattering technique in inverse kinematis by bombarding a proton-rich target (a thin polyethylene self-supporting foil) with a radioactive ^{18}Ne beam. The proton elastic scattering technique, a powerful analysis method of proton rich nuclei and a obliged investigation preceding a capture reaction measurement, has been widely applied for nuclear spectroscopy studies as well as for nuclear astrophysics experiments [7, 8, 9, 10, 11]. It makes use of the sensitivity of the protons to the presence of a

CP681, *Proton-Emitting Nuclei: Second International Symposium; PROCON 2003*,
edited by E. Maglione and F. Soramel
© 2003 American Institute of Physics 0-7354-0150-0/03/$20.00

resonant state in the compound nucleus. When the resonant state is scanned with the appropriate ion beam and target, the recoil proton spectra show spectacular changes when detected at forward laboratory angles. These spectra contain precise information on the resonance energy, angular momentum and width provided all experimental effects, mainly the energy resolution and angular resolution of the detectors, are properly taken into account [12].

The elastic scattering method has also some drawbacks. Ideally, all resonance widths could be measured using this technique, but in practice the energy resolution is limited to at least several keV (the exact value depends on the experimental conditions, mainly the angular resolution, target thickness, and beam energy resolution). For these reasons, the parameters of the ground and the first excited states of ^{19}Na, whose predicted widths are of the order of eV [2], cannot be obtained using this technique and their measurement was therefore beyond the scope of this work.

On the other hand, the energy straggling of the beam and the recoil protons on the target play also a non-negligible role. Therefore, even if the use of a thick target is convenient for covering a large energy range, for accurate measurements thin targets (of the order of a few hundreds of $\mu g/cm^2$) are preferred. Of course, the target thickness should be chosen such that the energy range of interest can be scanned (using one or several beam energies) and the expected interference pattern between the resonance and the Coulomb contributions can clearly show up. The elastic scattering method can obviously be extended to alpha scattering studies [13].

EXPERIMENTAL METHOD

The experiment was carried out at the Radioactive Beam Facility of the Centre de Recherches du Cyclotron at Louvain-la-Neuve, Belgium. Post-accelerated radioactive beams of ^{18}Ne^{2+} at $E_{lab} = 21$ and 23.5 MeV, and ^{18}Ne^{3+} at $E_{lab} = 28$ MeV, respectively, were provided by the CYCLONE110 cyclotron. The ^{18}Ne atoms were produced through the ^{19}F(p,2n)^{18}Ne reaction by bombarding a LiF target with an intense 30 MeV proton beam produced by the CYCLONE30 cyclotron and were ionised to the 2^+ or to the 3^+ state in a ECR source, before being post-accelerated. A detailed description of the production of the ^{18}Ne radioactive beam can be found in Ref. [14]. The average intensity of the ^{18}Ne beams on target was of the order of 4×10^6 pps, which was maintained for periods of several days, allowing us to obtain excellent statistics for each of the three beam energies (the statistical error of the proton events per energy bin of 25 keV at each angle ranges between 1% and 8% for all energies).

The target consisted of a thin polyethylene $(CH_2)_n$ self-supporting foil with a very thin Au coating evaporated on its upstream face. The Au layer was used for normalization (see below). The effect of this Au layer on the beam energy loss was negligible. Although polyethylene targets could suffer a steady decrease in hydrogen content, there was no sign of deterioration at the beam intensities used ($\sim 4 \times 10^6$ pps), which should have lead to a decreasing of the counting rate in energy spectra taken in successive runs at a given angle. The target thickness, determined by the energy loss of the ^{18}Ne beam [15], was (520 ± 10) μg/cm^2.

The target thickness and the three different beam energies have been chosen such that the energy range of interest could be scanned. A thicker target could be used to scan the whole energy range of interest using a single beam energy. However, as discussed above, the energy straggling of the beam and of the recoil protons would have been larger, introducing an additional uncertainty in the data. The thickness of the Au layer, used for normalization purposes (see below), was obtained by bombarding the targets with a 5.6 MeV ^{7}Li beam. The scattered ^{7}Li on Au and the recoil protons from the ^{1}H(^{7}Li,^{1}H) reaction were detected simultaneously in a two PIPS (Passivated Implanted Planar Silicon) detectors located at 60° with respect to the beam axis.

The recoil protons were detected in two LEDA (Louvain la Neuve-Edinburgh-Detector-Arrays) segmented silicon detector arrays [17] situated at 61.7 and 12.6 cm from the target (Figure 1). Due to the kinematical variation of the recoil proton energy with laboratory angle, only the 20 most forward angles were used at $\theta_{lab} = 4.9° - 11.7°$ ($\theta_{c.m} = 170.2° - 156.6°$) and $\theta_{lab} = 22.6° - 29.9°$ ($\theta_{c.m} = 131.0° - 120.2°$), with angular resolution $\Delta\theta_{lab} \simeq 0.2°$ and 0.9°, respectively. A 18 μm thick Mylar foil was situated in front of all LEDA sectors, except one, to stop the high energy scattered ^{18}Ne and most of the recoiled ^{12}C particles, and therefore to reduce the data acquisition dead time. The "uncovered" LEDA sector was used for monitoring the number of incoming beam particles by detecting the ^{18}Ne ions scattered on the thin Au layer, assuming a purely Rutherford elastic cross section for ^{197}Au(^{18}Ne,^{18}Ne)^{197}Au at the measured energies.

Owing to several factors (low beam intensity, distribution of charge states beyond the target), the determination of the total number of incident radioactive beam particles N_{Ne} cannot be done using standard techniques. Here, we have performed an independent determination of the factor $n_H N_{Ne} \Delta\Omega_{lab}$ (n_H is the number of H atoms in the target per cm^2, and $\Delta\Omega_{lab}$ is the solid angle covered by the strip detector situated an the laboratory angle ϕ_{lab}), from measurements of the ^{7}Li+p, ^{7}Li+Au, and ^{18}Ne+Au elastic scattering. The experimental conditions were the same for all the measurements. The advantage of this novel method is that one obtains an intrinsic normalization simultaneously to the recoil proton measurement. The total uncertainty in the cross section values due to the normalization procedure is 5.5%.

Moreover, the number of Au atoms in the target, N_{Au} could be calculated from the relation

$$N_{Au} = N_p \frac{C_p}{C_{Li}} \frac{(d\sigma/d\Omega)_{Li+p}}{(d\sigma/d\Omega)_{Li+Au}}, \tag{1}$$

where N_p is the number of H atoms in the target, C_p and C_{Li} are the number of detected recoil protons and scattered Li, respectively, and $(d\sigma/d\Omega)_{Li+p}$, $(d\sigma/d\Omega)_{Li+Au}$ are the

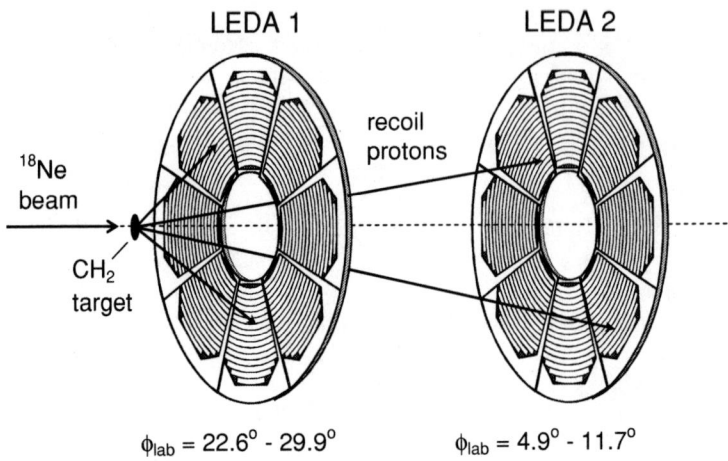

LEDA 1   LEDA 2

recoil
protons

^{18}Ne
beam

CH$_2$
target

ϕ_{lab} = 22.6° - 29.9° ϕ_{lab} = 4.9° - 11.7°

FIGURE 1. Schematic drawing of the experimental set up. The PIPS detectors used for normalization are not shown in the figure (see text).

differential cross sections for ^{1}H(^{7}Li,^{1}H)^{7}Li and ^{197}Au(^{7}Li,^{7}Li)^{197}Au, respectively. The ^{1}H(^{7}Li,^{1}H)^{7}Li cross section value was obtained from literature [16]. We have assumed a pure Rutherford cross section for ^{197}Au(^{7}Li,^{7}Li)^{197}Au. We derived a Au layer thickness of 8.0 ± 0.5 μg/cm^2 in excellent agrement with the nominal value obtained from monitoring during the Au evaporation.

For each LEDA strip, not only the energy but also the time-of-flight of the particle with respect to the cyclotron RF were recorded and displayed in a two-dimensional spectrum. Figure 2(a) shows a typical two dimensional spectrum obtained with a 23.5 MeV ^{18}Ne beam on a 0.52 mg/cm^2 polyethylene target during 2 days of running time. The spectra were obtained by adding the calibrated single spectra from several LEDA strips at the laboratory angle of $\theta_{lab} = 4.9°$ (total solid angle $\Delta\Omega_{lab} = 3.7$ msr). The energy calibration was performed by means of a 3-line α-source (^{239}Pu, ^{241}Am, ^{244}Cm) and a precision pulser. This spectrum corresponds also to the angle with lowest accumulated statistics (smallest solid angle). The recoil protons, marked by a selection on the figure, are well separated from the recoiled ^{12}C, not completely stopped in the Mylar foil, and the uncorrelated β background. In Figure 2(b), only the proton signal is shown as a function of the proton laboratory energy. The interference pattern between a s-wave resonance and the Rutherford contribution is clearly present. Similar spectra were obtained at all measured angles and at beam energies of 21.0, 23.5 and 28.0 MeV, covering laboratory proton energies from 2.5 to 5.5 MeV. Data obtained at the overlapping energy region near the $1/2^+$ resonance are in very good agreement within the experimental errors.

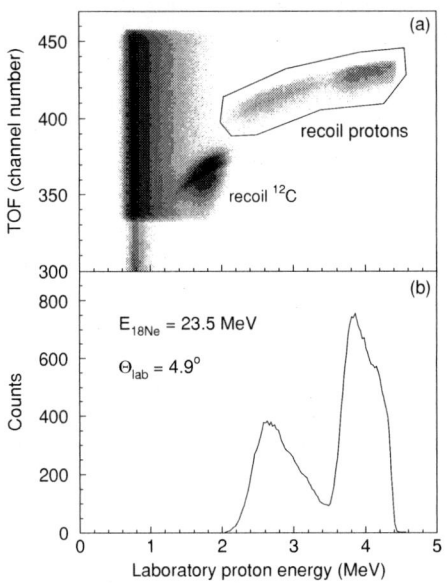

FIGURE 2. Summed TOF spectrum at $\theta_{lab} = 4.9°$ obtained with a 23.5 MeV ^{18}Ne beam on a 0.52 mg/cm^2 polyethylene target. The recoil proton energy spectrum, marked by a selection on the figure, are displayed on (b) as a function of the laboratory proton energy.

DATA ANALYSIS

The ^{18}Ne+p elastic scattering cross section

In reverse kinematics, the center of mass energy $E_{c.m.}$ of the system ^{18}Ne+p is related to the energy E_p of the recoil protons detected at a laboratory angle θ_{lab} by

$$E_{c.m.} = E_p \times \frac{M_b + M_t}{4 \, M_b \cos^2 \theta_{lab}}, \qquad (2)$$

where M_b and M_t are the mass number [18] of the beam and the target.

The detected proton energy E_p must be corrected by the energy loss [15] of the protons in the Mylar foil situated in front of the detectors and the energy loss inside the target. On the other hand, the proton spectra are slightly degraded by two effects that cannot be avoided:

- the opening angle of the detectors ($\Delta\theta_{lab}$), introducing an uncertainty in the proton energy ΔE_θ, given by [8]:

$$\Delta E_\theta = 2E_p \tan \theta_{lab} \Delta\theta_{lab}, \qquad (3)$$

- the energy straggling of the protons ΔE_s in the target and in the Mylar foil used to stop the heavier ions. A good estimate of ΔE_s is obtained by using the Bohr formula [19] (see, however, the discussion below).

The electronic noise introduces an additional energy broadening ΔE_e of about 15 keV FWHM in the laboratory system (about 4 keV in the c.m. system), which is significantly smaller than the kinematic broadening at backward angles and the straggling. The total energy broadening ΔE^1 is obtained by adding quadratically all the contributions,

$$\Delta E = (\Delta E_\theta^2 + \Delta E_s^2 + \Delta E_e^2)^{1/2} \tag{4}$$

Typical values of ΔE are 11 keV for a laboratory angle of $\theta_{lab} = 4.9°$, 16 keV for $\theta_{lab} = 11.7°$ (ΔE_{strag} is the most important contribution in both cases) and 40 keV for $\theta_{lab} = 29.9°$ (ΔE_θ is the most important contribution).

Although ΔE_θ and ΔE_e are well determined, one of the main sources of uncertainty is the value of ΔE_s. In particular, when the total energy broadening ΔE is dominated by ΔE_s (as it is the case for $\theta_{lab} = 4.9° - 11.7°$), we have found that ΔE_s can be underestimated. In order to limit the uncertainties related to the ΔE values, we have proceeded as follows. We have first performed an analysis using ΔE_s values as estimated by the Bohr formula. Once the best fit was obtained, several fits were performed by varying the ΔE values by steps of 0.25 keV. We have observed that for the larger laboratory angles, the best fit was found for ΔE equal to the calculated one. This is expected as ΔE is dominated by the opening angle uncertainty, that can be calculated exactly. The difference occurs mainly in the smaller laboratory angles, $\theta_{lab} = 4.9° - 11.7°$, where the fitted ΔE values are a bit larger than the estimated one. For example, for $\theta_{lab} = 4.9°$, the estimated value is 11 keV, while the best fit corresponds to $\Delta E = 12$ keV. We have adopted the values corresponding to the best fits.

From the raw recoil proton spectra for each of the three energies and the 20 laboratory angles, we have obtained ^{18}Ne+p elastic scattering differential cross sections in the c.m. system by the procedure explained below.

The laboratory differential cross section (in units of cm^2) at energy E_{lab} and angle θ_{lab} is calculated from

$$\frac{d\sigma}{d\Omega_{lab}}(E_{lab}, \theta_{lab}) = \frac{C_p}{n_H N_{Ne} \Delta\Omega_{lab}}, \tag{5}$$

where C_p is the number of detected protons of energy E_{lab} at an angle θ_{lab} in a detector covering a solid angle $\Delta\Omega_{lab}$ (in sr) summed in an energy interval defined by the data binning (here 25 keV for data taken at beam energies of 23.5 and 28.0 MeV, and 50 keV for data taken at 21.0 MeV), n_H is the number of H atoms per cm^2 in the target, N_{Ne} is the number of incident ^{18}Ne beam particles. As explained in the previous Section, a special attention has been taken to the determination of the factor $n_H N_{Ne} \Delta\Omega_{lab}$ in

---

[1] The θ dependence of ΔE is not explicitly written.

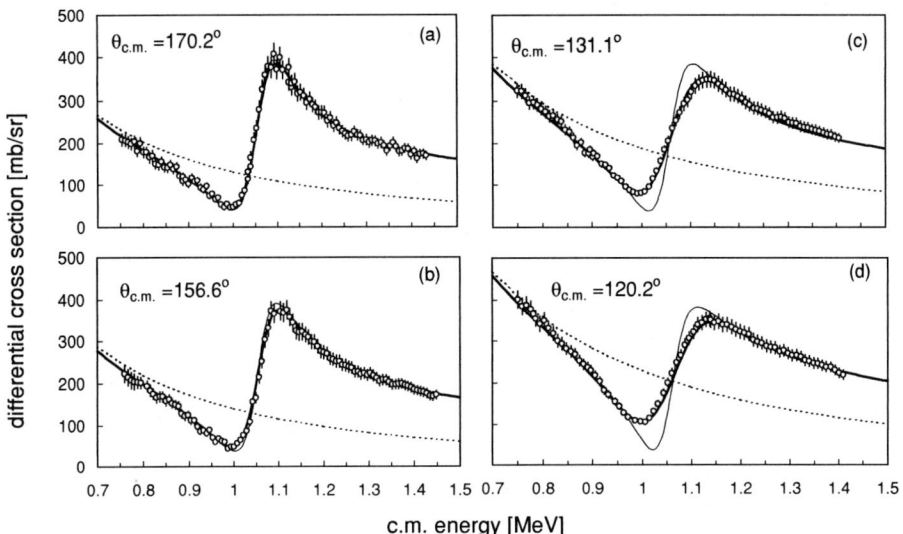

FIGURE 3. Elastic cross section (open circles) versus c.m. energies for the system ^{18}Ne+p for c.m. angles (a) 170.2°, (b) 156.6°, (c) 131.1° and (d) 120.2°. The thick solid curve is the best R-matrix fits that include the correction for energy resolution. The thin solid curve does not include this correction. The dotted curves are the Rutherford cross sections.

Eq. (5), which settles the normalization of the experimental yield.

The transformation of the laboratory angle and the differential cross section to the c.m. frame is given by

$$\theta_{c.m.} = \pi - 2\theta_{lab} \tag{6}$$

and

$$\frac{d\sigma}{d\Omega_{c.m.}}(E_{c.m}, \theta_{c.m.}) = \frac{1}{4\cos\theta_{lab}} \times \frac{d\sigma}{d\Omega_{lab}}(E_{lab}, \theta_{lab}) \tag{7}$$

The resulting ^{18}Ne+p cross sections are presented in Figure 3 for four typical c.m. angles $\theta_{c.m.} = 170.2°$, 156.6°, 131.1°, and 120.2° as a function of the c.m. energy. The solid curves are R-matrix fits as explained in next Section. The dotted curves are the Rutherford cross sections. The error bars are only statistical errors.

The $1/2^+$ state in ^{19}Na

We have used the R-matrix formalism [20] to fit the differential cross sections in the 120.2° − 170.2° angular range and the 0.7 − 1.5 MeV c.m. energy range. In the R-matrix theory, the nuclear phase shift is defined by

$$\delta^\ell = \delta^\ell_{HS} + \delta^\ell_R, \tag{8}$$

where δ^ℓ_{HS} and δ^ℓ_R are the hard-sphere and the R-matrix phase shifts, respectively. Here, we have used hard-sphere phase shifts for all partial waves (ℓ values up to $\ell_{max} = 2$ have been taken into account) except for $\ell = 0$, where we have applied the R-matrix phase shift given by

$$\delta^0_R = \arctan \frac{P_0 R^0}{1 - S_0 R^0}, \tag{9}$$

where P_0 and S_0 are the $\ell = 0$ penetration and shift factors, respectively, and R^0 is the R-matrix defined with a single pole. The fitted parameters are the pole parameters converted to the resonance energy E_R and the proton width Γ_p of the $1/2^+$ state in ^{19}Na (in R-matrix notation, E_R and Γ_p are "observed" parameters). In order to account for the experimental effects described above, we have convoluted the calculations with a Gaussian distribution, with a full width at half maximum ΔE, which was set equal to the total energy broadening as calculated above.

We have tested the sensitivity of the fit to the R-matrix channel radius a by performing fits for three different values $a = 4, 5$ and 6 fm. The results of these fits are presented in Table 1. We have adopted the values for $a = 5$ fm, which provides the lowest χ^2/N value, $\chi^2/N = 0.44$ for $E_R = 1.066 \pm 0.003$ MeV and $\Gamma_p = 101 \pm 3$ keV. The small errors are due to the strong sensitivity of χ^2 with respect to E_R and Γ_p. The thicker solid curve in Figure 2 is the R-matrix fit ($a = 5$ fm) for the adopted values (Table 1) and includes the correction for opening angle and beam straggling discussed above. The thinner solid curve (unresolved for (a)) corresponds to $\Delta E = 0$. R-matrix fits using other spin assignments strongly disagree with the experimental data. The dimensionless reduced width θ^2_p represents more than 20% of the Wigner limit. This value is similar to or even larger than the reduced width of $\ell = 0$ states in other nuclei such as ^{13}N ($E_x = 2.37$ MeV, $\theta^2_p = 21.8$ %) or ^{14}O ($E_x = 5.17$ MeV, $\theta^2_p = 21.4$ %).

TABLE 1.   Results of the R-matrix fits ($N = 367$)

	$a = 4$ fm	$a = 5$ fm	$a = 6$ fm
E_R (MeV)	1.067 ± 0.003	1.066 ± 0.003	1.064 ± 0.003
$\Gamma_p (keV)$	104 ± 3	101 ± 3	95 ± 3
χ^2/N	0.53	0.44	0.49
θ^2_p (%)	29.8	22.9	21.6

The isobar diagram for the mirror nuclei ^{19}O and ^{19}Na is shown in Figure 4, including the new $1/2^+$ state in ^{19}Na. The large θ^2_p value is also consistent with the important Coulomb shift (shown by an arrow in the figure). In ^{19}O, the excitation energy of the mirror state is $E_x = 1.472$ MeV [2] whereas it is $E_x = 0.745 \pm 0.013$ MeV for ^{19}Na. This Coulomb shift (~ 0.73 MeV) is among the largest values observed in light nuclei. It is typical of single-particle states, found in nuclei near the drip lines.

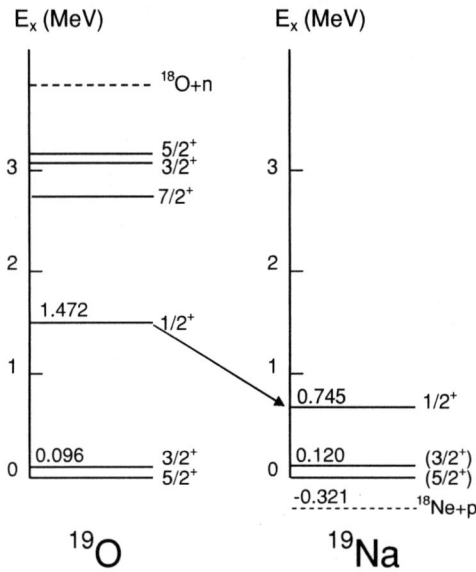

FIGURE 4. Isobar diagram for ^{19}O and ^{19}Na. The Coulomb shift between the $1/2^+$ mirror states is indicated with an arrow.

CONCLUSIONS

Using a low-energy intense ^{18}Ne radioactive beam and a thin polyethylene target the elastic scattering reaction ^{1}H(^{18}Ne,^{1}H)^{18}Ne has been studied. From a R-matrix fit of the elastic cross sections for c.m. energies $E_{c.m.} = 0.7 - 1.5$ MeV and for 20 c.m. angles in the range $\theta_{c.m} = 120.2° - 170.2°$, we have determined the energy and the width of the $1/2^+$ state in ^{19}Na, $E_R = 1.066 \pm 0.003$ MeV and $\Gamma_p = 101 \pm 3$ keV. This state corresponds to the mirror of the $E_x = 1.472$ MeV state in ^{19}O. We find a very large reduced width, typical of single-particle states. Consequently the Coulomb shift of this state with respect to the mirror one in ^{19}O is among the largest values observed in light nuclei.

According to the mirror nucleus ^{19}O, several states should be present in ^{19}Na in the c.m. energy range $2 - 3$ MeV. In order to obtain a global picture of the low-energy level scheme of ^{19}Na, further investigations are planned in the near future at Louvain-la-Neuve.

ACKNOWLEDGMENTS

I thank B.A. Brown and C.N. Davids for useful discussions during the PROCON03 conference.

The experimental work reported here was carried out in collaboration with: M. Couder, M. Gaelens, P. Leleux, A. Ninane, G. Tabacaru, F. Vanderbist (Université catholique de Louvain), P. Descouvemont (Université Libre de Bruxelles), T. Davinson, Ph.J. Woods (University of Edinburgh), J.S. Schweitzer (University of Connecticut), P. Himpe (University of Leuven), F. de Oliveira Santos (GANIL), E. Berthoumieux (CEA-Saclay), N.L. Achouri, J.-C. Angélique (LPC-Caen).

This work has been supported by the Belgian PAI program P5/07 on interuniversity attraction poles of the Belgian-state Federal Services for Scientific, Technical and Cultural Affairs and by the European Community-Access to Research Infrastructure action of the Improving Human Potential Program, contract N° HPRI-CT-1999-00110.

REFERENCES

1. Angulo., C., et al., Phys. Rev. C **67**, 014308 (2003)
2. Tilley, D.R., et al, Nucl. Phys. **A595**, 1 (1995).
3. Cerny, J., et al., Phys. Rev. Lett. **22**, 612 (1969).
4. Benenson, W., et al., Phys. Lett. **58B**, 46 (1975).
5. Görres, J., et al., Phys. Rev. C**51**, 392 (1995).
6. Görres, J., private communication.
7. Decrock, P., et al., Phys. Rev. Lett. **67**, 808 (1991).
8. Coszach, R., et al., Phys. Lett. B **353**, 184 (1995).
9. Bardayan, D.W., et al., Phys. Rev. Lett. **83**, 45 (1999).
10. Graulich, J.-S., et al., Phys. Rev. C **63**, 011302(R) (2000).
11. Angulo, C., et al., Nucl. Phys. **A716** (C), 211 (2003).
12. Graulich, J.S., et al., Eur. Phys. J. **A13**, 221 (2002).
13. Goldberg, V.Z., Pakhomov, A.E., Phys. At. Nucl. **56**, 1167 (1993).
14. Gaelens, M., et al., Proc. 15th Int. Conf. on Application of Accelerators in Research and Industry, Denton, 1998, ISBN-1-56396-825-8, p305.
15. Ziegler, J.F., Biersack, J.P., SRIM2000 program, version 39, 1999 IBM Co.
16. Warters, W.D., et al., Phys. Rev. **91**, 917 (1953).
17. Davinson, T., et al., Nucl. Inst. Meth. **A454**, 350 (2000).
18. Audi, G., Bersillon, O., Blachot, J., Wapstra, A.H., Nucl. Phys. **A624**, 1 (1997).
19. Bohr, N., Kgl. Dan. Vidsk. Sels. Mat. Fys. Medd. **18**, no. 8 (1948).
20. Lane, A.M., Thomas, R.G., Rev. Mod. Phys. **30**, 257 (1958).

Proton unbound states in ^{21}Mg and their astrophysical significance

Alexander Murphy, on behalf of the TUDA collaboration

School of Physics, The University of Edinburgh, JCMB, The King's Buildings, Mayfield Road, Edinburgh, SCOTLAND

Abstract. Recently, the world's first radioactive ^{20}Na beams have been accelerated at the TRIUMF facility. Using the TRIUMF-UK Detector Array facility we have performed elastic scattering of this beam from protons in a CH_2 target with the aim of determining resonance parameters of states in ^{21}Mg. The region of excitation probed in ^{21}Mg has astrophysical importance because resonant contributions to the ^{20}Na$(p,\gamma)^{21}$Mg reaction rate are thought to be part of the breakout path from the hot CNO cycle to the rp-process in x-ray bursters and perhaps novae. The data taken also serve a wider purpose in that comparisons may be made to isobaric analogue states in the mirror ^{21}F nucleus, and thus a study of Coulomb energy shifts may be undertaken.

An overview of the motivation, of the experiment, and a preliminary-analysis of the data taken, is presented.

EXPLOSIVE NUCLEAR BURNING IN CATACLYSMIC VARIABLES

It is believed that at least half of all stars are part of binary systems. As these stars evolve, mass transfer from one star to its companion can lead, in certain circumstances, to the development of a layer of degenerate hydrogen rich material above a compact object composed of evolved elements such as carbon, oxygen or neon. The accretion rate is largely responsible for the subsequent evolution of the compact object. High accretion rates lead to steady state burning such that material builds up until the total mass exceeds the Chandrasekhar limit, at which point the core cannot support its own weight and thus begins contracting. This leads to heating and internal, off-centre, carbon ignition. It is believed a flame front develops, with deflagration converting to detonation and eventual disruption of the entire star. This is a type-Ia supernova.

Continued accretion at lower rates however can lead to a thermonuclear runaway in which explosive hydrogen burning occurs. Such events are observed as, for example, novae and x-ray bursts. Accurate knowledge of the reaction rates involved in the explosion are required so that reliable models of these objects may be developed. One such reaction is ^{20}Na$(p,\gamma)^{21}$Mg, which is believed to be important in ONe novae explosions and in the process of breakout from the hot CNO cycle to the rp-process that occurs in x-ray bursters.

CP681, *Proton-Emitting Nuclei: Second International Symposium; PROCON 2003*,
edited by E. Maglione and F. Soramel
© 2003 American Institute of Physics 0-7354-0150-0/03/$20.00

FIGURE 1. An artists impression of the accretion from a Red Giant on to the surface of a White Dwarf.

Novae

Novae occur when hydrogen rich matter accretes onto the surface of a white dwarf (WD). This can occur in, for example, a binary system containing a white dwarf and a main sequence (MS) star. The MS star evolves as it would in an isolated system. As the central hydrogen fuel is used up and the core contracts, the released gravitational potential leads to an expansion of the outer layers and the star progresses to a red giant (RG) phase. The surface layers of the RG can expand beyond the Roche Lobe, and flow through the L1 Lagrange point into the gravitational well of the WD. This material forms an accretion disc around the WD which is then slowly deposited onto the WDs surface. This is illustrated in Figures 1 and 2.

Accretion rates of the order of $1 \times 10^{-9} M_\odot \mathrm{yr}^{-1}$, result in a (partially) degenerate layer of accreted material. The degenerate condition means that any rise in temperature is not accompanied by a change in density (i.e. expansion). Thus, when later accretion raises the temperature, the conditions for an explosive thermonuclear runaway (TNR) exist.

The white dwarf may be composed predominantly of either carbon and oxygen (CO) or oxygen and neon (ONe). The TNR is initiated when hydrogen is explosively burned on predominantly carbon, nitrogen and oxygen seeds in the C(arbon)-N(itrogen)-O(xygen) and H(ot) CNO cycles [1] It is believed that under conditions of sufficiently high temperature, or if a substantial number of heavy seed nuclei exist, then the explosive hydrogen burning is likely to continue to a variety of heavier nuclei as well. Such a scenario exists with ONe WDs [2]. In this case, the observed luminosity of novae cannot be achieved by simple burning of the CNO and HCNO cycles, yet the temperatures ($T_9 < 0.3$GK) are not high enough to allow breakout to the rp-process that could provide the required energy liberation. Instead, it is believed that the heavier seed nuclei present in the ONe WD allow the initiation of the NeNa and MgAl cycles. The NeNa cycle is illustrated in Figure 3.

The cycle begins with a ^{20}Ne seed which then undergoes the usual sequence of two

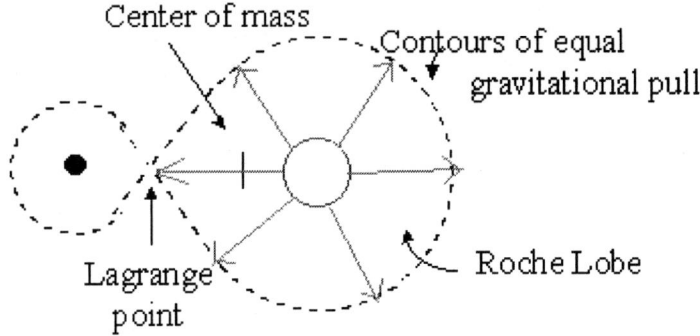

FIGURE 2. The main features of a binary stellar system. The expansion of the Red Giant leads to transfer of hydrogen rich material on to the surface an evolved compact star.

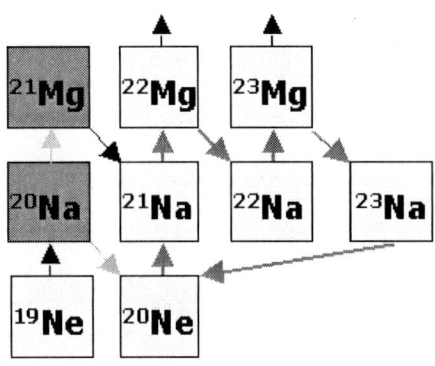

FIGURE 3. The ONe cycle of hydrogen burning.

(p,γ) reactions, followed by a $\beta+$ decay, another (p,γ) reaction, another $\beta+$ decay, and returns to ^{20}Ne via a (p,α) reaction. One of the sources of ^{20}Ne is from the beta-decay of ^{20}Na ($\tau_{1/2}$=446 ms), and thus the abundance of ^{20}Na influences the speed at which material can be processed by this cycle, and therefore how much energy this cycle may liberate. The abundance of ^{20}Na is determined by the competition between it's beta decay into ^{20}Ne and its transmutation via the ^{20}Na(p,γ)^{21}Mg reaction. This reaction rate is dominated by the resonant contributions of various proton unbound states in ^{21}Mg, as calculations [3] suggest only marginal contributions from direct capture to the total reaction rate in this energy region.

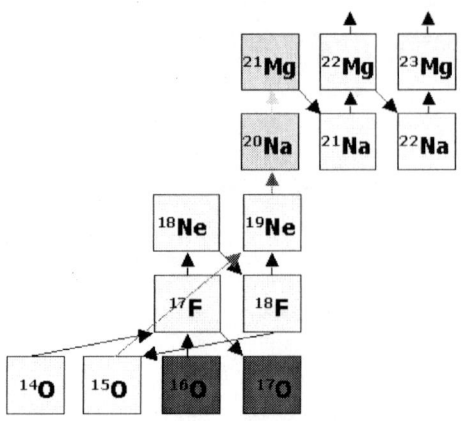

FIGURE 4. The reaction path sequence for breakout from the Hot CNO cycle to the rp-process at about T_9=0.6 GK.

X-ray bursters

If instead of a WD the evolved companion is a neutron star, the significantly deeper gravitational potential well will mean that the explosive hydrogen burning takes place in a more thoroughly degenerate, higher density environment. Consequently, the burning will involve both higher temperatures and more unstable, higher mass nuclei than in the novae counterpart [4]. Breakout from the HCNO cycle is achieved, and material then progresses into the rp-process [5]. Figure 4 illustrates the reaction pathway at about T_9=0.6 GK.

The initial path towards breakout from the HCNO cycles proceeds via the reaction sequence: $^{15}O(\alpha,\gamma)^{19}Ne(p,\gamma)^{20}Na(p,\gamma)^{21}Mg(\beta+)^{21}Na$. Thus, the $^{20}Na(p,\gamma)^{21}Mg$ reaction is an integral link in the energy generation of x-ray bursters, and a better determination of the properties of the states in the compound nucleus will lead to a reduction in the uncertainties in its rate.

PREVIOUS DATA AND NUCLEAR SIGNIFICANCE

Very little experimental data exist for proton rich nuclei in this mass region, and thus any new data are of significant interest. Spectroscopic information regarding the proton unbound states in ^{21}Mg are available only from the data of Kubono and collaborators [6] using the $^3He(^{24}Mg,^6He)^{21}Mg$ reaction. The spectroscopic properties they discovered are presented in Table 1. The resonance energies have quoted uncertainties of ±15 keV, while the spin and parity determinations are only tentative. The proton decay widths are calculated [3] assuming spectroscopic factors of 0.1. These uncertainties result in corresponding uncertainty in the $^{20}Na(p,\gamma)^{21}Mg$ reaction rate.

The three lowest levels in the data presented in Table 1 have been linked with levels in

TABLE 1. Spectroscopic data on states in ^{21}Mg, reproduced from reference 6 ©1992 with permission from Elsevier.

$E_x(^{21}\text{Mg})$ (MeV)	E_r (MeV)	J^π	Γ_p (MeV)
3.752	0.536	3/2-	9.94×10^{-5}
3.901	0.684	7/2-	8.72×10^{-7}
4,010	0.794	1/2+	4.72×10^{-3}
4.261	1.045		
4.824	1.608		

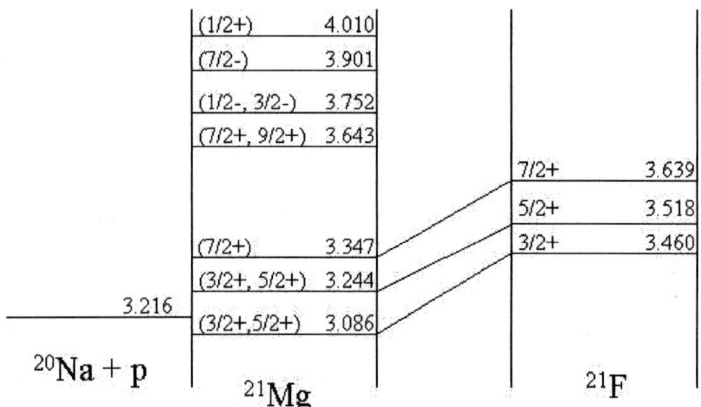

FIGURE 5. The experimental set up used.

the mirror nucleus ^{21}F, as indicated in Figure 5. As can be seen, if these isobaric analogue assignments are correct, that would imply rather large Coulomb energy shifts. This effect is understood within the formalism of Thomas and Ehrman [7], and arises because of the effect of the differing numbers of protons in the mirror nuclei resulting in different spatial distributions of charge and consequent breaking of the isospin symmetry. Such shifts are interesting not only for this specific case, but, since rather little similar data exist for other proton-rich nuclei in this mass region, these new data also have a wider significance. In general, for sd-shell nuclei, one could expect a shell model calculation to be fairly reliable, but the interaction used is largely untested on the proton-rich side of stability. Thus, for several reasons, confirming the location and nature of these states remains a high priority.

As mentioned, the only previous data collected are from a 3 neutron transfer reaction [6, 8]. These were interpreted with the aid of DWBA theory which is known to be not particularly reliable at low energies due to the poor description of compound nuclear effects. The use of such a multi-particle transfer reaction to access the states and then interpretation using DWBA techniques has been widely used in the past, and so comparison of measurements in specific cases using a more transparent analysis may be of some importance in reassessing existing data sets.

EXPERIMENTAL DETAILS

In order to determine the parameters of proton unbound states in ^{21}Mg, resonant elastic scattering of protons was employed. Experimentally, this consisted of impinging a radioactive ^{20}Na beam on to a hydrocarbon foil, and then detecting the recoil protons. The inverse kinematics coupled to the use of a large, spectroscopic quality, silicon strip detector array achieved a high efficiency, while a thick target method meant that a significant excitation function could be generated quickly. Further details are provided below.

The world's first post accelerated ^{20}Na beam was provided by the ISAC facility at TRIUMF. The ISOL technique was employed, with a primary 500 MeV proton beam impinging on a silicon-carbide target. Driver beam currents of around 20 μA were maintained. ^{20}Na ions were extracted from this target and post accelerated to the TRIUMF-UK Detector Array (TUDA). Radioactive beam intensities of up to 5×10^7 pps were delivered in to the experimental chamber with the ions in charge state 5. Two beam energies were employed: E_{lab}=1.25 MeV/u and 1.60 MeV/u. Accurate confirmation of the beam energies was assured by making complimentary measurements in two calibrated instruments. Firstly, a calibrated NMR reading of the 'Prague' analyzing magnet was used. Secondly, the beam was delivered in to the DRAGON spectrometer and focussed on to the exit slits of the first magnetic dipole where the calibrated field was again measured. For the higher beam energy, where the maximum rigidity of the DRAGON MD1 magnet was insufficient to bend the beam, this required first introducing gas in to the DRAGON target volume to slow the beam. Measurements of the required magnetic field were performed with various pressures of gas, and an extrapolation to zero gas pressure made. In all cases, an agreement between the beam energy measured with the Prague magnet and with DRAGON of $\sim$1keV/u was observed.

The target consisted of $\sim$795μg/cm^2 of CH$_2$. Such a thickness results in a significant spread of energies at which elastic scattering reactions can take place, corresponding to the depth in the target at which the reactions occur. For the lower beam energy reactions can occur at $\sim$0.51–1.20 MeV in the centre-of-mass, and for the higher beam energy between $\sim$0.91–1.54 MeV, the overlap of these two excitation functions ensuring experimental consistency.

Detection of the recoiling protons was achieved in two LEDA-design silicon detector arrays [9]. These were placed downstream of the target at 19 and 62 cm. The ^{20}Na beam was bunched and chopped to an 86 ns time structure with each bunch having a width of around 1 ns. Time-of-flight vs energy spectra were used to identify recoil protons. Preliminary spectra of proton energies observed in detecting elements placed at a laboratory scattering angle of 15.6° are shown in Figure 7. This shows protons being detected with a range of energies in agreement with that predicted due to the width of the target. Rutherford scattering determines that the yield in general falls off with higher energy, but superimposed on this are three distinct resonances, their shapes determined by the relative phase shifts between the Coulomb, and hard sphere and compound

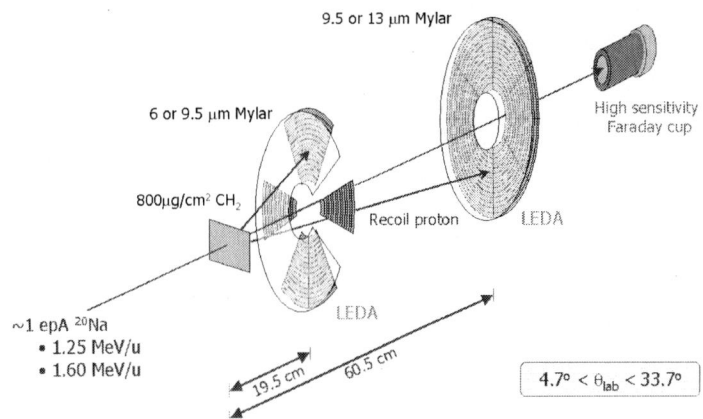

FIGURE 6. The experimental set up used.

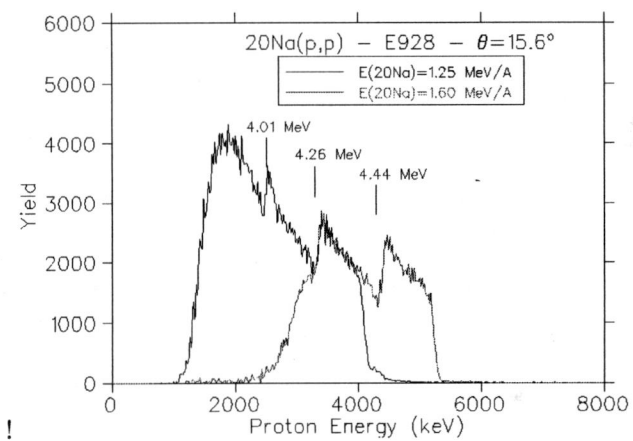

FIGURE 7. An example recoil proton energy spectrum for protons detected at 15.6° in the lab.

nuclear terms. These have approximate energies of 4.01, 4.26 and 4.44 MeV excitation energy in ^{21}Mg. The existence of states at 4.01 and 4.26 was already known, but the highest energy state is entirely new. Accurate determination of the energies, widths and spins of these states will require R-matrix analysis (multi-channel and allowing the possibility of non-zero angular momentum transfer), and this is currently in progress.

SUMMARY

The first measurement of the ^{20}Na(p,p)^{20}Na reaction has been made at the TRIUMF using the TUDA facility. Resonances observed in the compound nucleus will affect the

^{20}Na(p,γ)^{21}Mg reaction rate, which is thought to be of importance in both novae and x-ray bursters. Three distinct states have been observed, one of which for the first time, and R-matrix analysis of the data is currently underway.

ACKNOWLEDGMENTS

The authors would like to gratefully acknowledge the help and assistance of the TRI-UMF operations personnel and members of the DRAGON collaboration in the gathering of these data.

REFERENCES

1. M. Politano *et al.*, Ap. J. **448** (1995) 807
2. M.C. Wiescher, H. Schatz and A.E. Champagne, *Phil. Trans. R. Soc. Lond.*. **A 356** (1998) 2105
3. M.C. Wiescher, J. Görres, F.K. Thielemann and H. Ritter, Astron. & Astrophys. **160** (1986) 56
4. R. Wallace and S. Woosley, Ap. J. Suppl. **45** (1981) 389.
5. M.C. Wiescher, J. Görres and H. Schatz, J. Phys. **G. 25** (1999) 133
6. S. Kubono *et al.*, Nucl. Phys. **A537** (1992) 153
7. R.G. Thomas, Phys. Rev. **88** (1952) 1109.
8. S. Kubono *et al.*, Phys. Rec. C **43** (1991) 1821
9. T. Davinson *et al.* Nucl. Instrum. Methods. **454** (2000) 350

Spectroscopic studies of neutron-deficient light nuclei: decay properties of ^{21}Mg, ^{25}Si and ^{26}P

J.-C. Thomas*[†], L. Achouri**, J. Äystö[‡], R. Béraud[§], B. Blank*, G. Canchel*, S. Czajkowski*, P. Dendooven[¶], A. Ensallem[§], J. Giovinazzo*, N. Guillet*, J. Honkanen[‡], A. Jokinen[‡], A. Laird[||], M. Lewitowicz[††], C. Longour[‡‡], F. de Oliveira Santos[††] and M. Stanoiu[††][§§]

*Centre d'Etudes Nucléaires de Bordeaux-Gradignan, Allée du Haut-Vigneau, B.P. 120, 33170 Gradignan Cedex, France
[†]Instituut voor Kern-en Stralingsfysica, Celestijnenlaan 200D, 3001 Leuven, Belgium
**SUBATECH, Ecole des Mines, 4 rue A. Kastler, B.P. 20722, 44307 Nantes Cedex 3, France
[‡]Department of Physics, University of Jyväskylä, P.O. Box 35, 40351 Jyväskylä, Finland
[§]Institut de Physique Nucléaire de Lyon, 69622 Villerbanne Cedex, France
[¶]Kernfisich Versneller Instituut, Zernikelaan 25, 9747 AA Groningen, Nederland
[||]Department of Physics and Astronomy, University of Edinburgh, Edinburgh EH9 3JZ, United Kingdom
[††]Grand Accélérateur National d'Ions Lourds, B.P. 5027, 14076 Caen Cedex 5, France
[‡‡]Institut de Recherche Subatomique, 23 rue du Loess, B.P. 28, 67037 Strasbourg Cedex, France
[§§]Institut de Physique Nucleaire d'Orsay,15 rue G. Clemenceau, 91406 Orsay Cedex, France

Abstract. Neutron-deficient nuclei with T_z equals to $-3/2$ and -2 have been produced at the GANIL/LISE3 facility in fragmentation reactions of a 95 MeV/u ^{36}Ar primary beam in a ^{12}C target. For the first time, β-delayed proton and $\beta-\gamma$ emission has been simultaneously observed in the decay of ^{21}Mg, ^{25}Si and ^{26}P. The decay scheme of the latter is proposed and the Gamow-Teller strength distribution in its β decay is compared to shell-model calculations based on the USD interaction. The B(GT) values derived from the absolute measurement of the β-branching ratios are in agreement with the quenching factor of about 60% obtained for allowed Gamow-Teller transitions in this mass region. A precise half-life of 43.7 (6) ms was determined for ^{26}P, the $\beta-2p$ emission of which was studied. The expected contribution of spectroscopic studies of neutron-rich nuclei is discussed with respect to the mirror asymmetry phenomenon occuring in analogous β decays.

INTRODUCTION

The β decay of unstable nuclei toward a large set of daughter nuclear states is a powerful tool in nuclear structure studies and compilations of spectroscopic properties are available for most of the nuclei with A ranging from 17 to 39 [1, 2, 3, 4, 5]. Since the V-A description of the decay process involves quite simple operators, precise shell-model computations have been extensively done in the corresponding sd shell [6, 7, 8].

The link between experimental results and theoretical predictions is given by the reduced transition probabilities ft^+ and ft^- associated with the individual β^+ decay and the electronic capture ($E.C.$) and to the β^- decay, respectively. These parameters incorporate the phase space factor f and the reduced half-life $t = T_{1/2}/BR$ where $T_{1/2}$ is the total half-life of the decaying nucleus and BR the branching ratio of the considered

CP681, *Proton-Emitting Nuclei: Second International Symposium; PROCON 2003,*
edited by E. Maglione and F. Soramel
© 2003 American Institute of Physics 0-7354-0150-0/03/$20.00

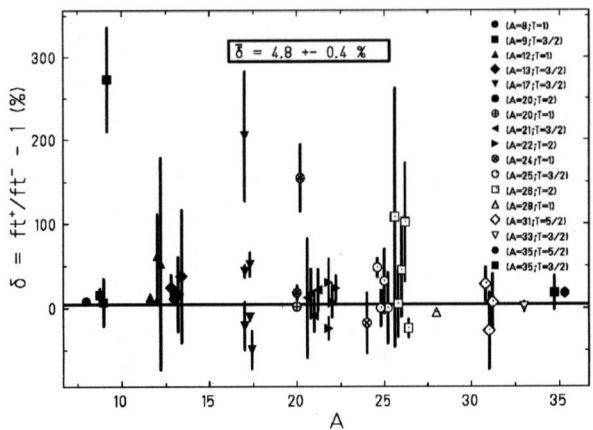

FIGURE 1. Systematics of the experimental values of the δ asymmetry parameter for nuclei with $A \leq 40$ and different isospins T. Only allowed Gamow-Teller transitions with $log(ft) \leq 6$ are considered.

β transition. In the frame of the V-A standard model of the β disintegration, the $ft^{\pm}$ parameters are given for allowed transitions by:

$$ft^{\pm} = \frac{\mathcal{K}}{g_V^2 |< f|\tau_{\pm}|i >|^2 + g_A^2 |< f|\sigma\tau_{\pm}|i >|^2} \tag{1}$$

where $\mathcal{K}$ is a constant and where g_V and g_A are the vector and axial-vector current coupling constants related to the Fermi and Gamow-Teller components of the β disintegration. Their strengths are derived from the calculation of the matrix elements $< f|\tau_{\pm}|i >$ and $< f|\sigma\tau_{\pm}|i >$.

To the first order, the agreement between the measured $ft^{\pm}$ values and the computed strengths appears to be a good test of nuclear wave functions built in the shell-model frame, stressing the role of the overlap between initial and final nuclear states as well as the configuration mixing occurring in parent and daughter states. However, a systematic deviation from theoretical predictions shows the limitation of our theoretical understanding and treatment of fundamental interactions. It is reported as the *mirror asymmetry anomaly in β decay* [9, 10, 11, 12, 13]. This phenomenon is related to the isospin non-conserving forces acting in the atomic nucleus. If nuclear forces were charge independent as it is admitted at a first approximation level, we would expect the β^+ (*E.C.*) and the β^- individual decays of analogous states belonging to mirror nuclei to be of equal strength. The deviation from this simple picture is characterized by the $\delta = ft^+/ft^- - 1$ asymmetry parameter, where the $+$ and $-$ signs are associated with the decay of the proton-rich and the neutron-rich members of the mirror nuclei couple.

An updated systematics of experimental δ values for nuclei with $A \leq 40$ is presented in figure 1 (see ref. [14] for details). A mean deviation of 5 % is found for these nuclei lying in the p and sd shells.

It was often attempted to explain the mirror asymmetry anomaly in the p shell either in terms of binding energy effects [11, 12, 15] or by introducing the concept of "second class currents" [16, 17, 18]. None of the theoretical approaches were able to reproduce the measured δ values. Nevertheless, some shell-model calculations are currently made to test the isospin non-conserving part of the interactions involved in β decay by studying the influence on the Gamow-Teller matrix elements of isospin mixing effects and radial overlap mismatches of nuclear wave functions [9]. From the experimental point of view, spectroscopic experiments should now be performed to measure the δ values of sd shell mirror nuclei for which reliable shell model calculations are now available [9].

With the parallel development of secondary radioactive beams and experimental techniques, a large set of neutron-deficient nuclei has been investigated since the β-delayed proton decay was first observed forty years ago [19]. Since Q_β values are increasing while nuclei become more exotic, the observation of $\beta - p$ and $\beta - \gamma$ decays of neutron-deficient nuclei gives the opportunity to probe the Gamow-Teller strength function up to more than 10 MeV in excitation energy. The whole energy window opened in the β decay can then be covered both by spectroscopic studies and charge exchange reactions [20]. Therefore, the theoretical description of the structure of atomic nuclei as well as our understanding of the weak interaction can be tested far from the stability line.

EXPERIMENTAL PROCEDURE

Fragment production and identification

In addition to ^{21}Mg, ^{25}Si and ^{26}P, the β-delayed proton and two-proton emitters ^{22}Al [21] and ^{27}S [22] have been studied during the same experimental campaign. All nuclei have been produced in the fragmentation of a 95 MeV/u ^{36}Ar primary beam delivered by the two cyclotrons of the GANIL facility. A $357.1\,mg/cm^2$ ^{12}C production target was placed in the SISSI device [23], the high angular acceptance and focusing properties of which increased the selectivity of the fragment separation operated by the LISE3 spectrometer. The latter included a shaped Be degrader at the intermediate focal plane and a Wien filter at the end of the line to refine the selection of the separated fragments. Ions of interest were stopped in the fourth element $E3$ of a silicon stack (figure 2).

The ion identification was performed by means of time of flight and energy loss measurements with the help of the silicon detectors $E1D6$ to $E4$ ($2 * 300\,\mu m$ and $2 * 500\,\mu m$ thickness, $600\,mm^2$ surface). It led to a precision in the counting rate better than 1% for ^{21}Mg and ^{25}Si and about 3% for the more exotic ^{26}P nucleus. The production rate of the selected ions varied from 60 to $100\,pps$, depending on their exoticity.

β-delayed proton-spectroscopy

In contrast to previous experiments [24, 25, 26] in which ions were deposited at the surface of an ion catcher, β-delayed protons are emitted inside the implantation detector E3. As a consequence, the proton spectrum rises on an important β background and the

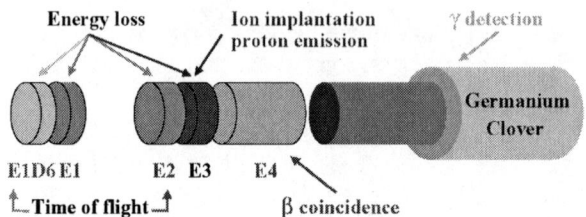

FIGURE 2. Identification and detection setup. The $E1D6$ and $E2$ silicon detectors were used for time of flight measurements, while the energy loss of the selected ions was determined with the first four detectors. Emitted charged particles were detected in the implantation detector E3. The last one, $E4$, was used as a β-coincidence detector. A segmented germanium-clover detector was finally employed to observe the β-delayed γ decay.

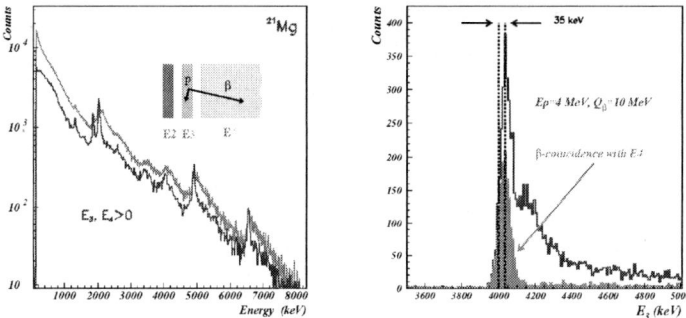

FIGURE 3. The left panel shows the influence of the β-coincidence condition ($E4 > 0$) on the energy spectrum delivered by the detector E3 for ^{21}Mg. The right panel exhibits a GEANT [28] simulation of the coincidence condition effect on the shape of a standard β-delayed proton emission event.

energy deposited by a proton is added to those of the recoiling ion and the emitted β particle. Consequently, low intensity proton groups are difficult to measure because of the small signal over noise ratio and proton peaks are deformed and energetically shifted due to the simultaneous β-particle energy deposit.

To minimize these effects, ions were implanted in the last $100\,\mu m$ of the detector $E3$ and a β coincidence with the large detector $E4$ (with $6\,mm$ thickness and $600\,mm^2$ surface) was requested. Therefore, as shown in figure 3 for ^{21}Mg, the β-energy deposit was strongly reduced and proton peaks were easy to identify and to fit by means of Gaussian distributions. The absolute intensities I_i^p of each individual proton group were then obtained as follows:

$$I_i^p = \frac{Sc_i^p}{Kc^p * N_{impl} * \varepsilon_i^p} \tag{2}$$

where Sc_i^p is the proton peak surface deduced from the β-coincidence spectrum, Kc^p the normalization factor to be taken into account due to the coincidence condition, N_{impl} the number of implanted ions and ε_i^p the proton detection efficiency of $E3$.

Proton detection efficiency. The proton detection efficiency was determined by means of simulations using the LISE [27] and GEANT [28] codes. For all ions, the detection efficiency was found to be larger than 50% for $8\,MeV$ protons, with a relative error lower than 6%.

Normalization factor Kc^p. It was estimated for each ion by fitting the corresponding proton peaks in the high energy part of the $E3$ spectrum, where they are well separated and where the β background is low. Kc^p values were found to vary from 10 to 13% with a relative error lower than 10%.

γ spectroscopy

As shown in figure 2, a segmented germanium detector was placed at 0 degree, a few centimeters behind the silicon stack. To reduce the dead time of the acquisition system, the output signals were not used as triggers for the data acquisition. As a consequence, the γ-detection efficiency depended on the kind of radioactivity event that triggered the acquisition. Since the energy loss of a proton in $E3$ is at least larger than a few hundred keV, $E3$ triggered the acquisition system each time a proton was emitted. On the other hand, most of the β particles emitted without accompanying protons did not lose enough energy in $E3$ to trigger the acquisition system. Therefore, the triggering rate for β-γ events is given approximately by the fraction of the solid angle under which the large silicon detector $E4$ is seen from $E3$. It was determined to be $35.0\,(45)\%$. The intrinsic γ-detection efficiency in the $300\,keV$ to $2000\,keV$ range was about 2 to 3%, with a relative error of about 20%.

EXPERIMENTAL RESULTS

Since the β-delayed proton emission of ^{21}Mg and ^{25}Si has already been investigated in the past [24, 26, 29, 30], only the experimental results obtained for ^{26}P are detailed in the following. The β-delayed proton spectrum of ^{26}P, conditioned by the detection of β particles in $E4$, is shown in the left panel of figure 4. The right panel shows the observed γ energy spectrum in the β decay of the same nucleus.

β-delayed proton emission of ^{26}P

The center of mass energy of the β-delayed one- or two-proton groups labeled in figure 4 is given in table 1. Their relative and absolute intensities are also reported. For each transition, the table indicates the populated daughter state and the deduced excitation energy of the emitting state in ^{26}Si. The latter were identified by means of proton-γ coincidences and using energy arguments. Only three proton groups (11, 15 and 24) could not be assigned.

The excitation energy of the I.A.S. $(E^*(I.A.S.) = 13015\,(4)\,keV)$ in ^{26}Si was derived from the individual center of mass energies of the one-proton groups emitted by this

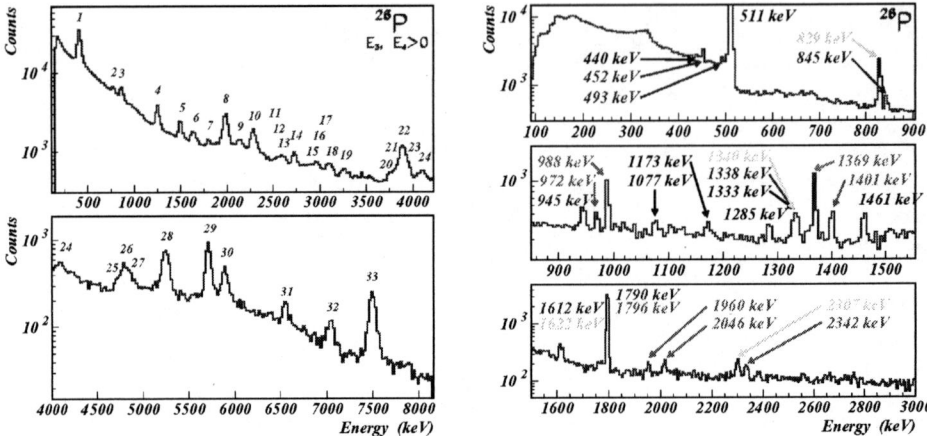

FIGURE 4. The left panel shows the β-delayed proton and two-proton spectrum obtained in coincidence with E4 in the decay of ^{26}P. The right panel presents the γ spectrum obtained for the same setting. The six γ lines at $972, 988, 1796, 1960, 2046$ and $2342\,keV$ are related to the β-delayed γ decay of ^{26}P.

state. The atomic mass excess of ^{26}P was then deduced using of the following relations [31]:

$$\Delta(^{26}P) = \Delta(^{26}Si) + E^*(I.A.S.) + \Delta E_c - \Delta_{nH} \text{ with } \Delta E_c = 1440.8 * \left(\frac{\bar{Z}}{A^{\frac{1}{3}}}\right) - 1026.3$$

(3)

where $\Delta(^{26}Si)$ is the atomic mass excess of ^{26}Si, ΔE_c the Coulomb isotopic shift between the I.A.S. in ^{26}Si and the ground state of ^{26}P and where Δ_{nH} is the mass excess difference between a neutron and a hydrogen atom. Taking $\bar{Z} = 14.5$ for the mean atomic number of the two $A = 26$ nuclei, the atomic mass excess of ^{26}P is equal to $11114\,(90)\,keV$. It leads to a $Q_\beta = \Delta(^{26}P) - \Delta(^{26}Si)$ value equal to $18258\,(90)\,keV$, and to a proton separation energy $S_p(^{26}P) = \Delta(H) + \Delta(^{25}Si) - \Delta(^{26}P) = 0\,(90)\,keV$ for ^{26}P. This latter value suggests that ^{26}P can hardly be a direct proton emitter since the available energy for such a disintegration would not exceed $100\,keV$.

Based on this mass excess determination, the $log(ft)$ value for the β feeding of the I.A.S. was deduced to be $3.13\,(5)$, which is close to the expected value of 3.188 [32].

β-delayed γ decay of ^{26}P

The γ spectrum obtained during the setting on ^{26}P is shown in the right panel of figure 4. Their measured absolute intensities were used to determine the absolute β-branching ratios towards the associated excited states in ^{26}Si. The new γ line at 1401 was assigned to the decay of the excited state of ^{26}Si at $4183\,(11)\,keV$ to the one at $1795.9\,(2)\,keV$ [4]. The β decay of the ground state of ^{26}Si is followed by the two observed γ rays at

TABLE 1. β-delayed proton and 2-proton emission of ^{26}P. The center of mass energy, the relative intensity and the absolute intensity of proton groups labeled in figure 4 are reported. The identification of the populated state is given in the fifth column. The last column gives the deduced excitation energy of the emitting state in ^{26}Si

1p	C.M. Energy (keV)	Relative intensity (%)	Absolute intensity (%)	populated state in ^{25}Al	Deduced emitting state in ^{26}Si *
1	412(2)	100.0(71)	17.96(90)	ground state	5929(5)
2	778(3)	4.3(5)	0.78(7)	ground state	6295(6)
3	866(2)	9.5(10)	1.71(15)	ground state	6384(5)
4	1248(2)	8.4(8)	1.51(12)	ground state	6765(5)
5	1499(2)	5.5(5)	0.99(7)	2nd excited state	7962(5)
6	1638(3)	3.6(4)	0.65(6)	1st excited state	7606(6)
7	1798(4)	1.1(3)	0.20(5)	2nd excited state	8254(5)
8	1983(2)	13.3(11)	2.39(16)	ground state	7501(5)
9	2139(4)	3.0(8)	0.54(14)	4th excited state	9433(4)
10	2288(3)	8.2(9)	1.47(12)	3rd excited state	9433(4)
11	2541(6)	0.5(2)	0.09(3)	not assigned	
12	2593(13)	1.5(3)	0.27(6)	1st excited state	8563(17)
13	2638(18)	0.6(2)	0.11(4)	ground state	8156(21)
14	2732(4)	2.6(4)	0.47(6)	ground state	8254(5)
15	2855(17)	< 0.8(2)	< 0.14(4)	not assigned	
16	2908(11)	0.3(3)	0.06(5)	2nd excited state	9370(15)
17	2968(5)	1.8(3)	0.32(5)	2nd excited state	9433(4)
18	3097(6)	1.7(4)	0.31(6)	4th excited state	10405(5)
19	3258(4)	1.9(2)	0.23(4)	2nd excited state	9725(5)
20	3766(9)	2.0(4)	0.36(7)	1st excited state	9725(5)
21	3817(6)	0.7(3)	0.13(5)	2nd excited state	10299(6)
23	3920(5)	6.7(9)	1.21(14)	ground state	9433(4)
24	4097(5)	< 2.1(3)	< 0.37(4)	not assigned	
25	4719(6)	1.3(2)	0.24(4)	1st excited state	10688(9)
26	4793(3)	3.0(4)	0.54(6)	ground state	10299(6)
27	4858(4)	2.5(3)	0.44(5)	1st excited state	10827(8)
29	5710(3)	7.8(7)	1.40(11)	4th excited state	13015(4)
30	5893(4)	4.1(8)	0.73(13)	3rd excited state	13015(4)
31	6551(4)	1.2(5)	0.21(8)	2nd excited state	13015(4)
32	7039(5)	1.0(1)	0.17(2)	1st excited state	13015(4)
33	7494(4)	3.4(3)	0.61(5)	ground state	13015(4)

2p	C.M. Energy (keV)	Relative intensity (%)	Absolute intensity (%)	populated state in ^{24}Mg	Deduced emitting state in ^{26}Si †
22	3879(3)	4.4(6)	0.79(12)	1st excited state	13015(4)
28	5247(3)	7.6(13)	1.37(22)	ground state	13015(4)

* The mean energy was taken when more than one decay is observed from the excited state

† Deduced from 1p transitions 29 to 33

830 and $1622 \, keV$. The two lines at 1341 and $2037 \, keV$ are due to the summing of the $511 \, keV$ γ rays with the intense γ transitions at 830 and $1796 \, keV$.

The other γ lines are associated to the β-delayed proton and two-proton decay towards excited states in ^{25}Al and ^{24}Mg, some of them being also due to the presence of ^{25}Si, ^{24}Al and ^{23}Mg as contaminants. Background γ lines are finally visible in the spectrum at $1461 \, keV$ (^{40}K), and at 1173 and $1333 \, keV$ (^{60}Co).

Measurement of ^{26}P half-life

The half-life of ^{26}P was determined by means of a time correlation procedure, based on the fact that decay events that are correlated to a selected implantation event follows a time-dependent distribution, whereas non-correlated events are randomly distributed. The half-life of ^{26}P was determined to be $43.7(6)\,ms$, in agreement with the value given by M.D. Cable *et al.* [25] ($20^{+35}_{-15}\,ms$).

β-decay scheme

The proposed decay scheme of ^{26}P is shown in figure 5. It is compared to calculations performed in the full sd shell by B.A. Brown [32] with the OXBASH code [33] and based on the USD interaction [34]. The Gamow-Teller strength distribution given by the summation of the individual $B(GT) = (g_A/g_V)^2 | < f|\sigma\tau_\pm|i > |^2$ parameters is compared to the one expected from the mirror β decay and from the calculated $log(ft)$ values.

The agreement between experimental results and theoretical calculations appears to be very satisfying, most of the observed excited states of ^{26}Si being reproduced by the model within a few hundred keV up to an excitation energy of $7\,MeV$.

The distributions of the summed Gamow-Teller strength as a function of the excitation energy of the populated states is presented in the insert. The experimental distribution is in good agreement with the theoretical calculations up to more than $10\,MeV$. Beyond, the model predicts the feeding of a lot of high energy excited states by low intensity β transitions that are not visible experimentally. Due to the small phase space factor f associated with such transitions, the related $B(GT)$ values are of importance, which explains the observed divergence. Nevertheless, the overall agreement below $10\,MeV$ indicates that the quenching of the Gamow-Teller strength for this sd shell nucleus is about 60%, which is the mean value that has been derived from previous studies in this mass region [8].

At low excitation energy, the summed Gamow-Teller strength is compared to the one deduced from the β decay of the mirror nucleus and assuming that nuclear forces are isospin independent ($\delta = 0$). Unfortunately, the error for the β-branching ratios to the proton-bound states is very large due to the poor precision on the γ efficiency and do not allow any constructive conclusion about the individual values of the δ asymmetry parameter.

Similar results were obtained for ^{21}Mg and ^{25}Si. The absolute measurement of the γ-ray intensities showed a large feeding of the $7/2^+_1$ excited state in ^{21}Na. The spin of the ^{21}Mg ground state was deduced to be $5/2^+$ rather than $3/2^+$ [4].

CONCLUSION

The β decay of the neutron-deficient ^{21}Mg, ^{25}Si and ^{26}P nuclei was studied at the LISE3 facility at GANIL. 60 to 300 ions per second were produced with a contamination rate of less than 1% for ^{21}Mg and ^{25}Si, and lower than 10% for ^{26}P. The decay scheme of the

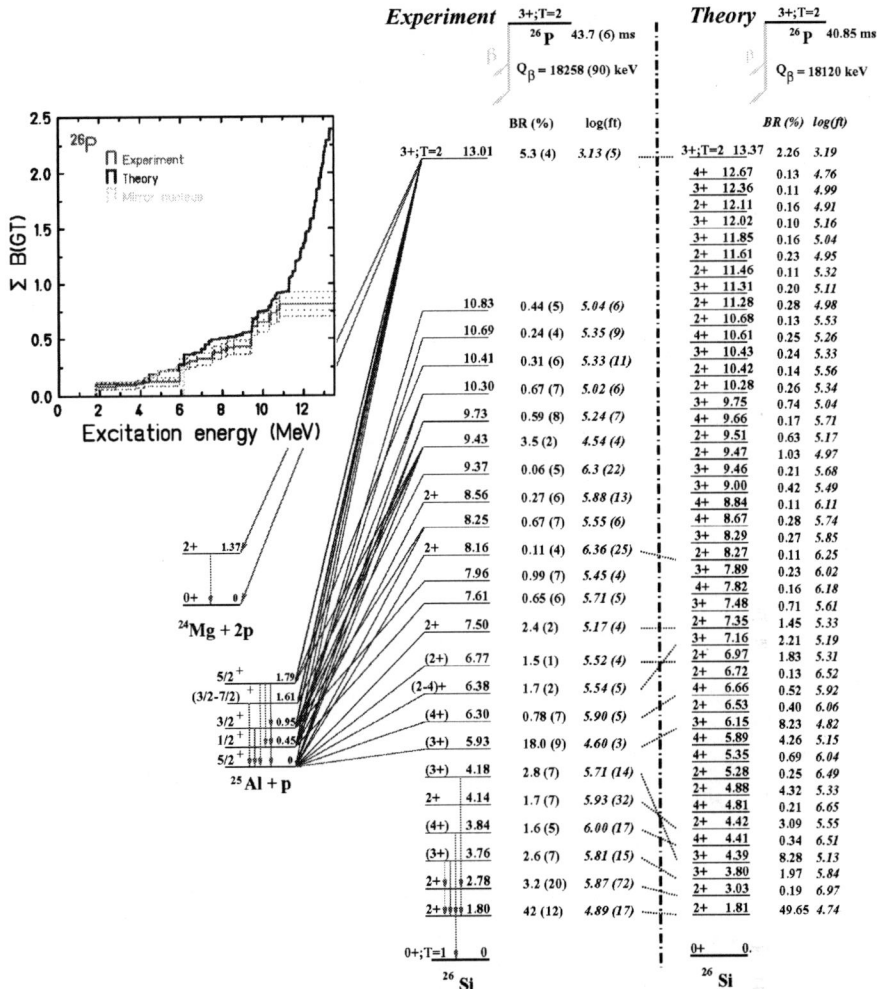

FIGURE 5. ^{26}P β-decay scheme. The experimental half-life and Q_β values are taken from this work. Results are compared to shell-model calculations performed by B.A. Brown [32]. The distribution of the summed Gamow-Teller strength ($\Sigma B(GT)$) is derived from the calculated $log(ft)$ values.

three nuclei was obtained, including for the first time the β decay towards proton-bound states. The comparison to shell-model calculation revealed two features: the reliability of such models when they are applied to middle-shell nuclei lying close to the proton drip-line, and the $\approx 60\%$ quenching of the Gamow-Teller strength of the individual β transitions.

Some nuclear properties were derived from the spectroscopic study of these nuclei: the half-life of ^{26}P was determined to be $43.7(6)\,ms$. Its proton separation energy as well as the maximum available energy in its β decay were estimated with a precision of $90\,keV$.

The two-proton emission of ^{26}P was observed and more than 30 one-proton transitions were identified. Based on γ-spectroscopy information, the spin of the ground state of ^{21}Mg was determined to be $5/2^+$.

The experiment presented here suffers from some deficiences that can be overcome. This would lead to the determination of accurate δ asymmetry parameter values, which are necessary to understand the origin of isospin non-conserving forces in nuclei.

ACKNOWLEDGMENTS

The authors would like to thank B.A. Brown for providing up-to-date shell-model calculations as well as C. Volpe and N.A. Smirnova for stimulating discussions on the mirror asymmetry question.

REFERENCES

1. D.R. Tilley, H.R. Weller, C.M. Chevez, and R.M. Chasteler, Nucl. Phys. A 564, 1 (1993)
2. D.R. Tilley, H.R. Weller, C.M. Chevez, and R.M. Chasteler, Nucl. Phys. A 595, 1 (1995)
3. D.R. Tilley, C.M. Chevez, J.h. Kelley, S. Raman, and H.R. Weller, Nucl. Phys. A 636, 247 (1998)
4. P.M. Endt, Nucl. Phys. A 590, 1 (1990)
5. P.M. Endt, Nucl. Phys. A 633, 1 (1998)
6. B.A. Brown and B.H. Wildenthal, At. Data Nucl. Data Tables 33, 347 (1985)
7. W.A. Landford and B. H. Wildenthal, Phys. Rev. C 7, 668 (1973)
8. B.H. Wildenthal, M. S. Curtin, and B.A. Brown, Phys. Rev. C 28, 1343 (1983)
9. N.A. Smirnova and C. Volpe, Nucl. Phys. A 714,441 (2003)
10. L. Axelsson et al., Nucl. Phys. A 634, 475 (1998)
11. I.S. Towner, Nucl. Phys. A 216, 589 (1973)
12. D.H. Wilkinson, Phys. Rev. Lett. 27, 1018 (1971)
13. D.H. Wilkinson and D. E. Alburger, Phys. Rev. Lett. 24, 1134 (1970)
14. J.-C. Thomas, Ph-D thesis, Centre d'Etudes Nucléaires de Bordeaux-Gradignan, Université de Bordeaux I, France, 2002, http://147.210.235.3/pdf/2002/THOMAS_JEAN_CHARLES_2002.pdf
15. F.C. Barker, Nucl. Phys. A 537, 147 (1992)
16. D.H. Wilkinson, Phys. Lett. 31 B, 447 (1970)
17. K. Kubodera, J. Delorme, and M. Rho, Phys. Rev. Lett. 38, 321 (1977)
18. D.H. Wilkinson, Eur. Phys. J. A 7, 307 (2000)
19. R. Barton et al., Can. J. Phys. 41, 2007 (1963)
20. B.D. Anderson et al., Phys. Rev. C 43, 50 (1991)
21. L. Achouri et al., in preparation
22. G. Canchel et al., Eur. Phys. J. A 12, 377 (2001)
23. A. Joubert et al., The SISSI Project, Proceedings of the Second Conference of the IEEE Particle Accelerator, San Francisco, May 1991, vol. 1, p. 594.
24. R.G. Sextro, R. A. Gough, and J. Cerny, Phys. Rev. C 8, 258 (1973)
25. M.D. Cable et al., Phys. Rev. C 30, 1276 (1984)
26. J.D. Robertson, D.M. Moltz, T.F. Lang, J.E. Reiff, and J. Cerny, Phys. Rev. C 47, 1455 (1993)
27. D. Bazin et al., Nucl. Instr. Meth. in Phys. Res. A 482, 307 (2002)
28. CERN web page at http://wwwinfo.cern.ch/asd/geant/
29. S. Hatori et al., Nucl. Phys. A 549, 327 (1992)
30. P.L. Reeder, A.M. Poskanzer, R.A. Esterlund, and R. McPherson, Phys. Rev. C 1, 781 (1966)
31. M.S. Antony, J. Britz, and A. Pape, At. Data Nucl. Data Tables 34, 279 (1986)
32. B.A. Brown, private communication
33. A. Echegoyan, W.M.D. McRae , and B.A. Brown, MSU-NSCL Report No. 524 (1985)
34. B.H. Wildenthal, Prog. Part. Nucl. Phys. 11, 5 (1984)

Study of ^{19}Na at SPIRAL

François de Oliveira Santos and the E400S collaboration

GANIL, BP 55027, F-14076 Caen Cedex 5, FRANCE
Oliveira@ganil.fr

Abstract. The excitation function for the elastic scattering p(^{18}Ne,p)^{18}Ne was measured. This study has been undertaken with the first radioactive beam from the SPIRAL facility at GANIL. A thick solid hydrogen cryogenic target was developed for this experiment. Several resonances have been observed, corresponding to excited states in the compound nucleus ^{19}Na.

INTRODUCTION

Sodium isotopes have been produced recently in a wide range of the neutron number, from the most neutron rich N=26: ^{37}Na identified in a recent experiment done at GANIL [1] with the new facility LISE2000, to the most proton rich, one step beyond the proton drip line, N=8: ^{19}Na. The study of the lightest sodium isotope has been the subject of the first SPIRAL experiment at GANIL.

The Motivations

There are several interests to study this nucleus. It is undoubtedly an important objective to undertake the study of this nucleus by the measurement of the ground and excited states properties. This allows us to compare them to the properties of the mirror nucleus ^{19}O states, and with theoretical calculations. In those light and proton rich nuclei several effects are sought (e.g. Thomas-Ehrman shift, breaking of isobaric symmetry, multi particle emission). Another application for the study of ^{19}Na is coming from the fact that the mass of the ^{19}Na ground state is a sensitive parameter in the calculation of the intensity for the reaction rate in the double proton capture: ^{18}Ne(2p,γ)^{20}Mg, one reaction that can have important effect in classical astrophysics since it starts from the ^{18}Ne, a nucleus already present in standard astrophysical models.

The isotope ^{19}Na was for the first time observed in 1969 by Cerny et al. [2] via a transfer reaction. They observed only one peak at the mass excess of 12.974 ± 0.070 MeV, a value close to that of 12.90 MeV predicted with the Isobaric Mass Multiplet Equation. This extracted mass of ^{19}Na implies that this nucleus is not bound for the proton emission, the ground state being only E_R = 366 ± 70 keV above the proton emission threshold. Since that first measurement only three new experiments have been dedicated to the study of ^{19}Na [3] [4] [5]. Today, only the ground state and two excited states have been seen experimentally. The ground state has been observed

CP681, *Proton-Emitting Nuclei: Second International Symposium; PROCON 2003*,
edited by E. Maglione and F. Soramel
© 2003 American Institute of Physics 0-7354-0150-0/03/$20.00

several times, the adopted value for its energy is $E_R = 320 \pm 10$ keV, but its spin has never been assigned, neither width measured. The first excited state has been observed only once and with a low statistics, 120 keV above the ground state. Nothing more is known about this state. The second excited state has been seen recently in a precise measurement (see Angulo et al. within this book), it is 1066 ± 2 keV above the proton emission threshold. Its spin is 1/2+ and its width $\Gamma = 101 \pm 3$ keV. This state corresponds without any doubt to the known second excited state in the mirror nucleus ^{19}O. It is 727 keV down shifted from its analogue. No more is experimentally known about ^{19}Na.

The Principle Of The Experiment

In the present experiment we have measured the excitation function of the elastic scattering reaction: $^{18}Ne(p,p)^{18}Ne$. It is well known that the excitation function for the elastic scattering at low energy is a combination of Rutherford contribution and anomalies (resonances) that correspond to states in the compound nucleus. There is a straightforward correspondence between those anomalies and the properties of the excited states of the compound nucleus. The energy of the state can be extracted from the position of the resonance, its width from the width and intensity of the resonance, its spin and parity from the shape and angular distribution of the resonance. The interest of using this kind of measurement for spectroscopy is manifest: this measurement is simple, give a lot of information about the compound nucleus states and the corresponding cross sections are high. Because ^{18}Ne is radioactive, we have measured this elastic scattering in inverse kinematics: $p(^{18}Ne,p)^{18}Ne$. The experiment has been done at GANIL with a radioactive beam produced by the new SPIRAL facility. The mean intensity of the beam was $5 \; 10^5$ pps. It is very time consuming to use a thin target and to change the energy of the incoming beam by small steps in order to measure the excitation function. To solve this difficulty we have used the thick target method. The idea of using thick target has been developed successfully in several experiments with stable and radioactive beams [6] [7]. If the implanted particles (^{18}Ne), at some point along its slowing down trajectory inside the target, has a center of mass energy that corresponds to a state in the compound nucleus ($^{19}Na = ^{18}Ne + p$), the probability for the elastic scattering change (resonance). The scattered proton can escape the target and can be detected at forward angles in the laboratory frame. We have used a simple silicon detectors telescope to measure the energy of the light charged particles. The thick target method makes it possible to obtain a complete and continuous excitation function over a wide energy range, by simply detecting the scattered protons and measuring their energies, without changing the energy of the incoming beam.

The Cryogenic Target

We have used a solid hydrogen cryogenic target for two important reasons. The first, the use of compound targets (e.g. $(CH_2)_n$) introduces atoms (e.g. Carbon) in which nuclear reactions can occur and pollute the measurement. The second, the use of a pure hydrogen target maximizes the counting rate because the higher stoechiometric ratio leads to the highest effective target thickness, this effect is developed hereafter. The main requirements imposed for solid hydrogen targets usable under vacuum with a particle beam are: thickness about 1 mm, very thin windows, and uniform thickness and density. A special cryogenic system has been designed to make the target used in the present experiment. It is composed of a H_2 target cell and a He cell on either side of the target. During the hydrogen target production phase, equivalent He pressure is maintained on either side of the target windows up to the complete formation of the solid H_2 target. Once the target is formed, the helium gas is evacuated.

TESTS AND CALIBRATIONS WITH $^{18}O(P,P)^{18}O$

The raw measured spectrum (number of counts as a function of the detected particles energies) has to be corrected with two functions in order to produce the final spectrum (differential cross section as a function of the center of mass energy). The first correction (Figure 1a) is applied for the energy loss of the protons in the hydrogen target, between the positions of the scattering events up to the silicon detector. The second correction (Figure 1b) is due to the fact that the effective target thickness depends on the energy loss of the incident ions inside the target. In fact the higher is the energy loss of the incoming ions, the lower is the effective target thickness, and so the lower is the counting rate. This effect explains why it is very important to use a pure hydrogen target. In that case the energy loss is only due to the interaction with the hydrogen atoms of the target. With the pure hydrogen target we get the lowest energy loss, and the highest proton density, which both increase the counting rate.

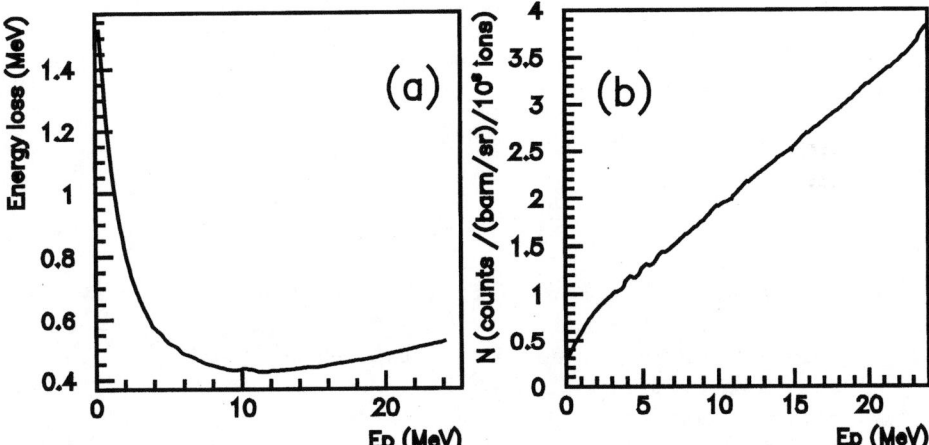

FIGURE 1. Those two figures are results of a simulation of the experiment. (a) The energy loss of the proton inside the hydrogen target is plotted as a function of the detected proton energy. This energy loss is a function of the energy of the proton when it is scattered, and of the distance the proton has to pass in the target. (b) The number of counts (per barn, per steradian and per 10^9 incident ions) is plotted as a function of the detected proton energy. This demonstrates the influence of the effective target thickness as a function of the proton energy.

We have performed a measurement with a stable beam to evaluate the quality of the setting (mainly the cryogenic target) and to calibrate the data. This test was accomplished with an [18]O stable beam, produced at the same energy as [18]Ne. The compound nucleus [18]O + p = [19]F is quite well known at the corresponding excitation energies. In the Figure 2 we have performed two comparisons: we have compared our results with experimental data from a very precise measurement published by Orihara et al. [9], and also with the results of a R-matrix calculation we have done with the code Anarki [8] and using the known states properties in [19]F (about 40 states). From those comparisons it was possible to determine the hydrogen target thickness of 1.15 mm (density of 88.5 mg/cm^3). It was also possible to determine the solid angle of the silicon detector, it is $d\Omega = 0.01$ sr.

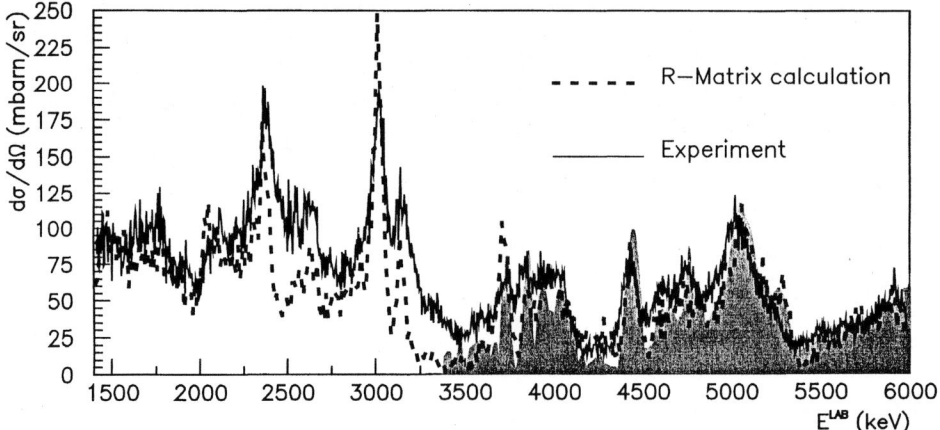

FIGURE 2. The reconstructed differential cross section (lab, direct kinematics, $\Theta^{ab} = 180°$) for the elastic scattering reaction [18]O(p,p)[18]O is plotted as a function of the proton energy in the laboratory system in direct kinematics. The continuous line represents the data reconstructed from our measurement, the dotted line shows the R-Matrix calculation using the known properties of ^{19}F, and the filled area corresponds to the experimental measurement from the reference [9]. The overall agreement is excellent. Our experimental resolution is $\sigma \approx$ 40keV.

RESULTS FOR [19]NA

The same analysis of the data has been applied to produce the [19]Na spectrum in figure 3. This result is still preliminary and doesn't represent the full statistics. The energy range of the spectrum is limited due to experimental cut-offs. The high value limit at about 6 MeV is close to the expected maximum excitation energy calculated using the full energy of the incident [18]Ne ions E = 7 MeVA. The low energy limit E $\approx$ 1.5 MeV is mainly due to the selection (banana in a E-ΔE silicon telescope) we have done to identify protons among different particles detected after the target (proton, alpha and electrons). It is also due to the fact that protons have to cross some distance inside the hydrogen target to exit, resulting in some energy loss as already shown.

In order to analyze the experimental data we have used the known properties of [19]Na, as well as the known properties of the mirror nucleus [19]O. Finally, we have done shell model calculations (in the spsdpf space and with the WBT interaction) with the program Oxbash. The first complete set of shell model calculations for the proton elastic and inelastic scattering has been performed for the first time in this work. The continuous line in figure 3 represents our R-matrix calculation. The first peak, labeled 1 in figure 3, corresponds to the known second excited state in ^{19}Na. Unfortunately, we were not able to observe this state because of the low energy cut-off. Even if it is not visible in the figure 3, this state also influences the general shape and normalization of the spectrum at higher excitation energy. If one takes this effect into account, our measurement is in a good agreement with the calculation. In a second step of the calculation we have introduced three new states in the R-matrix analysis: corresponding to the states 4, 5 and 6. In fact our shell model calculation has shown only three states with width larger than 5 keV for elastic scattering, they all have spin 3/2-. Starting from the predicted properties for those states we have fitted the experimental spectrum with the R-matrix analysis, the result is shown in the figure 3.

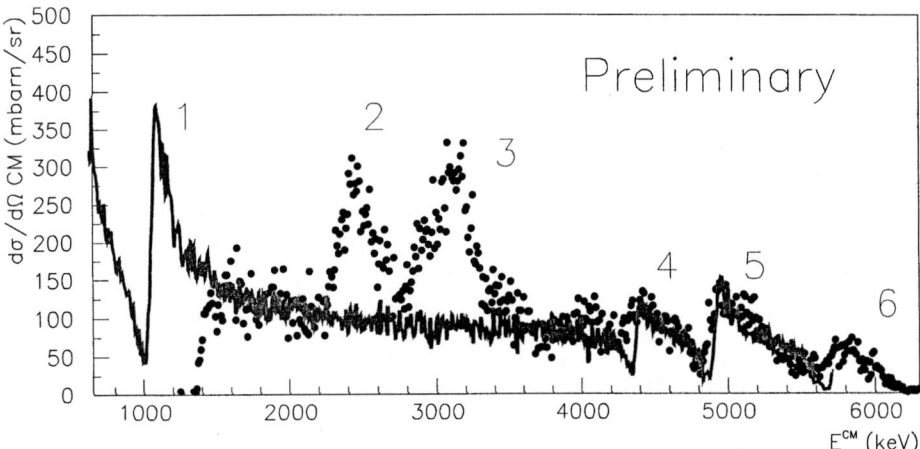

FIGURE 3. A preliminary, reconstructed differential cross section (CM, direct kinematics, Θ^{CM} = 180°) for the elastic scattering reaction ^{18}Ne(p,p)^{18}Ne as a function of the proton energy in CM is shown. The points are experimental data; the continuous line represents a calculation from R-Matrix model using the known and expected properties of ^{19}Na. The numbers correspond to the peaks described in the text.

The measured properties of the peak 4 E_x=4371 keV and Γ=30 keV and the peak 5 E_x=4903 keV and Γ=50 keV are very close to calculated values: peak 4 E_x=4258 keV and Γ=61 keV; peak 5 E_x=4667 keV and Γ=47 keV. The used value for the state 6: E_x=5466 keV and Γ=160 keV, are indicative. In fact, states at energy higher than 5.5 MeV are difficult to analyze because at those energies there is a high density of broad states not easy to disentangle.

Even if we have found a good agreement in those fits, the two main observed peaks 2 and 3 in the figure 3 couldn't be explained by the elastic scattering. In the same shell model calculations we have performed, several broad states have been predicted for the inelastic scattering. In that case the raw spectrum has to be re-analyzed with new correction functions, in order to take into account different kinematics and energy loss. This results in the following properties for the two peaks labeled 2 and 3, if interpreted as inelastic scattering: peak 2 E_x=3560 keV and $\Gamma \approx$ 130keV; peak 3 E_x=4540 keV and $\Gamma \approx$ 150 keV, in good agreement with our predictions: E_x = 3167 keV, Γ=187 keV and J^π=5/2+; E_x=3746 keV, Γ=305 keV and J^π=3/2+.

A full description of the experimental details, calculations and final results will be published in the following article [10].

ACKNOWLEDGMENTS

A large part of the success of this experiment is due to the quality of the SPIRAL beam, and we would like to thanks again all the GANIL staff for this result. This work has partially been supported by the EU Access to Large Scale Facilities program.

REFERENCES

1. S.M. Lukyanov, Yu. E. Penionzhkevich, R. Astabatyan, S. Lobastov, Yu. Sobolev, D. Guillemaud-Mueller, G. Faivre, F. Ibrahim, A.C. Mueller, F. Pougheon, O. Perru, O. Sorlin, I. Matea, R. Anne, C. Cauvin, R. Hue, G. Georgiev, M. Lewitowicz, F. de Oliveira Santos, D. Verney, Z. Dlouhy, J. Mrazek, D. Baiborodin, F. Negoita, C. Borcea, A. Buta, I. Stefan and S. Grevy, *J.Phys.G.* **28**, L41-L45 (2002)
2. J. Cerny, R.A. Mendelson, G.J. Wozniak, J.E. Esterl and J.C. Hardy, *Phys. Rev. Let.* **V22**, 612(1969)
3. W. Benenson, A. Guichard, E. Kashy, D. Mueller, H. Nann and L.W. Robinson, *Phys. Let.* **V58B** N1, 46(1975)
4. Thomas Zerguerras, Thesis IPNO T01-05 (2001) IPN Orsay France
5. C. Angulo, G. Tabacaru, M. Couder, M. Gaelens, P. Leleux, A. Ninane, F. Vanderbist, T. Davinson, P.J. Woods, J.S. Schweitzer, N.L. Achouri, J.C. Angélique, E. Berthoumieux, F. de Oliveira Santos, P. Himpe, and P. Descouvemont, *Phys. Rev.* **C67**, 014308(2003)
6. K. Markenroth, L. Axelsson, S. Baxter, M.J.G. Borge, C. Donzaud, S. Fayans, H.O.U. Fynbo, V.Z. Golberg, S. Grévy, D. Guillemaud-Mueller, B. Jonson, K.-M. Källman, S. Leenhardt, M. Lewitowicz, T. Lönnroth, P. Manngard, I. Martel, A.C. Mueller, I. Mukha, T. Nilsson, G. Nyman, N.A. Orr, K. Riisager, G.V. Rogatev, M.-G. Saint-Laurent, I.N. Serkov, N.B. Shul'gina, O. Sorlin, M. Steiner, O. Tengblad, M. Thoennessen, E. Tryggestad, W. H. Trzaska, F. Wenander, J.S. Winfield, and R. Wolski., *Phys. Rev.* **C62**, 034308(2000)
7. V.Z. Goldberg et al, *Yad. Fiz.* **60**, 1186(1997)
8. E. Berthoumieux, B. Berthier, C. Moreau? J.P. Gallien, A.C. Raoux, *NIM* **B136-138**, 55-59 (1998)
9. H. Orihara, G. Rudolf and Ph. Gorodetzky, *Nucl. Phys.* **A203**, 78 (1973)
10. F. de Oliveira Santos et al. in preparation

EXPERIMENTAL TECHNIQUES

The ISOLDE Silicon Ball

L.M. Fraile for the ISOLDE Si-ball collaboration

ISOLDE / EP Division, CERN, CH-1211 Geneva 23
Dpto. de Física Atómica, Molecular y Nuclear, Univ. Complutense, E-28040 Madrid, Spain
Department of Physics, FIN-40014,University of Jyväskylä, Finland
Instituto de Estructura de la Materia, CSIC, E-28006, Madrid, Spain
Helsinki Institute of Physics, FIN-00014, University of Helsinki, Finland
Institut for Fysik og Astronomi, Århus Univ., DK-8000 Århus, Denmark
Department of Physics, Charmers Univ. of Technology, S-41296 Götabor, Sweden
Department of Physics, Osaka University, Toxocara, Osaka 560-0043, Japan

Abstract. The investigation of weakly bound nuclei close to the particle driplines makes necessary the development of new spectroscopy devices with the capability of detecting charged particles and precisely determining their energy, angular distribution and nature. With this aim the ISOLDE Silicon Ball is under construction. It is a charged particle spectroscopy device with the requirements of high geometrical efficiency and broad energy range coverage, designed for the investigation of the exotic nuclei produced at ISOLDE and at other similar facilities. In order to allow for particle identification the simultaneous use of the Time of Flight (TOF) and Pulse Shape Discrimination (PSD) techniques is intended. Recoil tagging capabilities, suitable for transfer reactions to be performed at REX-ISOLDE, should be foreseen for a future development. The design and realization of the first prototype, together with the first tests are reported.

1. INTRODUCTION

The recent advances in the study of weakly bound nuclei close to the particle driplines make necessary the development of spectroscopy devices with high efficiency and granularity, in order to detect the outcoming charged particles and efficiently determine their nature, energy and angular distribution. With this aim the ISOLDE Silicon Ball for the detection of charged particles is being constructed. It is intended for β-decay investigation of exotic nuclei produced at ISOLDE and other facilities, and in the future should provide tagging capabilities for transfer reactions performed at REX-ISOLDE.

2. PHYSICAL MOTIVATION

Beta-decay plays an important role on the investigation of nuclear structure far from stability. Position sensitive devices with good granularity, efficiency and particle discrimination are essential in this studies, as for instance to disentangle the mechanism of β-2p decays in ^{31}Ar [1], or to distinguish the β-p and β-α branches in ^{17}Ne [2]. Exotic decay channels are also predicted to occur from highly excited states in nuclei at the neutron dripline, as for example the light di-neutron halo nucleus ^{11}Li [3]. This nucleus has a very complex decay with many different open multi-particle decay channels, and

CP681, *Proton-Emitting Nuclei: Second International Symposium; PROCON 2003*,
edited by E. Maglione and F. Soramel

is key in the understanding of the nuclear halo properties. The capacity of measuring β-p and p-p correlations is very valuable this type of studies. Precisely, the most interesting cases in which the ISOLDE Silicon Ball can show its capabilities are studies of beta-delayed multi-particle decays.

Other applications of the detector array are accurate measurements of β-delayed protons and the study of isospin symmetry via Gamow-Teller decays. Among those, one of the key cases is the ^{58}Zn → ^{58}Cu decay, which will be one of the first applications [4], see below. Other investigations include electron-neutrino correlations in Fermi decays, search for proton decay from the ground state in astrophysically relevant nuclei, as ^{69}Br or ^{73}Rb [5] and in the future particle transfer reactions in weakly bound nuclei, involving charged particles.

3. SI-BALL DESIGN

3.1. Main requirements

Position sensitivity, high granularity and broad angular range between emitted particles are required, with a geometrical efficiency as close to 4π as possible. The device should detect charged particles in a wide energy range with sufficient energy resolution, low detection threshold and the ability of determining angular correlations. Optimal energy resolution and a detector thickness sufficient to stop the particles of interest are needed. A very good identification of particles within the whole range is needed. The application of Pulse Shape Discrimination and Time of Flight is intended, which limits the choice of the size of the device. Moreover, good versatility is required to swap between the different configurations, and the possibility of combination with other detectors and spectrometers has to remain open.

3.2. Experimental techniques

In order to allow for particle identification the simultaneous use of the Time of Flight (TOF) and Pulse Shape Discrimination (PSD) is foreseen. The TOF for identification with Silicon detectors has been successfully tested and will be particularly used for low energy range identification. The use of this technique compels to increase the dimensions of the device in order to have a reasonable flying path. In the first stage of the project the starting time is provided by a 3-mm-thick BICRON BC-408 plastic scintillator placed as close as possible to the beam collection point, providing a flying path of 15 cm.

On the other hand, the PSD technique exploits the properties of the pulse signal of different particles in planar Si detectors [6] in order to extract information on their charge and mass. The signal shape is very sensitive to the density and length of the ionization track of the particles, due to the finite drift time of the electrons and holes produced in the ionization track and the plasma effect [7]. Nevertheless, the application of such method for thicker detectors and low energy particles is still a challenge. The particle discrimination can be achieved by means of the zero-crossing method, which is well known from neutron-gamma discrimination with scintillation detectors, by observing the zero-crossing time versus the energy deposition E in the detector. Alternative ways of identification are provided by the use of fast differentiating preamplifiers and low noise, wide band time filter amplifiers in conjunction with standard electronics [8] or by the use of digital electronics. The employ of the PSD technique requires the use of very homogeneous high resistivity Silicon detectors and dedicated electronics (PA, TFA) in the design. Suitable materials for the detectors are neutron transmuted doped (NTD) Silicon with resistivity in the order of 5 kΩ·cm or n-type silicon with <100>-crystal orientation and higher resistivity.

FIGURE 1. Partial assembly of the first phase of the ISOLDE Silicon Ball in its aluminum frame. The frame itself consists of interchangeable modules, similar to that shown in the bottom right corner, in which two or four segmented square Si detectors are fixed.

3.3. Geometrical design

The proposed design combines the advantages of a self-supporting structure in a compact design, with high geometrical efficiency and rather low mass. It has the shape of a rhombicuboctahedron, made of 18 square faces of 119 x 119 mm² inner area and 8 more faces made of equilateral triangles of 119 mm side, as shown in Figure 1. Each of the square faces is covered with 4 standard 51 x 51 mm² Si detectors, which in turn are segmented in 4 quadrants (25.5 x 25.5 mm²) for better granularity. The solid angle subtended by each of the segments is around 0.21%. In the final design 4 equilateral triangular detectors of 51 mm side will fill each of the triangular faces. The solid angle subtended by each of them is about 0.35%. The number of detectors adds up to 104 (72+32), and the channels to be read to 320. In the first stage of the project just one half of the 4π device including only square detectors (144 channels) has been built.

The detector system is placed inside vacuum in a spectroscopy chamber that allows the combination with other type of detectors (scintillators, HPGe). The device is cooled to a constant temperature of $-10\ ^oC$ by means of C_3F_8 liquid circulating through copper cooling pipes entering the vacuum chamber.

3.4. Detectors

With the aim of achieving a good configuration versatility, and flexibility of integration with other standard Si detectors, the design makes use of commercially available silicon detectors. Such detectors are provided by Micron Semiconductor Ltd. and manufactured on n-type silicon with <100>-crystal orientation and resistivity above 15 kΩ·cm. The active thickness is $988 \pm 6\ \mu$m, fully depleted at around 200 V. In this way the energy of fully stopped protons ranges from below 100 keV to 12 MeV. These MSQ25-1000 type detectors are mounted in a PCB transmission package specially designed to minimize dead areas. The dead layer and entrance window thickness has been reduced to lower the detection thresholds. Tests of the feasibility of the pulse shape discrimination technique with alpha particles and betas have been accomplished as well. Although time differences are observed between both types of particles, further tests with different ions are required.

3.5. Electronics

Low noise multichannel preamplifier modules, and integrated shaper, discriminator and TFA NIM modules from the company MESYTEC are used as front-end electronics for the Si-ball. The digitization electronics consists of CAEN V785 32 channel peak sensing ADCs and V775 32 channel multievent TDCs. The acquisition system is based on VME electronics and on the GSI multibranch system (MBS). The whole solution suits other ISOLDE Experimental Physics groups, which use multiple detector arrays or highly segmented detectors with a typical number of channels of the order of 100 to 300.

4. FIRST TESTS AND APPLICATIONS

After the completion of the device the first tests of the performance of the detectors, and of the detection techniques, have been carried out. The detectors behave in a satisfactory way in terms of resolution and time response (see Figure 2). Moreover, the ISOLDE Silicon Ball, the electronics chain and the DAQ system were fine tuned online by means of particle emitting isotopes. An example of an α spectrum after the β-decay of ^{20}Na is shown in Figure 2.

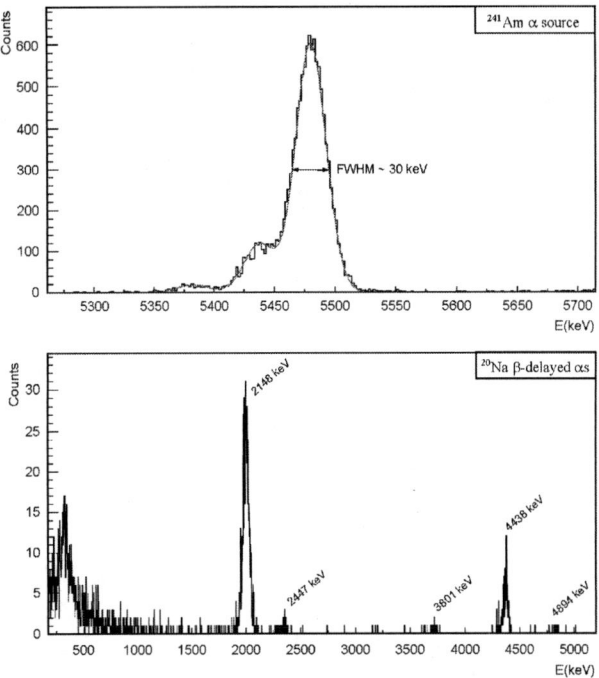

FIGURE 2. *Top*: Calibration spectra with ^{241}Am illustrating the energy resolution. The spectra is the sum of data from 28 detectors, corresponding to about 5.9% solid angle. *Bottom*: An example of an online ^{20}Na spectrum showing the main α lines, taken for to the same 28 detectors.

4.1. Isospin symmetry of transitions

If the nuclear interaction is charge symmetric, then the isospin should be a good quantum number. In particular, nuclear transitions from the ground state of the isobaric pair $<T,T_Z> = <1,\pm1>$ to excited states in the nucleus between the pair with T=0 should be symmetric, and the strength of analogue transitions (IAS) should be equivalent. This fact can be tested in the case of T = 1 → 0 for the symmetry pair with A = 58 [4]. On one side the transition ^{58}Ni ($T_Z = 1$) → ^{58}Cu ($T_Z = 0$) can be investigated via the (3He,t) reaction. On the other, ^{58}Cu is populated from ^{58}Zn (Tz = -1) via β-decay. In the latter, the detection of β-delayed protons with good energy resolution in a wide range is essential in order to obtain the β strength. The ISOLDE Si-ball is well suited for this type of study, which has been performed at ISOLDE in the autumn 2002 using the setup sketched in Figure 3. The analysis is in progress.

5. SUMMARY

The ISOLDE Si-ball is a 4π particle detection array under construction at ISOLDE, CERN. The detection and discrimination of charged particles with good E resolution and angular coverage is intended. In the first stage, a single 2π hemisphere with square 1-mm-thick <100>-Si detectors has been constructed. The first tests and experiments have taken place at the end of year 2002. Further β-decay experiments are scheduled for 2003.

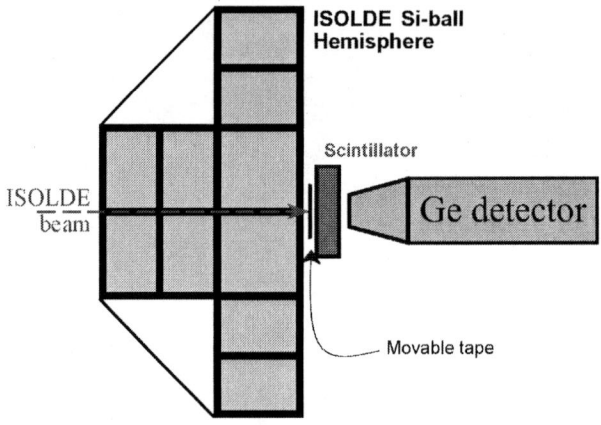

FIGURE 3. Sketch of the experimental setup used for the investigation of β-delayed protons following the ^{58}Zn decay.

ACKNOWLEDGMENTS

This work has been supported by the E.U. funded EXOTAG project, contract HPRI-1999-CT-50017.

REFERENCES

1. H.O.U. Fynbo et al., Nuc. Phys. A677 (2000) 38.
2. L.M. Fraile et al., Proposal to the ISOLDE n-TOF Comittee.
3. T. Nilsson, G. Nyman, K. Riisager, Hyperfine Interact. 129 (2000) 67.
4. Y. Fujita et al., CERN-INTC-2001-008; INTC-P-136, and Euro. Phys. J. A 13 (2002) 411.
5. M. Oinonen et al., CERN-INTC-2001-009 ; INTC-P-137
6. J.B.A. England et al., NIM A280 (1989) 291.
7. G. Pausch et al. NIM A337 (1994) 572, NIM A349 (1994) 281.
8. M. Mutterer et al., IEEE-TNS 47 (2000) 756.

Digital Spectroscopy for Proton Emitters

Robert Grzywacz

ORNL, Physics Division, Oak Ridge, TN 37830, USA
IFD, Warsaw University, Pl-00681 Warsaw, Hoża 69, Poland

Abstract. The detection system based on Digital Signal Processing has been developed at ORNL and applied to study proton radioactivity. Novel concepts of the Digital Spectroscopy are presented. They have been applied in experiments difficult or impossible to perform with standard systems which combine shaping amplifiers with analog-to-digital converters. The DSP features, such as, trigger-less operation, pulse shape recording and pulse shape based triggering schemes, led to successful measurements of the fine structure in the proton decay of ^{145}Tm and ^{141}Ho and to observation of the two-proton emission in ^{45}Fe.

DIGITAL PULSE PROCESSING

This paper presents the application of digital pulse processing technique to proton emission experiments. The primary motivation for this development came from the need of the detection of very short lived proton emitters, which may have lifetimes as short as microseconds. However, the versatility of digital pulse processing allows for much broader applications in particle and gamma-ray spectroscopy [1].

The general concept of real time Digital Pulse Processing (DPP) will be introduced, followed by a brief description of the electronic board [2], implementing the DPP concepts. Finally, the on-line experiments on proton emitting nuclei which take advantage of DPP are presented. The ability of directly storing the detector pulse shapes and the real time "proton catcher" algorithm are two main properties of the system which allowed the discovery of the fine structure in ^{145}Tm [3, 4]. A real time processing algorithm which can determine the arrival time and amplitude of the detector signal enabled the measurement of the fine structure in a decay of a longer lived ^{141}Ho [4] using the same system.

Digital Pulse Processing is a fully numerical analysis of the detector pulse signals. The numerical algorithms replace the analog shaping and timing circuitry to derive properties of the pulse such as its amplitude and arrival time, see fig. 1. The advantage of DPP arises from flexibility in implementing the algorithms. This enables much more complex pulse analysis necessary for example in gamma-ray tracking [5], or particle identification [6]. One of DPP features which is exploited in the proton emission experiments is new approach of handling pileup pulses.

The first necessary step for the DPP analysis scheme is the signal digitization. This process has to be done with sufficient accuracy to preserve the information which is carried by the pulse shape. Once the pulse is transformed into digital form it becomes immune to distortions caused by electronic noise and temperature instabilities. Compu-

CP681, *Proton-Emitting Nuclei: Second International Symposium; PROCON 2003*,
edited by E. Maglione and F. Soramel
© 2003 American Institute of Physics 0-7354-0150-0/03/$20.00

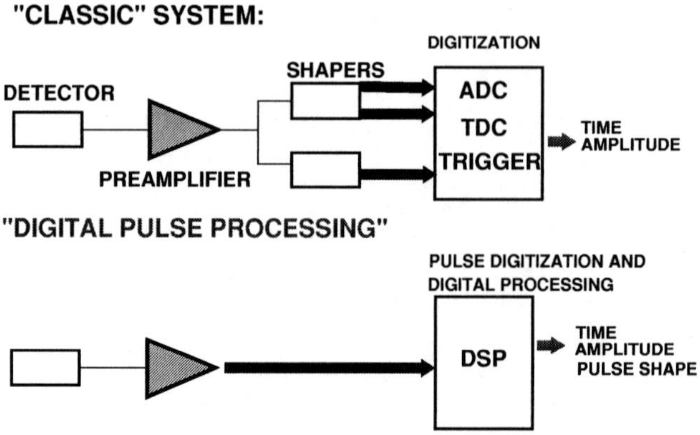

FIGURE 1. The system employing Digital Pulse Processing will be have a simplified architecture due to the integration of various functions in the pulse processing real time algorithms.

tational operations can be performed on identical copies of the pulse and are limited only by the processor performance and programmer's skill and inventiveness. The numerical analysis of the pulse shape can be done either in 'real-time', as soon as the pulse is digitized, or in a post-processing step, on the stored data. Both operation modes have been used in described experiments. The post-processing solution requires the output and storage of a large number of data reduced later to only a few numbers classifying the pulse such as energy and time. Storing and synchronizing large data streams becomes inconveniently difficult for even small size multi-parametric experiments. This is the obvious advantage of the fast, real time processing algorithms which can reduce the data stream by providing the functionality of analog shaping units. It leads to the main difficulty, to develop the real-time algorithms for any given detector system.

THE ELECTRONIC BOARD - DGF4C

Digital Pulse Processing techniques are not unknown in experimental physics. However the accuracy, overall stability and robustness of the widely used analog circuit boards developed for nuclear physics applications were achieved only recently.

A number of DPP ideas have been successfully implemented in the design of the DGF-4C board [2, 7]. It combines the capability for real time pulse-shape analysis which provides the necessary functionality with optional waveform storage for post-processing.

The DGF4C is a single-width Camac board featuring four signal input channels. A detailed description can be found in refs [2, 7]. The system comprises of 4 analog signal conditioning units (ASC), 4 fast sampling Analog to Digital Converters (ADC), 4 Field Programmable Gate Arrays (FPGA), and a Digital Signal Processor (DSP). The role of the analog circuit (ASC) is to map the signal amplitude to the ADC range and to provide an anti-aliasing analog filter. It may change the signal's amplitude and DC-

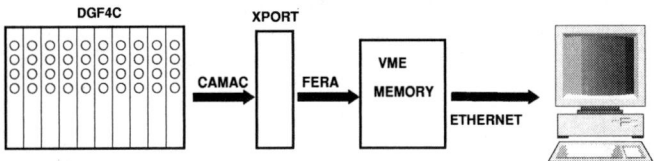

FIGURE 2. The readout scheme. The DGF memory buffers are read into XPORT memory. After this the DGFs can resume the digitization while the data are being shipped to VME and the to the workstation for storage.

offset, but will preserve the overall pulse shape. The digitization is performed at a rate of 40 MSPS, i.e. in 25 ns intervals, using 12-bit ADCs. After the digitization all operations on the signal are performed using discrete computations. The data from the ADCs are streamed into the real time processors implemented in the FPGAs. They continuously perform pileup inspection and compute arrival times and pulse heights using trapezoidal filters.

A trapezoidal filter with short time constants is used to provide a local trigger, which is used for signal time stamping, and to feed the pileup inspector. A trapezoidal filter with long and adjustable time constants is used for the energy measurement. At the same time a digital copy of the pulse can be put into a memory and stored for later readout or analysis. For more elaborate filtering the time/amplitude or trace data are transferred to the DSP processor. Finally, the DSP formats time/amplitude and waveform data, writes them into a buffer and increments spectra in its histogram memory.

The data for each board are stored in the 16kB output buffers. In the multi-module system the readout of all buffers is initiated by the module which first reports its buffer to be full. The input digitizing is halted during readout to avoid electronic interference. The data readout is executed under CAMAC protocol. Because data are organized into large banks, the readout process can be done using block readout which reduces the time overhead associated with the event-by-event readout mode. Recently, the CAMAC to FERA converter XPORT [8] has been added to the system. This module, equipped with an FPGA, can be programmed as a DGF readout controller and FIFO buffer memory. The additional buffering ability can decouple the VME processing from the DGF readout thus reducing the overall readout dead time, partially curing the problems related to the slowness of the CAMAC protocol.

The DGF is a self-triggering device but is also equipped with external triggering and gating capabilities.

The achieved energy resolutions of the DGF coupled with the silicon DSSD, 65 μm thick, 40x40 mm with 1 mm wide strips, was 26 keV FWHM for the 5.586 MeV ^{241}Am external source and 18.5 keV FWHM measured with 1.17 MeV ^{141}Ho implanted protons.

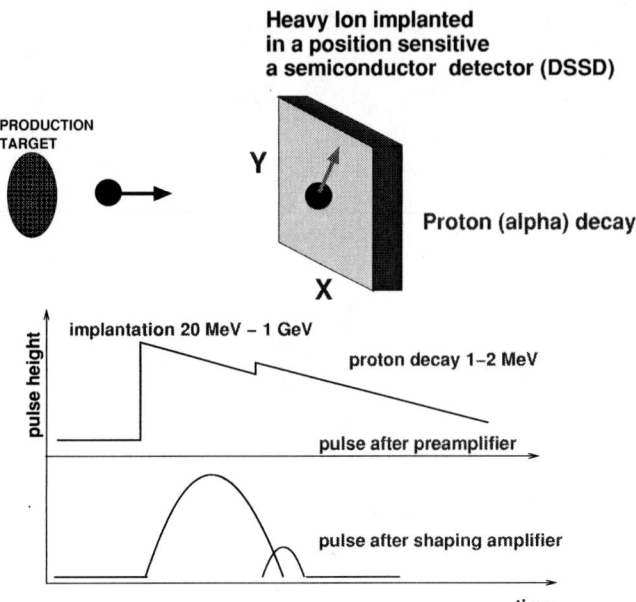

FIGURE 3. The proton emission experimental scheme. Upper graph: Proton emitting nuclei are produced in a thin target. They are transmitted to the detection setup and separated from the unwanted recoils and primary beam using an electromagnetic separator. After the implantation in a silicon detector, the proton decay process can be observed. Lower graph: The high amplitude of the recoil induced signal is followed by the small proton decay pulse. Schematically are presented the pulse shapes after the preamplifier and shaping amplifier. For the very short proton emitters in the microsecond range, the small decay pulse will not be distinguishable from the large implant pulse on the amplifier output.

PROTON EMISSION STUDIES

The primary motivation for the implementation of DSP was to handle the cases of closely overlapping pulses, which analog electronics have difficulty to resolve with good resolution in time and energy, see Fig. 3. Proper determination of the energy of overlapping pulses remains difficult when using fixed-length trapezoidal filters. Here the ability to store pulse shapes and use post-processing has been exploited. The interesting pileup pulse waveforms were recognized, selected and stored together with the time, amplitude data and pulse shape from other detectors.

The DGF based acquisition system was developed to measure microsecond particle radioactivities at Holifield Radioactive Ion Beam Facility (HRIBF) in Oak Ridge. In such experiments, proton- or alpha-emitting exotic nuclei are produced with heavy-ion fusion reactions on thin targets. Recoiling fusion products are mass separated in a large acceptance recoil mass spectrometer, passed through a position sensitive detector for M/Q determination, and implanted in a Double-sided Silicon Strip Detector (DSSD), like those described in Ref. [9]. The HRIBF DSSD setup and Recoil Mass Spectrometer (RMS) are discussed in Ref. [10]. The recoiling heavy ion is slowed down in a degrader foil and implanted in the DSSD, where it deposits an energy of the order of 15-50 MeV. An emitted proton carries a much smaller kinetic energy, of the order of 1 MeV. For the short-lived proton radioactivity the main difficulty is to detect two closely spaced pulses with a large energy difference. This problem was solved through the use of DGFs [11].

The system developed at HRIBF consists of nineteen DGF modules in a single CAMAC chassis. Eighteen of them are dedicated to digitizing the signals from the preamplifier [12] of the Double-sided Silicon Strip Detector (DSSD) where the ions and protons are stopped. When searching for, or studying, short microsecond activities these modules are used to record the full waveform of the preamplifier signal associated with the recoil implantation and subsequent decay. For each channel containing a possibly interesting waveform we store the 25 μs long 'trace'. Each trace has an assigned time stamp, with 48 bit precision in units of 25 ns, and the signal amplitude is coded with 12 bit resolution (4096 channels).

The four channels of the nineteenth module are used to digitize data from Position Sensitive Detectors (PSD) [10], which provide independent information on the ion arrival.

To reduce the data stream a fast self-trigger system was implemented for each channel of the DGF. It is programmed in the real time processing part of the DGF and its role is to recognize the pile-ups occurring within 10 microseconds; all other signals are ignored (i.e. they are not sent to the output data buffer). Such pileups are very likely to be ion-ion random coincidences but also real ion-proton decay events, thus the algorithm has been called a 'proton catcher'.

Proton emission events have been observed from about 500 nanoseconds after the ion implantation into the DSSD. An event qualified as a valid proton event should have identical amplitudes of the second pulse from both the front and back strips of the DSSD, identical absolute time (within the resolution of the algorithm), and have no PSD signal. They should be preceded by the heavy ion induced pulse which has the same absolute time from both front and back strips and be in coincidence with a PSD signal. The proton amplitudes were extracted from the stored wave-forms with a resolution of 70 keV at 1.7 MeV by our algorithm [13], while the resolution of the DSSD-DGF signal observed for long-lived proton radioactivity was determined to be 20 keV. The worsening of the resolution is related to the short integration time available for the piled-up pulse. The advantage of the DPP-based solution over the previously used system was demonstrated in the [145]Tm experiment [3, 11]. The efficiency was increased by a factor of 7 over that obtained with the conventional electronics, thus allowing the observation of the fine structure in the decay of this 3-μs activity because of the increased detection sensitivity at short times.

The DGFs were used successfully in an experiment aiming at a discovery of the fine structure in the decay of the deformed [141]Ho nucleus [4]. In this experiment the DGF's were used in the so-called 'standard mode', which means that only the time and amplitudes of the pulses have been stored. Because of the long lifetime of the proton emitting state, the use of the 'proton catcher' was not required. Currently, 'standard mode' and 'proton catcher' mode cannot be simultaneously invoked in the same module.

An attractive possibility is to employ this system in Recoil Decay Tagging experiments [14] using e.g. Clarion [10] gamma-ray array. The main difficulty arises from the readout incompatibility of the existing Clarion electronics and DGF based system.

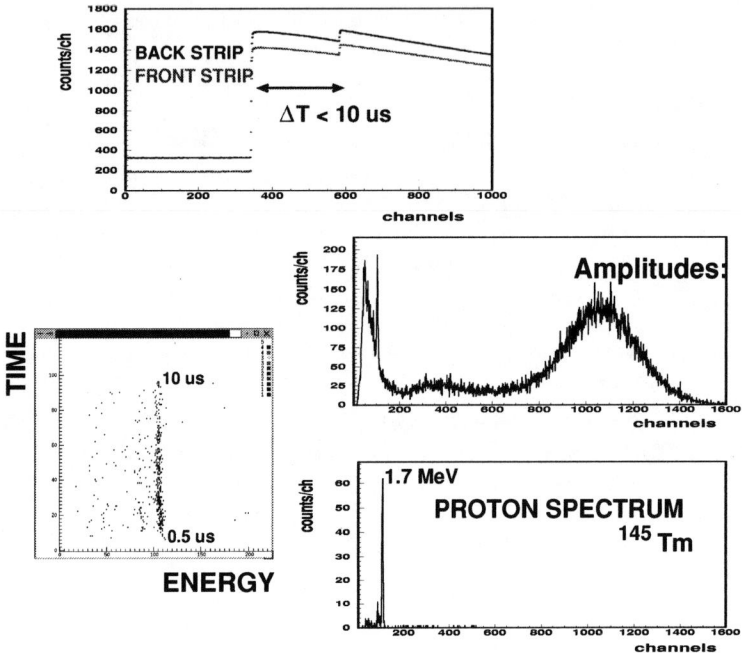

FIGURE 4. The analysis of the traces in the proton decay experiment, leading to the observation of the fine structure in the decay of ^{145}Tm [11, 3]. The pileup-type signals occurring within 10 μs are stored using DGF and "proton catcher" algorithm. The captured front and back strip pulses are shown (top figure). One channel corresponds to 25 ns. The amplitude of pileup pulses is determined (middle right). The time-difference vs. amplitude between the first and second pulse shows clearly the proton decay pattern of ^{145}Tm (left). A projection on the energy axis reveals the main peak at 1.73 MeV and a fine structure peak at 1.40 MeV.

Clarion operates using standard shapers and analog and digital converters and has event-by-event mode readout. DGF based system works in quasi-singles mode with a block readout. This problem can be overcome by using a time stamping option. DGF provide the time stamping capability also for external logic signal. Upon Clarion trigger the DGF will generate the time signal which will then be read in the event by event mode. The time stamped Clarion and DGF data streams can be than combined by the event builder algorithm and proton-recoil-gamma correlation can be recovered.

Search for two proton decay

Some candidates for proton and two proton emitters cannot be produced in a fusion reaction with stable beams and targets due to the very small reaction cross section. They can be produced only at heavy-ion fragmentation facilities. The general scheme of an experiment which uses this production method is rather similar (production target, frag-

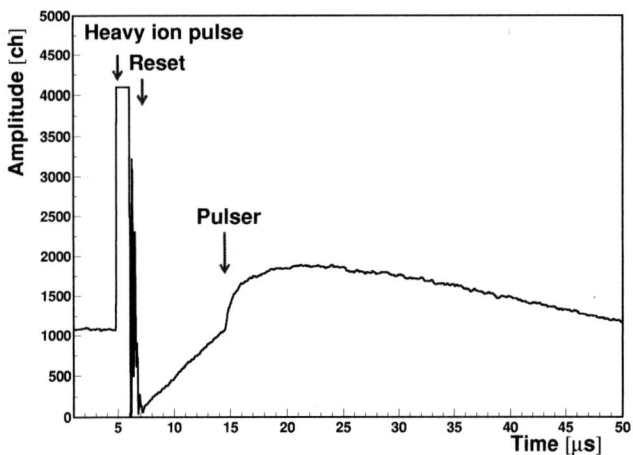

FIGURE 5. A example of a signal from the 'reset'-preamplifier. The first out-of-scale pulse is induced by the heavy-ion signal which is followed by reseting action. The second pulse at 15 μs is due to a pulser simulating a decay event.

ment spectrometer, silicon detector) to the setup for proton emission studies described above. However, in fragmentation the primary beam energy and the resulting fragment energy are much higher ($\approx$1000 MeV) than in the fusion reaction ($\approx$10 MeV). In addition the range straggling problem requires the use of multiple or very thick stopping silicon detectors. The development described below was aimed at the discovery of a new type of nuclear decay - two proton (2p) radioactivity. The promising candidates for such an observation are ^{48}Ni and ^{45}Fe; both nuclei were recently identified in fragmentation experiments [15, 16]. The 2p-decay energy of these nuclei was expected to be 1-1.5 MeV [15]; the predicted partial lifetime for such an energy range vary widely, from 10^{-2} to 10^{-6} s. Thus, the experimental system has to be sensitive to decays with microsecond and millisecond lifetimes. The experimental signature of the 2p-decay will be the measured total energy deposit of both protons in a silicon detector.

We performed experiments to test the possibility of using DGFs to discover 2p-radioactivity. In an approach similar to that used at HRIBF, the waveform of the signal induced by implantation was stored. However, for completeness we stored the 'traces' for each ion of interest, not just pileups, thus the 'proton catcher' mode was not used. Signals which were necessary for the event-by-event ion identification from the secondary-beam time-of-flight, energy loss and position detectors were also digitized and processed in real time on board.

The first attempt to use the DPP in 2p-decay search experiment was performed at GANIL - with intermediate energy (tens of MeV per nucleon) primary beam. The DGF system was placed in parallel with the main 'analog' data acquisition system. It consisted of six DGFs, four of them used for 'trace' storage from 16 strips and two for recording the ion detector signals for time-of-flight and energy loss measurement. This allowed for independent storage of the DGF data necessary for ion identification as well as traces.

The energy of implanted ions was degraded until 900 MeV was deposited in the stopping DSSD detector. It was large compared to the expected signal from the 2p decay (1-1.5 MeV). With 4096 channels of the flash ADC to cover the whole range of signals emerging from the preamplifiers, namely from 1 GeV down to 0, the expected 2p-proton decay signal would have an amplitude of only a few ADC channels. Several events of ^{45}Fe registered in this experiment, but they did not show any evidence for fast 2p-decay, however we deemed it inconclusive because of the small expected proton pulse amplitude. The DGF acquisition in this experiment was not sensitive to the lifetimes longer then 100 μs.

To overcome the problem of large amplitude difference between recoil and decay pulses, a new approach has been developed [17]. It requires the use of a new kind of a preamplifier [18]. The preamplifier is 'blocked' for about 1 microsecond at the beginning of the recoil-induced signal. This solution ensures that the large charge induced by the recoil in the silicon detector does not overload the preamplifier circuit, which can then have the large amplification required for the correct definition of the small decay pulse. In principle it should eliminate the necessity of acquiring the full waveform of the signal. However, the charge injected in the preamplifier upon reactivation (as soon as one microsecond after implantation) still induces a signal large enough to make it difficult to properly derive the decay pulse amplitude by an on-board algorithm, see Fig. 5. Thus, it was necessary to store the 'trace' and analyze it 'off-line'.

A DGF based system using reset-preamplifiers was successfully tested and applied at the FRS separator at the GSI heavy ion laboratory. In the experiment with high energy primary beam of ^{58}Ni at 650 MeV/u, aiming at the measurement of the decay mode of ^{45}Fe, the detection setup consisted of 7 planar silicon detectors placed inside a NaI barrel with very high gamma detection efficiency. Eleven DGF modules were used in this experiment. All signals required ion identification and the decay occurring within a 3 μs to 10 ms window after the ^{45}Fe implantation were digitized. In order to identify emission occurring within microseconds after implantation, the DGFs stored signal waveforms for 50 μs after implantation from the reset preamps, and another two were used to digitize the decay branch of the silicon signals from 30 microseconds to 10 milliseconds after implantation of the valid heavy ion. Within the 10 ms after implantation, DGFs were used to digitize the NaI signals.

In this experiment six events of ^{45}Fe have been identified [19]. From the analysis of the traces it has been concluded that no 2p-emission in the microsecond range has been observed. However, analyzing the longer time correlation yielded 4 well-correlated decay events at 1.1 MeV with a 3 ms decay lifetime [19] consistent with the interpretation as 2p-radioactivity. This finding has been confirmed by the data collected during a GANIL experiment [20]

SUMMARY

The particle detection setup combining the use of double-sided silicon strip detector and digital pulse processing front-end electronic has been developed and successfully used in proton emission experiments. Various approaches including real time and post-processing algorithms have been applied. The DPP approach is particularly useful in handling pileup signals. The measurements of microsecond particle emitters is inefficient or impossible without such system. Time stamping and self-triggering greatly simplify the data acquisition design in decay experiments. The real-time analysis is a very useful tool for implementing pulse-shape dependent triggering schemes as demonstrated by the 'proton catcher' algorithm.

There are still so far unexplored capabilities offered by application of digital pulse processing in particle decay spectroscopy. The pulse shape analysis can be employed in particle tracking and can be used to differentiate between particle type (e.g. positrons and protons) because of their different energy losses in the detector material.

The author would like to acknowledge Michael Momayezi for the invention of the "proton catcher", and also for his continuous advisory support during experiments, Marek Karny for the contribution to trace analysis, and Jim McConnell for the effort to implement the digital system at HRIBF.

ORNL is managed by UT-Battelle, LLC for the U.S. DOE under contract DE-AC05-00OR22725. This work has been supported also by Polish Committee of Scientific Research under grant KBN 2 P03B 036 15, and University of Tennessee under grant DE-FG02-96ER40983.

REFERENCES

1. R. Grzywacz Nucl. Instr. Meth. **B204** (2003) 649
2. B. Hubbard-Nelson, M. Momayezi and W.K. Warburton, Nucl. Instr. Meth. **A422** (1999) 41
3. M. Karny *et al.*, Phys. Rev. Lett. **90** (2003)012502
4. K. Rykaczewski *et al.*, in contribution to Nuclear Structure Conference, May 22-25, 2002 and this proceedings
5. K. Vetter *et al.*, Nucl. Instr. Meth. **A452** (2000) 223
6. M. Mutterer *et al.*, IEEE Trans. Nucl. Scie. **47** (2000) 756
7. W.K. Warburton, M. Momayezi, B. Hubbard-Nelson and W. Skulski Conference on Industrial Radiation and Radioisotope Measurement Applications IRRMA-99, October 3-7, 1999, Raleigh, NC, Applied Radiation and Isotopes (2000), available also at *www.xia.com*
8. J. Toke, *private communication*, see also *www.jtec-instruments.com*
9. J. C. Batchelder *et al.*, Phys. Rev. **C57**, R1042 (1998)
10. C. Gross *et al.*, Nucl. Instr. Meth. **A450** (2000) 12
11. K. Rykaczewski *et al.*, Nucl. Phys. **A682** (2001) 270c
12. P.J. Sellin *et al.*, Nucl. Instr. Meth. **311** (1992) 217
13. M. Karny, private communication
14. R. Simon *et al*, Z. Phys. **A325** (1986) 197 E. S. Paul *et al*, Phys. Rev. **C51** (1995) 78
15. B. Blank *et al.*, Phys. Rev. Lett. **84** (2000) 1116
16. B. Blank *et al.*, Phys. Rev. Lett. **77** (1996) 2893
17. M. Pfützner *et al.* Nucl. Instr. Meth. **A493** (2002) 155
18. E. Badura, *private communication*
19. M. Pfützner, Eur. Phys. Jour. **A14** (2002) 279
20. J. Givinazzo, Phys. Rev. Lett. **A89** (2002) 102501

List of Participants

Corina ANDREOIU
Oliver Lodge Lab. - Liverpool
University of Liverpool
Oxford Street
L69 7ZE Liverpool
UK
ca@ns.ph.liv.ac.uk

Andrei ANDREYEV
Oliver Lodge Lab. - Liverpool
University of Liverpool
Oxford Street
L69 7ZE Liverpool
UK
aan@ns.ph.liv.ac.uk

Carmen ANGULO
UCL - Louvain la Neuve
Centre de Recherches du Cyclotron
2, Chemin du Cyclotron
B-1348 Louvain-la-Neuve
Belgium
angulo@cyc.ucl.ac.be

Jon BATCHELDER
JIHIR/ORNL
(ORNL), Physics Division
P.O. Box 2008, Bldg 6008
Oak Ridge, TN 37831-6374
USA
batcheld@mail.phy.ornl.gov

Dino BAZZACCO
INFN, Padova
via Marzolo, 8
I-35131 Padova
Italy
bazzacco@pd.infn.it

Maria J.G. BORGE
CSIC - Madrid
Instituto de Estructure de la Materia
Serrano 113 bis
E-28006 Madrid
Spain
imtmj07@pinar2.csic.es

Alex B. BROWN
Michigan State University
MSU, Physics and Astronomy Bld.
W.Side Physics Road
East Lansing, MI 48824-1116
USA
brown@nscl.msu.edu

Laura CONFALONIERI
Hamamatsu Photonics Italia
Str. della Moia 1/E
I-20020 Arese
Italy
confalonieri@hamamatsu.it

Dave M. CULLEN
University of Manchester
Dep. of Physics and Astronomy
Manchester M13 9PL
UK
dmc@mags.ph.man.ac.uk

Cary DAVIDS
Argonne National Laboratory
Bldg 203
9700 South Cass Avenue
Argonne IL 60439-4843
USA
davids@anl.gov

Thomas DAVINSON
University of Edinburgh
James Clerk Maxwell Building
The King's Buildings
Edinburgh EH9 3JZ
UK
t.davinson@ph.ed.ac.uk

Giacomo de ANGELIS
INFN, Legnaro
Laboratories of Legnaro
Viale dell'Università, 2
I-35020 Legnaro
Italy
deangelis@lnl.infn.it

François DE OLIVEIRA SANTOS
GANIL - Caen
Bld. Henri Bequerel - B.P. 5027
14076 Caen Cedex 5
France
oliveira@ganil.fr

Lidia S. FERREIRA
IST - Lisboa
Centro Fisica Int. Fundamentais
Av. Rovisco Pais
1049-001 Lisboa
Portugal
flidia@ist.utl.pt

Luis M. FRAILE
CERN - Genève
EP Division ISOLDE Group
CH-1211 Geneve 23
Switzerland
luis.fraile@cern.ch

Jerome GIOVINAZZO
CEN - Bordeaux
Le Haut Vigneau
Boite Postale 120
F-33175 Gradignan Cedex
France
giovinaz@cenbg.in2p3.fr

Leonid GRIGORENKO
GSI - Darmstadt
Gesellschaft für Schwerionenforschung
Plankstraße 1, Postfach 11 05 52
D-64291 Darmstadt
Germany
l.grigorenko@gsi.de

Robert GRZYWACZ
Oak Ridge National Laboratory
ORNL, Physics Division
P.O. Box 2008, Bldg 6000 MS 6371
Oak Ridge, TN 37831-6371
USA
grzywacz@mail.phy.ornl.gov

Antonio INSOLIA
University of Catania and INFN
Department of Physics
Via S.Sofia 64
I-95123 - Catania
Italy
insolia@ct.infn.it

Ari JOKINEN
CERN - Genève
PS Division
CH-1211 Geneve 23
Switzerland
ari.jokinen@cern.ch

Stanislav KADMENSKY
Voronezh State University
1, University square
Voronezh, 394693
Russia
kadmensky@cd.vsu.ru

Heikki KETTUNEN
University of Jyväskylä
Department of Physics
P.O. Box 35 (YFL)
FIN-40014 Jyväskylä
Finland
Heikki.Kettunen@phys.jyu.fi

Wojciech KROLAS
Institute of Nuclear Physics
Radzikowskiego 152
PL-31-342 - Krakow
Poland
wojciech.krolas@ifj.edu.pl

Andras Tibor KRUPPA
ATOMKI - Debrecen
Institute of Nuclear Research
Bem tér 18/C
H-4026 Debrecen
Hungary
atk@chaos.atomki.hu

Miguel LOPES
IST - Lisboa
Centro Fisica Int. Fundamentais
Av. Rovisco Pais
1049-001 Lisboa
Portugal
guelcl@hotmail.com

Sergiy LUKYANOV
IST - Lisboa
Centro Fisica Int. Fundamentais
Av. Rovisco Pais
1049-001 Lisboa
Portugal
sergiy@gtae3.ist.utl.pt

Santo LUNARDI
University of Padova and INFN
Department of Physics
Via F. Marzolo 8
I-35131 Padova
Italy
lunardi@pd.infn.it

Enrico MAGLIONE
University of Padova and INFN
Department of Physics
Via F. Marzolo 8
I-35131 Padova
Italy
maglione@pd.infn.it

Nicu MARGINEAN
INFN, Legnaro
Laboratories of Legnaro
Viale dell'Università, 2
I-35020 Legnaro
Italy
nicu@lnl.infn.it

Chiara MAZZOCCHI
GSI - Darmstadt
Gesellschaft für Schwerionenforschung
Plankstraße 1, Postfach 11 05 52
D-64291 Darmstadt
Germany
c.mazzocchi@gsi.de

Marco MAZZOCCO
University of Padova and INFN
Department of Physics
Via F. Marzolo 8
I-35131 Padova
Italy
mazzocco@pd.infn.it

Giovanna MONTAGNOLI
University of Padova and INFN
Department of Physics
Via F. Marzolo 8
I-35131 Padova
Italy
montagnoli@pd.infn.it

Ivan MUKHA
GSI - Darmstadt
Gesellschaft für Schwerionenforschung
Planckstraße 1, Postfach 11 05 52
D-64291 Darmstadt
Germany
i.mukha@gsi.de

271

Alex MURPHY
University of Edinburgh
James Clerk Maxwell Building
Mayfield Road
Edinburgh EH9 3JZ
UK
amurphy@ph.ed.ac.uk

Alessandro PASCOLINI
University of Padova and INFN
Department of Physics
Via F. Marzolo 8
I-35131 Padova
Italy
pascolini@pd.infn.it

Marek PFÜTZNER
Warsaw University
Stefan Pienkowski Institute of
Experimental Physics
ul. Hoza 69
PL-00-681 Warsaw
Poland
pfutzner@mimuw.edu.pl

Jirina RIKOVSKA STONE
University of Oxford
Department of Physics
Parks Road
Oxford OX1 3PU
UK
j.stone1@physics.ox.ac.uk

Peter RING
Technische Universität München
Physik-Department T30
James-Franck-Straße 1
D-857478 Garching
Germany
peter_ring@physik.tu-muenchen.de

Andrew P.ROBINSON
University of Edinburgh
James Clerk Maxwell Building
Mayfield Road
Edinburgh EH9 3JZ
UK
apr@ph.ed.ac.uk

Dirk RUDOLPH
Lund University
Department of Physics
S-22100 Lund
Sweden
dirk.rudolph@kosufy.lu.se

Krzysztof RYKACZEWSKI
Oak Ridge National Laboratory ORNL
Physics Division
P.O. Box 2008, Bldg 6000 MS 6371
Oak Ridge TN 37831-6371
USA
rykaczew@mail.phy.ornl.gov

Fernando SCARLASSARA
University of Padova and INFN
Department of Physics
Via F. Marzolo 8
I-35131 Padova
Italy
scarlassara@pd.infn.it

Cath SCHOLEY
University of Jyväskylä
Department of Physics
P.O. Box 35 (YFL)
FIN-40014 Jyväskylä
Finland
cs@phys.jyu.fi

Paolo SCOPEL
University of Padova
Department of Physics
Via F. Marzolo 8
I-35131 Padova
Italy
scopel@pd.infn.it

Dariusz SEWERYNIAK
Argonne National Laboratory
Bldg 203
9700 South Cass Avenue
Argonne, IL 60439-4843
USA
seweryniak@anl.gov

Cosimo SIGNORINI
University of Padova and INFN
Department of Physics
Via F. Marzolo 8
I-35131 Padova
Italy
signorini@pd.infn.it

Francesca SORAMEL
Univeristy of Udine and INFN
Department of Physics
Via delle Scienze 208
I-33100 Udine
Italy
soramel@pd.infn.it

Alberto M. STEFANINI
INFN, Legnaro
Laboratories of Legnaro
Viale dell'Università 2
I-35020 Legnaro
Italy
alberto.stefanini@lnl.infn.it

Jean-Charles THOMAS
University of Leuven
Instituut voor Kernen Stralingsfysica
(IKS)
Celestijnenlaan 200 D
B-3001 Leuven
Belgium
Jean-Charles.Thomas@fys.kuleuven.ac.be

Monica TROTTA
INFN, Legnaro
Laboratories of Legnaro
Viale dell'Università 2
I-35020 Legnaro
Italy
trotta@lnl.infn.it

Phil WOODS
University of Edinburgh
James Clerk Maxwell Building
Mayfield Road
Edinburgh EH9 3JZ
UK
pjw@ph.ed.ac.uk

Chang-Hong YU
Oak Ridge National Laboratory
ORNL, Physics Division
P.O. Box 2008, Bldg 6000 MS 6371
Oak Ridge TN 37831-6371
USA
chy@mail.phy.ornl.gov

Liu ZHONG
University of Edinburgh
James Clerk Maxwell Building
Mayfield Road
Edinburgh, EH9 3JZ
UK
zliu@ph.ed.ac.uk

275

281

A

Abu Saleem, K., 187
Achouri, L., 235
Ahmad, I., 187
Angulo, C., 217
Äystö, J., 235

B

Badura, E., 105
Barker, F. C., 118
Batchelder, J. C., 11, 144, 172, 183
Batist, L., 139, 209
Belleguic, V., 139
Béraud, R., 235
Bingham, C. R., 11, 105, 172, 183
Blank, B., 105, 111, 235
Blazhev, A., 209
Bonetti, R., 18
Borcea, C., 111
Borge, M. J. G., 193
Brown, B. A., 111, 118
Bruce, A., 187

C

Canchel, G., 235
Carpenter, M. P., 187
Cerny, J., 144
Chartier, M., 105, 111
Cullen, D. M., 187
Czajkowski, S., 111, 235

D

Davids, C. N., 41, 187
Delion, D. S., 149
Dendooven, P., 235
de Oliveira Santos, F., 111, 235, 245
Döring, J., 139, 209

E

Eeckhaudt, S., 26
Enqvist, T., 26
Ensallem, A., 235
Esbensen, H., 41

F

Ferreira, L. S., 50, 77
Fletcher, A. M., 187
Fleury, A., 111
Fong, D., 11, 183
Fraile, L. M., 253
Freeman, S. J., 187

G

Geissel, H., 105
Gernetti, M., 18
Gierlik, M., 139
Ginter, T. N., 11, 144
Giovinazzo, J., 105, 111, 235
Grahn, T., 26
Grawe, H., 209
Greenlees, P. T., 26
Gregorich, K. E., 144
Grigorenko, L. V., 105, 126
Gross, C. J., 11, 172, 183
Grzywacz, R., 11, 105, 111, 172, 183, 259
Guglielmetti, A., 18
Guillet, N., 235
Guo, F. Q., 144

H

Hagino, K., 11
Hamilton, J. H., 11, 183
Hartley, D. J., 11, 183
Heinz, A., 187
Hellström, M., 105
Hoffman, C., 209
Honkanen, J., 235
Hwang, J. K., 11, 183

Related Titles from AIP Conference Proceedings

To learn more about these titles, or the AIP Conference Proceedings Series, please visit the
webpage **http://proceedings.aip.org/proceedings**